Recent Trends in Cancer Biology: Spotlight on Signaling Cascades and microRNAs

Sundas Fayyaz · Ammad Ahmad Farooqi
Editors

Recent Trends in Cancer Biology: Spotlight on Signaling Cascades and microRNAs

Cell Signaling Pathways and microRNAs in Cancer Biology

Springer

Editors
Sundas Fayyaz
Department of Biochemistry
Rashid Latif Medical College (RLMC)
Lahore, Pakistan

Ammad Ahmad Farooqi
Institute of Biomedical and Genetic
Engineering (IBGE)
Islamabad, Pakistan

ISBN 978-3-030-10074-2 ISBN 978-3-319-71553-7 (eBook)
https://doi.org/10.1007/978-3-319-71553-7

Softcover re-print of the Hardcover 1st edition 2018

Printed on acid-free paper

This Springer imprint is published by Springer Nature
The registered company is Springer International Publishing AG
The registered company address is: Gewerbestrasse 11, 6330 Cham, Switzerland

Preface

Cell signaling is one of the most intricate and fascinating areas of molecular biology, and it will not be wrong to say that we have witnessed groundbreaking discoveries in the era of molecular biology. It was in the late 1970s when scientists started to discover interwoven network of proteins. Genetic, genomic, and proteomic studies have improved our understanding of these highly synchronized, hierarchically assembled arrays of proteins, which transduce the signals linearly and also cross-talk with other signaling cascades.

This book comprehensively reviews most recent advancements in our understanding of the involvement of intracellular signaling cascades in cancer development and progression. Detailed information presented in this book will help younger oncologists find their way in the labyrinth of signal transduction cascades and how these dysregulated pathways can be therapeutically exploited to improve clinical outcomes. Detailed mechanistic insights of the signaling cascades are described but without ignoring the multiplicity of interconnections and cross-talks. After description of modulators and effectors of different pathways, brief description is also provided related to the alterations found in cancers as well as of the targeted pharmacological approaches that can be used to inhibit different proteins in a signaling pathway. Additionally we have also emphasized the role of miRNAs in different cancers.

The first chapter is focused on the essential role of signal transduction cascades in endometriosis and how these pathways promote transformations from benign to premalignant endometriosis. Dr. Talha Abdul Halim and his group comprehensively discussed mechanistic insights. The next chapter is written by the distinguished scientist Dr. George Calin whose team has contributed substantially in putting together the missing pieces of an incomplete jigsaw puzzle related to the role of noncoding RNAs in human cancers. Dr. Maria Ciccone and Dr. George Calin extensively elaborated how microRNAs, tyrosine kinases, and epigenetic modifications played key roles in leukemogenesis. Dr. Carolina Ruivo and Dr. Sonia Melo summarized the knowledge related to the rapidly emerging role of exosomes in tumor development, metastatic spread, and drug resistance. Dr. Yi Lim and his team provided an update related to ATM kinase, a multitalented regulator of DNA damage signaling. G protein-coupled estrogen receptor (GPER) induced signaling has been reported in prostate cancer. Dr. Sílvia Socorro and coworkers shed light on GPER-induced intracellular pathway and different agonists and antagonists to modulate GPER pathway. An overview of the aberrantly expressed miRNAs in bone cancer was provided by Dr. Janaina Dernowsek. Dr. Massimo Mallardo and his team emphasized on intricately controlled target genes by miR-25 in different cancers followed by a presentation by Dr. Kayla Lewis and Dr. Liu Yi who described how sonic hedgehog signaling pathway can be therapeutically exploited to treat lung cancer. Another highly investigated signaling cascade in molecular oncology is TGF/SMAD signaling cascade. Dr. Mohadeseh Hasanpourghadi and Dr. Mohammad Rais Mustafa conceptually portrayed most recent updates about regulation of TGF/SMAD pathway by different natural products. Dr. Eun Sohn impressively presented the use of natural products to modulate oncogenic and tumor suppressor miRNAs in different cancers. Dr. Armando Luis Garcia and colleagues shared most current knowledge related to Wnt signaling and miRNAs in sebaceous carcinoma of the eyelids and how clinicians will benefit from a greater understanding of the disease. Dr. Maria Luisa Gasparri and her team exclusively focused on the potential use of miRNAs as biomarkers in breast cancer. Dr. Chiara Martinelli comprehensively reviewed the underlying mechanisms of AML. Dr. Krassimira Todorova and Dr. Soren Hayrabedyan shared expert opinion about miRNA regulation of different genes in prostate cancer. This chapter gives a smart analysis of different proteins, which are deregulated in prostate cancer. Dr. Bayraktar Oznur and Dr. Gozuacik Devrim provided an in-depth analysis of miRNA regulation of autophagy. The last chapter, contributed by Dr. Ilhan Yaylim and her team, provided a conceptual framework of the contributory role of CEACAMs in different cancers.

We would like to offer our sincere gratitude to all the contributing authors. Without their help this book would not have been possible. Finally, we would like to dedicate this book to our three wonderful babies, Burhan, Ibrahim, and Jibran, for their understanding, unconditional love, and sacrifice to enhance our scientific career.

Lahore, Pakistan — Sundas Fayyaz
Islamabad, Pakistan — Ammad Ahmad Farooqi

Contents

Contributors

Rukset Attar Department of Obstetrics and Gynecology, Yeditepe University Hospital, Istanbul, Turkey

Zein Mersini Besharat Department of Experimental Medicine, "Sapienza" University of Rome, Rome, Italy

Aris Raad Besharat Department of Gynecology and Obstetrics, "Sapienza" University of Rome, Rome, Italy

Shahzad Bhatti Department of Human Genetics, University of Health Sciences, Lahore, Pakistan

Ghazala Butt Department of Botany, GCU, Lahore, Pakistan

Carmen Caiazza Department of Molecular Medicine and Medical Biotechnology, University of Naples Federico II, Napoli, Italy

George A. Calin Hematology Section, Azienda Ospedaliero-Universitaria Arcispedale S Anna, Ferrara, Italy

Henrique J. Cardoso CICS-UBI, Health Sciences Research Centre, Faculty of Health Sciences, University of Beira Interior, Covilhã, Portugal

Maria Ciccone Department of Experimental Therapeutics and Leukemia and the Center for RNA Interference and Non-coding RNA, University of Texas M.D. Anderson Cancer Center, Houston, TX, USA

Janaina de Andréa Dernowsek Center for Information Technology Renato Archer (CTI), 3D Technologies Research Group (NT3D), Campinas, SP, Brazil

Gozuacik Devrim Molecular Biology, Genetics and Bioengineering Program, Faculty of Engineering and Natural Sciences, Sabanci University, Istanbul, Turkey

Ravi Doddapaneni Dr. Nasser Al-Rashid Orbital Vision Research Center, Bascom Palmer Eye Institute, Department of Ophthalmology, University of Miami Miller School of Medicine, Miami, FL, USA

Cristina Donfrancesco Department of Experimental Medicine, University of Rome "Sapienza", Rome, Italy

Department of Obstetrics and Gynecology, "San Tommaso dei Battuti" Hospital, Portogruaro (VE), Italy

Sander Dubovy Dr. Nasser Al-Rashid Orbital Vision Research Center, Bascom Palmer Eye Institute, Department of Ophthalmology, University of Miami Miller School of Medicine, Miami, FL, USA

Ammad Ahmad Farooqi Institute of Biomedical and Genetic Engineering (IBGE), Islamabad, Pakistan

Maria Paula Fernandez Department of Ophthalmology, University of Miami School of Medicine, Miami, FL, USA

Elisabetta Ferretti Department of Experimental Medicine, "Sapienza" University of Rome, Rome, Italy

Neuromed Institute, Pozzilli, IS, Italy

Marília I. Figueira CICS-UBI, Health Sciences Research Centre, Faculty of Health Sciences, University of Beira Interior, Covilhã, Portugal

Armando Luis Garcia Dr. Nasser Al-Rashid Orbital Vision Research Center, Bascom Palmer Eye Institute, Department of Ophthalmology, University of Miami Miller School of Medicine, Miami, FL, USA

Maria Luisa Gasparri Department of Gynecology and Obstetrics, "Sapienza" University of Rome, Rome, Italy

Surgical and Medical Department of Translational Medicine, "Sapienza" University of Rome, Rome, Italy

Department of Gynecology and Obstetrics, University Hospital of Bern, Bern, Switzerland

Talha Abdul Halim Department of Obstetrics & Gynaecology, RLMC, Lahore, Pakistan

Mohadeseh Hasanpourghadi Department of Pharmacology, Faculty of Medicine, University of Malaya, Kuala Lumpur, Malaysia

Soren Hayrabedyan Institute of Biology and Immunology of Reproduction "Acad. Kiril Bratanov", Bulgarian Academy of Sciences, Laboratory of Reproductive OMICs Technologies, Sofia, Bulgaria

Catherine Jeeyun Choi Dr. Nasser Al-Rashid Orbital Vision Research Center, Bascom Palmer Eye Institute, Department of Ophthalmology, University of Miami Miller School of Medicine, Miami, FL, USA

Ke Jin The Sheila and David Fuente Graduate Program in Cancer Biology, Sylvester Comprehensive Cancer Center, University of Miami Miller School of Medicine, Miami, FL, USA

Sumbul Khalid Department of Bioinformatics and Biotechnology, International Islamic University, Islamabad, Pakistan

Kayla C. Lewis Department of Cancer Biology, Mayo Clinic, Jacksonville, FL, USA

Yi Chieh Lim Translational Brain Cancer Research, QIMR Berghofer Medical Research Institute, Brisbane City, QLD, Australia

Yi Liu Department of Cancer Biology, Mayo Clinic, Jacksonville, FL, USA

Massimo Mallardo Department of Molecular Medicine and Medical Biotechnology, University of Naples Federico II, Napoli, Italy

Chiara Martinelli Department of Biosciences, University of Milan, Milan, Italy

Sónia A. Melo Instituto de Investigação e Inovação em Saúde, Universidade do Porto, Porto, Portugal

Institute of Pathology and Molecular Immunology of the University of Porto (IPATIMUP), Porto, Portugal

Medical Faculty of the University of Porto (FMUP), Porto, Portugal

Michael David Mueller Department of Gynecology and Obstetrics, University Hospital of Bern, Bern, Switzerland

Mohd. Rais Mustafa Department of Pharmacology, Faculty of Medicine, University of Malaya, Kuala Lumpur, Malaysia

Konstantinos Nirgianakis Department of Gynecology and Obstetrics, University Hospital of Bern, Bern, Switzerland

Bayraktar Oznur Faculty of Medicine, Medical Biology and Genetics Department, Okan University, Istanbul, Turkey

Pierluigi Benedetti Panici Department of Gynecology and Obstetrics, "Sapienza" University of Rome, Rome, Italy

Andrea Papadia Department of Gynecology and Obstetrics, University Hospital of Bern, Bern, Switzerland

Hwan Tae Park Peripheral Neuropathy Research Center, Department of Physiology, College of Medicine, Dong-A University, Busan, South Korea

Daniel Pelaez Dr. Nasser Al-Rashid Orbital Vision Research Center, Bascom Palmer Eye Institute, Department of Ophthalmology, University of Miami Miller School of Medicine, Miami, FL, USA

Department of Biomedical Engineering, University of Miami Miller School of Medicine, Miami, FL, USA

Palmiro Poltronieri CNR-ISPA, Institute of Sciences of Food Productions, National Research Council, Lecce, Italy

Carolina F. Ruivo Instituto de Investigação e Inovação em Saúde, Universidade do Porto, Porto, Portugal

Institute of Pathology and Molecular Immunology of the University of Porto (IPATIMUP), Porto, Portugal

Ilary Ruscito Department of Gynecology and Obstetrics, "Sapienza" University of Rome, Rome, Italy

Sílvia Socorro CICS-UBI, Health Sciences Research Centre, Faculty of Health Sciences, University of Beira Interior, Covilhã, Portugal

Eun Jung Sohn College of Korean Medicine, Kyung Hee University, Seoul, Republic of Korea

Peripheral Neuropathy Research Center, Department of Physiology, College of Medicine, Dong-A University, Busan, South Korea

Wensi Tao Dr. Nasser Al-Rashid Orbital Vision Research Center, Bascom Palmer Eye Institute, Department of Ophthalmology, University of Miami Miller School of Medicine, Miami, FL, USA

Krassimira Todorova Institute of Biology and Immunology of Reproduction "Acad. Kiril Bratanov", Bulgarian Academy of Sciences, Laboratory of Reproductive OMICs Technologies, Sofia, Bulgaria

David Tse Dr. Nasser Al-Rashid Orbital Vision Research Center, Bascom Palmer Eye Institute, Department of Ophthalmology, University of Miami Miller School of Medicine, Miami, FL, USA

Ilhan Yaylim Aziz Sancar Institute of Experimental Medicine, İstanbul University, İstanbul, Turkey

Farrukh Zaman Department of Obstetrics & Gynaecology, RLMC, Lahore, Pakistan

Chapter 1
From Endometriosis to Cancer: Spotlight on Intracellular Signaling Cascades and MicroRNAs

Talha Abdul Halim, Rukset Attar, Cristina Donfrancesco, Ammad Ahmad Farooqi, and Farrukh Zaman

Abstract Increasingly sophisticated information has started to shed light on essential role of signal transduction cascades in endometriosis and how these pathways promote transformations from benign to premalignant endometriosis. It is becoming progressively more understandable that genetic/epigenetic mutations, inactivation of tumor suppressors, aberrant expression of different microRNAs play decisive role in malignant transformation of endometriosis.

Keywords Cancer · Endometriosis · MicroRNA · Therapy

Introduction

Endometriosis represents a benign, chronic gynecological condition, affecting about 6–10% of young fertile women. The prevalence rises to 30–80% in patients with pelvic pain and to 20–50% in those with infertility [1, 2]. The most frequent disease localization is the pelvis, and the most affected organs are ovaries, fallopian tubes, bladder, rectosigmoid colon and myometrium (i.e. adenomyosis). Possible extrapelvic foci of disease can be found in abdomen, appendix, abdominal anterior wall, lungs, urinary tract and nervous system. The classic triad of

T. A. Halim · F. Zaman
Department of Obstetrics & Gynaecology, RLMC, Lahore, Pakistan

R. Attar (✉)
Department of Obstetrics and Gynecology, Yeditepe University Hospital, Istanbul, Turkey
e-mail: ruksetattar@hotmail.com

C. Donfrancesco
Department of Experimental Medicine, University of Rome "Sapienza", Rome, Italy

Department of Obstetrics and Gynecology, "San Tommaso dei Battuti" Hospital, Portogruaro (VE), Italy

A. A. Farooqi
Institute of Biomedical and Genetic Engineering (IBGE), Islamabad, Pakistan

S. Fayyaz, A. A. Farooqi (eds.), *Recent Trends in Cancer Biology: Spotlight on Signaling Cascades and microRNAs*, https://doi.org/10.1007/978-3-319-71553-7_1

symptoms is composed by dysmenorrhea, dyschezia and dyspareunia, but patients may have atypical manifestations or may be completely asymptomatic, making diagnosis difficult [3]. However, the severity of symptoms does not always correlate with the severity of disease [4]. Different authors have emphasized the role of emotional and psychological responses to stressful factors, which may contribute in perception of pain [4].

The definitive diagnosis is done histopathologically which is possible only at the time of surgery. Typical lesions are made of ectopic endometrial glands and stroma, while inflammation and fibrosis are usually present in varying degrees. Nevertheless endometriosis can be considered an inflammatory disease, as the related pain is due to an increase in inflammatory mediators, neuromodulation of the peripheral sensory neurons by estrogens and neurological dysfunction [2].

To date various serum markers have been investigated to help in the initial diagnosis and disease recurrences. Furthermore in the blood and peritoneal cavity of affected patients, an increase in concentration of cytokines and growth factors has been documented. CA125 represents the most widely studied factor to help in diagnosis of first presentation and recurrences of disease, considering that its levels can be high [5]. However, the biomarker has a low specificity in premenopausal women. Its values can be found elevated not only in ovarian cancers, but also in non-ovarian gynecological cancers, in non-gynecological malignancies, in women with pelvic inflammatory disease, in those with myomas, in pregnancy or during menstruation-and in systemic diseases too, as cirrhosis or tuberculosis [6]. Thus, for its low specificity, CA125 is not so helpful in distinguish endometriosis and ovarian cancer [7].

In patients with endometriosis, CA125 peritoneal fluid concentrations have been evaluated in order to express a risk of developing cancer, but we do not have conclusive data. In the study by Mckinnon et al. patients with endometriosis who developed ovarian cancer presented high peritoneal concentrations of another marker, the Human epididymis protein 4 (HE4). This molecule was detected as the best for ovarian cancer identification [8]. When investigated in the serum, sensitivity of HE4 and CA125 was of 65.5% and 58.6%, respectively, while when combining HE4+CA125 the sensitivity reached 68.9% with the same specificity for ovarian cancer detection [9].

It has to be stressed that endometriosis has competency typical of invasive cancer, as angiogenesis, apoptosis, abnormal cell proliferation and, as mentioned above, invasion of distant organs [10, 11] even if it is a benign disease. The most accepted theory for endometriosis origin remains Sampson's theory: endometrial elements, after retrograde menstruation, graft and proliferate at ectopic sites [12, 13]. This ability to fix on a surface with subsequent infiltration, proliferation and vascularization is typical of cancer. Various authors have described an association between endometriosis and ovarian cancer [14, 15].

Patients affected by endometriosis present a 2.5–4 fold increase in the risk of developing ovarian cancer in the next 10 years, especially of the endometrioid or clear cell subtypes. The risk is particularly elevated in women with diagnosis at young age or with over 10 years disease history [16]. It seems to be a real strong association between endometriosis and invasive endometrioid and clear-cell ovarian cancers [17, 18] if we also consider that women affected by these histotypes of cancer present more occurrence of endometriosis respect to women with other histotypes [19].

The diagnosis of ovarian cancer associated to endometriosis is usually made at younger age; cancer tends to be of earlier stage, of lower grade, and prognosis is usually better [20]. In the study by Bonous et al. conducted on 203 patients affected by ovarian cancer, endometriosis-associated cancer incidence is 22.2%. In the same study ovarian cancer related to endometriosis appears to be diagnosed at an earlier stage and to confer a better overall survival, but, when stratified by stage, its advantage in survival disappears [21].

Nowadays, with the advent of molecular biology, many investigations have been focused on the link between ovarian cancer and its precursors - as endometriosis - in order to better understand the signaling pathways which underlie its development and progression.

Signaling Pathways Involved in Endometriosis

Immunohistochemistry assay provided evidence of presence of HMGB1 (High mobility group box-1) in endometrial cells. Furthermore, there was a significant increase in expression of HMGB1 during secretory phase in endometriosis group [22]. Passively released HMGB1 interacted with TLR4 and induced sterile inflammation through a signaling pathway which involved NFκB in human endometrium. HMGB-1 mediated effects were blocked by co-inhibition of TLR4 and NFκB which highlighted possible involvement of NFκB pathway. HMGB1-mediated NFκB signal transduction may be central mechanism during earlier phase of endometrial inflammation in patients with endometriosis [22].

There was a notable loss of PTEN (Phosphatase and tensin homolog) in both endometriosis and invasive tumor tissues [23]. Whereas, ER (estrogen receptor) expression was lost in OCCC relative to endometriosis. Significant overexpression of XRCC5, eEF1A2, PPP1R14B and PTCH2 was noted in OCCC and associated-endometriosis. However, characteristically unique changes in expression levels of genes were not observed in benign endometriosis [23]. Immunohistochemistry confirmed the loss of expression of Wilms' tumour protein 1 (WT1) in OCCC, while significant co-expression of ER and WT1 was reported in endometriotic tissues [23]. Expression of lineage-specific genes in endometriosis but their repression in OCCC pointed towards an epigenetically 'reprogrammable' state that transformed differentiated endometriotic cells into pluripotent state in OCCC [23].

Patients were classified according to different types of endometriosis: ovarian endometrioma (OMA), deep infiltrating endometriosis (DIE) and superficial peritoneal endometriosis (SUP) [24]. Major histocompatibility complex class I homologues MICA and MICB are NKG2D ligands noted to be substantially enhanced in endometriosis. Significantly higher MICA ratio levels were recorded in women suffering from endometriosis in comparison to controls. Similarly, there was a rise in

MICB levels in peritoneal fluid of endometriosis-affected women in comparison to disease-free women. ULBP2, MICA and MICB ratio levels were also noted to be considerably increased in DIE in comparison to controls [24].

HIF-1α was frequently overexpressed in ectopic endometriotic tissues and repressed expression of DUSP2 (dual-specificity phosphatase-2) [25]. DUSP2, a phosphatase is centrally involved in inactivation of mitogen-activated protein kinase (MAPK). Markedly downregulated IL-6 expression was noticed in DUSP2 overexpressing endometrial stromal cells. Hypoxic conditions significantly suppressed DUSP2 levels and simultaneously induced the expression of IL-6 in eutopic endometrial stromal cells. IL-6 enhanced cellular proliferation in eutopic endometrial stromal cells. Hypoxic conditions protected endometrial stromal cells from apoptotic cell death via activation of IL-6/STAT3 signaling cascade [25].

Loss of KLF11 was noted in human endometriotic lesions as compared to eutopic endometrium. Loss of KLF resulted in epigenetically dysregulated target genes [26]. There is evidence of association of Dopamine receptor 2 (DRD2) with vascularity and fibrosis in endometriosis.Klf11/Drd2 cascade played a role in the development and progression of endometriosis. Drd2 expression in lesions from wild-type and Klf11−/− mice was investigated. Endometriotic model was generated by autologously transplanted everted uterine segments on to flank parietal peritoneum that enabled peritoneal exposure of eutopic endometrium. Wild-type models displayed near complete regression of lesions, Klf11−/− models developed progressive disease. Dopamine receptor 2 expression was increased in the region of regressed wild-type lesions compared to Klf11−/− lesions [26].

Correlation between the drop in Nuclear factor erythroid-derived 2-like 2 (NRF2) and Glutamate Cysteine Ligase (GCL) levels and endometriotic lesion growth has recently been reported. Uterine horns from Nrf2−/− mice were implanted into syngeneic animals [27]. Weight and volume of Nrf2−/− ectopic implants were higher significantly. Another murine model of endometriosis was studied for role of Nrf2 induction on endometriosis in vivo using dimethyl-fumarate (DMF). Significant reductions in weight and volume of lesions were noted in DMF-treated mice. Moreover, there was an increase in gene expression of both GCLC and Nrf2 in ectopic implants of DMF-treated mice. It was concluded that Nrf2 induction prevented endometriosis development [27].

Significantly upregulated levels of OCT4 (octamer-binding transcription factor-4) and TGF-β receptor-I (TGFRI) were observed in the high-migratory ectopic endometriotic tissues as compared to hyperplastic or low-migratory normal endometrium [28]. Positive correlations between OCT4 and TGFRI and either OCT4 or TGFRI with migration associated-genes (SLUG, TWIST and SNAIL) were noticed in endometriotic tissues. TGFβI dose-dependently increased OCT4, SNAIL and N-Cadherin. However, OCT4 inhibition drastically impaired TGFβI triggered increase in expression levels of SNAIL and N-Cadherin in endometrial carcinoma cells and endometriotic stromal cells. TGFβI remarkably enhanced migratory capacity of endometriotic cells but OCT4 silencing abolished TGFβI induced cellular migration. Data clearly presented vital role of TGFβ in modulation of different target genes in presence of OCT4 which contributed to ectopic endometrial growth by promoting migration of endometrial cells [28].

Natural Products Mediated Regulation of Proteins Network in Endometriosis

Glycyrrhizin, a triterpene isolated from *Glycyrrhiza glabra* was noted to effectively suppress interleukin (IL-1β), tumor necrosis factor-α (TNF-α), NO and prostaglandin-E2 production in Lipopolysaccharide (LPS) stimulated mouse endometrial epithelial cells (MEEC) [29]. Glycyrrhizin dose-dependently inhibited LPS-induced NF-kB activation and Toll-like receptor 4 (TLR4) expression in MEEC cells [29].

Transformation of Endometriosis into Cancer

In 1988, La Grenade and Silverberg presented a landmark finding and reported a direct association between atypical endometriosis and ovarian cancer [30]. Combined effect of KRAS mutations and p53 loss were studied recently in a transgenic animal model [31]. Recently reported high-impact research highlighted that K-ras mutations and conditionally deleted p53 within the ovarian surface epithelium triggered formation of ovarian lesions which had proliferation potential and endometrioid glandular morphology. Moreover, double mutant ovaries formed high-grade ovarian carcinosarcomas which were poorly differentiated. K-rasG12D effects were also tested in ovarian cancer cells (MCAS, PA-1) and results revealed that K-rasG12D overexpressing cells had significantly higher proliferation, migration and invasive potential [31].

AT-Rich Interactive Domain 1A (ARID1A), a chromatin remodeler was noted to be a key player in the regulation of phenotypic and molecular alterations that potentially contributed to the malignant transformation of endometriotic cells to ovarian clear cell carcinoma (OCCC) [32]. Significant increased anchorage-independent colony formation was noted in ARID1A silenced endometriosis cells. Additionally, adhesive and invasive properties of ARID1A silenced endometriosis cells were also noted to be remarkably enhanced. ARID1A downregulation reconfigured architecture of chromatin through extensively increased H3K27ac (acetylation at the 27th lysine of histone H3) and a modest, context-dependently altered H3K27me3 (Histone 3 lysine 27 trimethylation) at gene promoters. There data clearly suggested that ARID1A inhibition can be a "trigger" for transformation of endometriotic cells into cancerous cells [32].

mRNA expression levels of myostatin and myostatin receptors ALK4 (activin receptor-like kinase 4), ALK5 and activin receptor type-IIB (ActRIIB) were studied in endometrium of healthy women during menstrual cycle, in benign (endometriosis, polyps) and malignant (endometrial adenocarcinoma) conditions [33]. There was a 19-fold increase in ALK5 mRNA in deep infiltrating endometriosis (DIE), whereas 14-fold increase in ActRIIB mRNA was noticed in DIE lesions as compared to control endometrium. Upregulated expression levels of ALK5 and ActRIIB were detected in ovarian endometrioma (OMA). ALK4 and ALK5 levels were markedly upregulated in DIE than in OMA. Expression levels of ALK4, ALK5 and ActRIIB were high in endometrial adenocarcinomas as compared to polyps and control endometrium [33].

MicroRNA Mediated Regulation of Endometrioid Endometrial Carcinoma

Dysregulations of microRNAs have been frequently noted in endometrial endometrioid adenocarcinoma (EEC). qRT-PCR results revealed lower expression level of miR-206 in $ER\alpha^+$ EECs as compared to ERα-negative samples [34]. MiR-206 negatively regulated ERα and affected its downstream target genes in $ER\alpha^+$ EECs cells. miR-206 overexpression time-dependently inhibited growth of $ER\alpha^+$ EEC cells. There was a significant reduction in invasive properties of Ishikawa and RL95-2 cells transfected with miR-206 precursors. miR-206 overexpression significantly reduced levels of matrix metalloproteinases (MMP2, MMP9). miR-206 exerted inhibitory effects on secretion of MMP2 and MMP9, particularly MMP9 in EEC cells [34].

It is noteworthy that in comparison to adjacently located tissues and normal endometrium, miR-199a-3p is significantly downregulated in EEC tissues [35]. miR-199a-3p quantitatively controlled mTOR (mammalian target of rapamycin) by binding to the mTOR-3′ UTR. Data clearly suggested that miR-199a-3p inhibited cellular proliferation by negatively regulating mTOR in EEC cells [35].

MiR-23a directly targeted and downregulated SMAD3 in HEC-1-A cells [36]. Reduction in SMAD3 was noted after treatment with miR-23a agomir however, treatment with a miR-23a antagomir induced an increase in SMAD3 levels in HEC-1-A cells. Furthermore, miR-23a overexpression increased E-cadherin and decreased vimentin and α-smooth muscle actin in HEC-1-A cells [36].

Multiple lines of evidence increasingly linked MALAT1 (metastasis associated lung adenocarcinoma transcript 1) to diverse cancers [37]. miR-200c binding sites have been identified in MALAT1. MiR-200c mimics markedly repressed cellular growth and there was a reduction in the proportion of cells in S-phase and an increment in cellular proportion in G1-phase. Tumor weight and volume were markedly reduced in the mice xenografted with miR-200c expressing RL-952 cells [37].

miRNA-200a, miRNA-200b and miRNA-429 are oncogenic miRNAs which directly targeted PTEN gene in endometrioid endometrial adenocarcinoma [38–40]. MiRNA-370 suppressed cellular proliferation and sensitized endometrioid ovarian cancer cells to cisplatin (CDDP) by directly targeting endoglin (ENG) [41].

Conclusion

Endometriosis and endometriosis-associated malignancies have attracted considerable attention. Recent advancements in technological tools for massively parallel, high-throughput sequencing of DNA have enabled us to comprehensively characterize mutations in different diseases. In a recently published study in New England Journal of Medicine, cancer driver mutations were identified only in the epithelium but not the stroma of the same endometriosis lesions [42]. Therefore, presumably those mutations made endometriotic epithelial cells selectively advantageous! This seemingly effective selective pressure in the epithelial compartment might be helpful in the emergence of distinct clonal populations within the same lesion [42].

A systematic analysis of the data obtained from high-throughput technologies is necessary to help us in developing a better understanding of miRNA-target networks in endometriosis and endometriosis associated malignancies. Using miRNA and mRNA data from the **publically available repository** of cancer genomic and proteomic data (TCGA), miRNA and mRNA networks can be re-interpreted in detail in endometriosis and endometriosis associated malignancies. Databases such as TCGA will prove to be helpful in developing a deeper knowledge related to novel miRNA pathways involved in cancer. CLIP-seq (cross-linked immunoprecipitation followed by next generation sequencing) methodologies such as HITS-CLIP35, iCLIP37 and PAR-CLIP36 have helped us to put missing pieces of the jigsaw-puzzle together for identification of miRNA targets with higher confidence.

Identification of miRNA–mRNA associations is possible through biochemical pulldown assays of specific miRNA and associated mRNA targets, followed by sequencing. Near to complete landscape of the miRNA targetome, that clearly defines the number of tumor suppressors and/or oncogenes targeted by a particular miRNA has yet to be portrayed. The capability of miRNAs to target multiple genes is scientifically appealing, as this feature may be instrumental in the targeting of multiple compensatory pathways. However, we still have outstanding questions related to efficacy of a specific miRNA because a particular miRNA targetome might include both oncogenes and tumor suppressors, as well as a number of targets not involved in cancer.

References

1. Missmer SA, Hankinson SE, Spiegelman D, Barbieri RL, Marshall LM, Hunter DJ. Incidence of laparoscopically confirmed endometriosis by demographic, anthropometric, and lifestyle factors. Am J Epidemiol. 2004;160:784–96.
2. Rafique S, Decherney AH. Medical management of endometriosis. Clin Obstet Gynecol. 2017;60:485. https://doi.org/10.1097/GRF.0000000000000292. [Epub ahead of print]

3. Porpora MG, Koninckx PR, Piazze J, Natili M, Colagrande S, Cosmi EV. Correlation between endometriosis and pelvic pain. J Am Assoc Gynecol Laparosc. 1999;6(4):429–34.
4. Szendei G, Hernádi Z, Dévényi N, Csapó Z. Is there any correlation between stages of endometriosis and severity of chronic pelvic pain? Possibilities of treatment. Gynecol Endocrinol. 2005;21(2):93–100.
5. Cavaggioni G, Lia C, Resta S, Antonielli T, Benedetti Panici P, Megiorni F, Porpora MG. Are mood and anxiety disorders and alexithymia associated with endometriosis? A preliminary study. Biomed Res Int. 2014;22:786830.
6. Low WY, Edelmann RJ. Psychosocial aspects of endometriosis: a review. J Psychosom Obstet Gynaecol. 1991;12(1):3–12.
7. Gupta S, Agarwal A, Sekhon L, Krajcir N, Cocuzza M, Falcone T. Serum and peritoneal abnormalities in endometriosis: potential use as diagnostic markers. Minerva Ginecol. 2006;58(6):527–51.
8. Anastasi E, Manganaro L, Granato T, Benedetti Panici P, Frati L, Porpora MG. Is CA72-4 a useful biomarker in differential diagnosis between ovarian endometrioma and epithelial ovarian cancer? Dis Markers. 2013;35(5):331–5.
9. Bast RC Jr, Klug TL. St. John E. A radioimmunoassay using a monoclonal antibody to monitor the course of epithelial ovarian cancer. N Engl J Med. 1983;309(15):883–7.
10. Mckinnon B, Mueller MD, Nirgianakis K, Bersinger NA. Comparison of ovarian cancer markers in endometriosis favours HE4 over CA125. Mol Med Rep. 2015;12(4):5179–84.
11. Kadija S, Stefanovic A, Jeremik K, Radojevic MM, Nikolic L, Markovic I, Atanackovic J. The utility of human epididymal protein 4, cancer antigen 125, and risk for malignancy algorithm in ovarian cancer and endometriosis. Int J Gynecol Cancer. 2012;22(2):238–44.
12. Chung SY, Kim SJ, Kim TH, Ryu WG, Park SJ, Lee DY, Paik HC, Kim HJ, Cho SH, Kim JK, Park KJ, Ryu YH. Computed tomography findings of pathologically confirmed pulmonary parenchymal endometriosis. J Comput Assist Tomogr. 2005;29:815–8.
13. Fraser IS. Recognising, understanding and managing endometriosis. J Reprod Sci. 2008;1:56–64.
14. Halme J, Hammond M, Hulka J, Raj SG, Talbert LM. Retrograde menstruation in healthy women and in patients with endometriosis. Obstet Gynecol. 1984;64:151–4.
15. Sampson J. Peritoneal endometriosis due to the menstrual dissemination of endometrial tissue into the peritoneal cavity. Am J Obstet Gynecol. 1927;14:422–69.
16. Brinton LA, Gridley G, Persson I, Baron J, Bergqvist A. Cancer risk after a hospital discharge diagnosis of endometriosis. Am J Obstet Gynecol. 1997;176:572–9.
17. Swiersz L. Role of endometriosis in cancer and tumour development. Ann N Y Acad Sci. 2002;955:281–92.
18. Heidemann LN, Hartwell D, Heidemann CH, Jochumsen KM. The relation between endometriosis and ovarian cancer – a review. Acta Obstet Gynecol Scand. 2014;93:20–31.
19. Kobayashi H, Sumimoto K, Moniwa N, Imai M, Takakura K, Kuromaki T, Morioka E, Arisawa K, Terao T. Risk of developing ovarian cancer among women with ovarian endometrioma: a cohort study in Shizuoka, Japan. Int J Gynecol Cancer. 2007;17:37–43.
20. Rossing MA, Cushing-Haugen KL, Wicklund KG, Doherty JA, Weiss NS. Risk of epithelial ovarian cancer in relation to benign ovarian conditions and ovarian surgery. Cancer Causes Control. 2008;19:1357–64.
21. Vigano P, Somigliana E, Parazzini F, Vercellini P. Bias versus causality: interpreting recent evidence of association between endometriosis and ovarian cancer. Fertil Steril. 2007;88:588–93.
22. Guo SW. The association of endometriosis with ovarian cancer: a critical review of epidemiological data, Endometriosis, vol. 22. Berlin: Springer; 2014. p. 363–84.
23. Bonous V, Ferrero A, Fuso L, Ravarino N, Ceccaroni M, Menato G, Biglia N. Endometriosis-associated ovarian cancer: a distinct clinical entity? Anticancer Res. 2016;36(7):3445–9.
24. Yun BH, Chon SJ, Choi YS, Cho S, Lee BS, Seo SK. Pathophysiology of endometriosis: role of high mobility group box-1 and toll-like receptor 4 developing inflammation in endometrium. PLoS One. 2016;11(2):e0148165. https://doi.org/10.1371/journal.pone.0148165

25. Worley MJ Jr, Liu S, Hua Y, Kwok JS, Samuel A, Hou L, Shoni M, Lu S, Sandberg EM, Keryan A, Wu D, Ng SK, Kuo WP, Parra-Herran CE, Tsui SK, Welch W, Crum C, Berkowitz RS, Ng SW. Molecular changes in endometriosis-associated ovarian clear cell carcinoma. Eur J Cancer. 2015;51(13):1831–42. https://doi.org/10.1016/j.ejca.2015.05.011
26. González-Foruria I, Santulli P, Chouzenoux S, Carmona F, Batteux F, Chapron C. Soluble ligands for the NKG2D receptor are released during endometriosis and correlate with disease severity. PLoS One. 2015;10(3):e0119961. https://doi.org/10.1371/journal.pone.0119961
27. Hsiao KY, Chang N, Tsai JL, Lin SC, Tsai SJ, Wu MH. Hypoxia-inhibited DUSP2 expression promotes IL-6/STAT3 signaling in endometriosis. Am J Reprod Immunol. 2017;78. https://doi.org/10.1111/aji.12690.
28. Richards EG, Zheng Y, Shenoy CC, Ainsworth AJ, Delaney AA, Jones TL, Khan Z, Daftary GS. KLF11 is an epigenetic mediator of DRD2/dopaminergic signaling in endometriosis. Reprod Sci. 2017;24(8):1129–38. https://doi.org/10.1177/1933719117698582
29. Marcellin L, Santulli P, Chouzenoux S, Cerles O, Nicco C, Dousset B, Pallardy M, Kerdine-Römer S, Just PA, Chapron C, Batteux F. Alteration of Nrf2 and glutamate cysteine ligase expression contribute to lesions growth and fibrogenesis in ectopic endometriosis. Free Radic Biol Med. 2017;110:1–10. https://doi.org/10.1016/j.freeradbiomed.2017.04.362
30. Au HK, Chang JH, Wu YC, Kuo YC, Chen YH, Lee WC, Chang TS, Lan PC, Kuo HC, Lee KL, Lee MT, Tzeng CR, Huang YH. TGF-βI regulates cell migration through pluripotent transcription factor OCT4 in endometriosis. PLoS One. 2015;10(12):e0145256. https://doi.org/10.1371/journal.pone.0145256
31. Wang XR, Hao HG, Chu L. Glycyrrhizin inhibits LPS-induced inflammatory mediator production in endometrial epithelial cells. Microb Pathog. 2017;109:110–3. https://doi.org/10.1016/j.micpath.2017.05.032
32. LaGrenade A, Silverberg SG. Ovarian tumors associated with atypical endometriosis. Hum Pathol. 1988;19(9):1080–4.
33. Tang FH, Hsieh TH, Hsu CY, Lin HY, Long CY, Cheng KH, Tsai EM. KRAS mutation coupled with p53 loss is sufficient to induce ovarian carcinosarcomas in mice. Int J Cancer. 2017;140(8):1860–9. https://doi.org/10.1002/ijc.30591
34. Lakshminarasimhan R, Andreu-Vieyra C, Lawrenson K, Duymich CE, Gayther SA, Liang G, Jones PA. Down-regulation of ARID1A is sufficient to initiate neoplastic transformation along with epigenetic reprogramming in non-tumorigenic endometriotic cells. Cancer Lett. 2017;401:11–9. https://doi.org/10.1016/j.canlet.2017.04.040
35. Carrarelli P, Funghi L, Ciarmela P, Centini G, Reis FM, Dela Cruz C, Mattei A, Vannuccini S, Petraglia F. Deep infiltrating endometriosis and endometrial adenocarcinoma express high levels of myostatin and its receptors messenger RNAs. Reprod Sci. 2017;24:1577. https://doi.org/10.1177/1933719117698579
36. Chen X, Yan Q, Li S, Zhou L, Yang H, Yang Y, Liu X, Wan X. Expression of the tumor suppressor miR-206 is associated with cellular proliferative inhibition and impairs invasion in ERα-positive endometrioid adenocarcinoma. Cancer Lett. 2012;314(1):41–53. https://doi.org/10.1016/j.canlet.2011.09.014
37. Wu D, Huang HJ, He CN, Wang KY. MicroRNA-199a-3p regulates endometrial cancer cell proliferation by targeting mammalian target of rapamycin (mTOR). Int J Gynecol Cancer. 2013;23(7):1191–7. https://doi.org/10.1097/IGC.0b013e31829ea779
38. Liu P, Wang C, Ma C, Wu Q, Zhang W, Lao G. MicroRNA-23a regulates epithelial-to-mesenchymal transition in endometrial endometrioid adenocarcinoma by targeting SMAD3. Cancer Cell Int. 2016;16(1):67. https://doi.org/10.1186/s12935-016-0342-1
39. Li Q, Zhang C, Chen R, Xiong H, Qiu F, Liu S, Zhang M, Wang F, Wang Y, Zhou X, Xiao G, Wang X, Jiang Q. Disrupting MALAT1/miR-200c sponge decreases invasion and migration in endometrioid endometrial carcinoma. Cancer Lett. 2016;383(1):28–40. https://doi.org/10.1016/j.canlet.2016.09.019
40. Yoneyama K, Ishibashi O, Kawase R, Kurose K, Takeshita T. miR-200a, miR-200b and miR-429 are onco-miRs that target the PTEN gene in endometrioid endometrial carcinoma. Anticancer Res. 2015;35(3):1401–10.

41. Chen XP, Chen YG, Lan JY, Shen ZJ. MicroRNA-370 suppresses proliferation and promotes endometrioid ovarian cancer chemosensitivity to cDDP by negatively regulating ENG. Cancer Lett. 2014;353(2):201–10. https://doi.org/10.1016/j.canlet.2014.07.026.
42. Anglesio MS, Papadopoulos N, Ayhan A, Nazeran TM, Noë M, Horlings HM, Lum A, Jones S, Senz J, Seckin T, Ho J, RC W, Lac V, Ogawa H, Tessier-Cloutier B, Alhassan R, Wang A, Wang Y, Cohen JD, Wong F, Hasanovic A, Orr N, Zhang M, Popoli M, McMahon W, Wood LD, Mattox A, Allaire C, Segars J, Williams C, Tomasetti C, Boyd N, Kinzler KW, Gilks CB, Diaz L, Wang TL, Vogelstein B, Yong PJ, Huntsman DG, Shih IM. Cancer-associated mutations in endometriosis without cancer. N Engl J Med. 2017;376(19):1835–48. https://doi.org/10.1056/NEJMoa1614814

Chapter 2
Tyrosine Kinases, microRNAs, Epigenetics: New Insights in the Mechanisms of Leukemogenesis

Maria Ciccone and George A. Calin

Abstract Haematological malignancies include a broad spectrum of diseases ranging from indolent disorders up to very aggressive leukemias. Recently, numerous studies have contributed to deepen the knowledge of the mechanisms underlying leukemogenesis. Interestingly, different types of leukemias may share the same molecular abnormality (for example, the loss of the *TP53* gene). Conversely, only few haematological malignancies harbour a single and specific aberration (for example the *BCR/ABL* gene fusion in chronic myeloid leukemia). Rather, they represent the final step of a complex transformation process starting from a normal cell that acquires multiple genetic abnormalities because of an intrinsic genetic "frailty" along with stimuli from the cellular microenvironment triggering clonal evolution. Furthermore, either the damage of the genes that are critical in cell growth and death pathways or the disruption of the check-machinery that tunes and supervise the expression of the genome inside the cell (epigenetics), may occur during the clonal evolution. The knowledge of the mechanisms underlying leukemogenesis has addressed the scientific community to test molecules that are able to target specific proteins or genes to verify whether they could replace or integrate the conventional chemotherapy in order to either spare in terms of unneeded toxicity or improve in terms of disease remission and survival. In many cases, the survival improvement and a more acceptable therapy-related toxicity were achieved following the spread of the "target" therapy. This chapter aims to discuss the new insights in the mechanisms of leukemogenesis and their consequences on therapeutic goals.

Keywords MicroRNA · Tyrosine kinase · Hematological malignancies

M. Ciccone (✉)
Department of Experimental Therapeutics and Leukemia and the Center for RNA Interference and Non-coding RNA, University of Texas M.D. Anderson Cancer Center, Houston, TX, USA
e-mail: m.ciccone@ospfe.it

G. A. Calin
Hematology Section, Azienda Ospedaliero-Universitaria Arcispedale S Anna, Ferrara, Italy

S. Fayyaz, A. A. Farooqi (eds.), *Recent Trends in Cancer Biology: Spotlight on Signaling Cascades and microRNAs*, https://doi.org/10.1007/978-3-319-71553-7_2

Introduction

Hematological malignancies include a broad spectrum of diseases ranging from indolent disorders up to very aggressive leukemias [1]. The knowledge of the mechanisms underlying leukemogenesis has contributed to the achievement of several goals. At first, it provided new tools for the diagnostic work up within the group of the hematological malignancies as the last World Health Organization classification has reaffirmed [1]. Second, it improved the prognostication systems that discriminate low risk from intermediate or high risk disease within a specified group of haematological malignancies [2–4]. Third, the possibility to better refine the prognosis has the advantage to modulate the treatment plan according to risk category, with more intensified schedule for patients who present with more aggressive disease while low dose therapies are reserved to low risk disease [5–7]. Forth, it revolutionized the core of anti-leukemic treatment moving from high toxic and low specific chemo-based treatments to less toxic and targeted therapies [8–11]. Nevertheless, the disclosure of the molecular basis of leukemogenesis has raised several questions that future clinical trials and basic science research may eventually solve. Why patients with apparently same genetic features will respond better than others to the same treatment? Which are the primary genetic abnormalities that trigger the multi-step process to leukemia onset? And at which time of life they will appear at first? In fact, the availability of new and more accurate techniques (for example, the next genome sequencing) has allowed the detection of very small clones with genetic abnormalities supposed to foster malignant transformation; and has allowed to detect them at their very early appearance along the life of a subject, and to monitor the clone expansion until the disease become evident ([12, 13]. Furthermore, numerous studies have documented the influence of extrinsic factors, i.e. microenvironment, on the cell which harbour one or more genetic abnormalities that in presence of negative stimuli could unmask its intrinsic potential to transform into malignant cell [14, 15].

Importantly, the pile of details we acquired regarding leukemogenesis is not deprived of few concerns. It has been debated whether the high production costs of the new targeted molecules balance their well proven efficacy. The financial sustainability of these new treatments has significant relevance in those health care systems where the patient has to co-pay for the treatment affecting in particular low income people, or developing countries. Recently, some authors have raised some concerns regarding the adherence and timely initiation of tyrosine kinase inhibitors (TKIs) among patients with chronic myeloid leukemia due to the impossibility to contribute to therapy costs as requested by insurance companies [16, 17]. For doctors and researchers who spend their lives to come out with new weapons to defeat leukemia is deeply disappointing to know there are patients who might not be cured depending exclusively on their low income.

Finally, the possibility to test own "genetic susceptibility" to leukemia as early as possible has some ethical implications. A subject with a "known" susceptibility may commit suicide or give up breeding.

In conclusion the aim of the knowledge of the molecular basis of leukemogenesis should be the improvement in the odds to be cured from leukemia without any significant financial, social or ethic impeachments.

Tyrosine Kinases and Tyrosine Kinase Inhibitors

Tyrosine Kinases: The Lessons from the Ableson and the Bruton's Kinase

Among haematological malignancies, the first disease that was recognized to be strictly related with a specific genetic abnormality was chronic myeloid leukemia (CML) that became a model for the future generation of researchers who aimed to understand the molecular mechanisms underlying leukemia [18]. CML is a proliferative myeloid leukemia characterized by leucocytosis, thrombocytosis, splenomegaly and anemia. The cell of origin is represented by a very immature stem cell that acquires the translocation between the long arms of the chromosome 9 and 22, originating a shorter 22 chromosome, called Philadelphia chromosome (or Ph). The t(9;22) translocation results in the juxtaposition of the oncogene Ableson (*ABL*) at the 9 chromosome with the *BCR* (*break cluster region*) gene located at 22 chromosome, the *BCR-ABL* gene [18, 19]. The BCR-ABL protein works as a tyrosine kinase that is persistently activated thus contributing to the upregulation of cell division in myeloid progenitors [20]. Interestingly, the t(9;22) is not only the hallmark of CML but it can be detected in a form of acute lymphoblastic leukemia (ALL) with a very aggressive behaviour [21]. Although the Ph+ ALL and the CML share the same type of reciprocal translocation, they represent two completely distinct entities with CML being a chronic, indolent myeloproliferative disease while ALL is an acute, aggressive, B-cell derived leukemia [21, 22]. Indeed, the BCR-ABL protein in Ph+ ALL has in most cases a lower molecular weight (p190) than the BCR-ABL product of patients with CML because of a different break point translocation [23]. However the different molecular weight doesn't seem to explain the fact that the same genetic abnormality may be associated with two deeply distinguished diseases. Recently, it has been demonstrated that in patients with Ph+ ALL, in addition with *BCR-ABL* fusion gene, the deletion of the transcription factor Ikaros (IKZF1) at 7p12 contribute to leukemic transformation. Furthermore, the Src-kinase family may be involved in the pathogenesis of Ph+ ALL as it is part of the intracellular signaling pathway under the control of the BCR-ABL protein [24]. So far, we can speculate that rarely an hematological malignancy derives from a single genetic abnormality and a single step process, as in CML. On the contrary, in most cases the disease is the result of a multi step process and multiple genetic abnormalities affecting a single or several cellular pathways suggesting that the inhibition of a single signaling may not be sufficient to leukemic burden control and to cure the disease. The presence of Ph chromosome was initially detected by conventional cytogenetics analysis, showing the presence of a shorter 22 chromosome [19]. However, by time new and more accurate techniques were developed, particularly the polymerase chain reaction (PCR) that allows the detection of the *bcr-abl* transcript at very low level and that for this reason is particularly useful for the monitoring of the disease in patients on treatment [25].

Another interesting model of how the tyrosine kinases are crucial in leukemic transformation is chronic lymphocytic leukemia (CLL) which represent the most common leukemia among adult population in western countries [26]. Patients with CLL may have an aggressive disease requiring treatment early after diagnosis or may display an indolent course with life expectation overlapping persons without the disease and no need of treatment [27]. Numerous studies have contributed to define which are the biological markers that at diagnosis may predict the need of treatment, the likelihood to respond to treatment and therefore the survival [28]. In the last years, Bruton's tyrosine kinase (Btk) emerged as a key-role in the pathogenesis of CLL [8, 29–31]. In B-cells, the BCR (B-cell receptor) engagement by the antigen determines the activation of a cascade involving the BTK tyrosine kinase which ends up with the activation of NF-κB [14, 32, 33]. Similarly, the tyrosine kinase PI3Kδ may participate with the signaling transduction of the BCR. It has been speculated that in the majority of B-cell malignancies the BCR-derived pathway is deregulated as suggested by the demonstration of recurrent genetic abnormalities of the genes (CARD11, CD79B, MYD88) that are involved in BCR signal transduction [34–37]. In diffuse large B-cell lymphoma (DLBCL) the CARD11 mutations are associated with the constitutionally activation of CARD11 protein, the spontaneous and BCR-independent dimerization of CARD11, and the formation of complex resulting in the upregulation of the NF-κB pathway [36]. Similarly, in 90% of patients affected by Waldenstrom's Macroglobulinemia a mutation of MYD88 has been observed [37]. The mutant protein binds with and stabilizes the active form of the BTK, followed by the upregulation of the BCR and the NF-κB pathways [34, 35]. Nevertheless, the BCR activation may derive from the microenvironment as it has been shown in CLL where at the lymph-nodes within "active" areas defined proliferation centers (PCs) a not well-known antigen triggers the B-cell activation through the BCR, promotes B-cell division and genetic instability [38, 39].

In the last years, the data from gene expression profiles have revealed a great number of mutations involving tyrosine kinases, thus suggesting potential targets for disease treatment. Furthermore, the presence of tyrosine kinase recurrent abnormalities in hematological malignancies and the availability of more accurate technique for the detection of specific genetic aberration (like the real-time PCR, RT-PCR) has allowed the spread of the molecular tests to monitor disease response to treatment or disease progression [25, 40, 41]. For example, in patients with CML or Ph+ ALL the efficacy of treatment is measured by means of RT-PCR and the persistence of BCR-ABL transcript at pre-defined time points predicts for shorter survival and disease-free progression [41, 42].

In conclusion, the identification that tyrosine kinases play a key role in hematological malignancies has deepened the knowledge of the biology of the disease; it has contributed to better define the prognostic risk within a same disease category;

it has suggested targetable proteins by new inhibitor molecules; and finally has offered a new tool for disease monitoring.

Tyrosine Kinase Inhibitors

The first target therapy that claimed the attention of scientific and social communities attention all over the world was Imatinib (Gleevec, Novartis) that won the cover of the Time newspaper as the "magic bullet" [43]. Indeed, at that time it represented a true novelty in the scenario of cancer therapies as it brought out that leukemia (and thus cancer) could be treated (and we know now, cured) avoiding chemotherapy and more intense and toxic strategies like bone marrow transplantation. It taught that patients with leukemia may achieve either deep response to treatment or good quality of life. It introduced the concept that a chronic and curative therapy may exist for leukemia as well as for diabetes or arterial hypertension. It fostered the born of a new field of medicine that provides proofs of the beneficial effects of a continuous anti-leukemic treatment over increased expenses derived from a potentially endless medicine intake.

Imatinib works by binding at the ATP active site of BCR-ABL inhibiting the enzyme activity of the protein. The ATP active site may exist in an "opened" (active) or "closed" (inactive) conformation. The imatinib binds at the closed conformation locking it at inactive state and thus blocking the BCR-ABL activity [44]. The occurrence of mutations at the binding site of the imatinib is associated with resistance and disease progression in CML and ALL [45]. The mutations may occur in the course of the disease (secondary mutations) or may be present at the diagnosis (primary) [40, 45]. However, new generations of BCR-ABL TKIs have been developed which differ from imatinib because of an increased affinity for the binding site and the ability to overcome in most cases drug resistance due to mutations [46–48]. Unfortunately, few mutations have been associated with multiple TKIs resistance and very aggressive disease [49].

Importantly, BCR-ABL TKIs have off-target effects, meaning that they may inhibit also other, not leukemia-specific, kinases. The off-target effects of BCR-ABL TKIs account for a part of toxicity that has been observed in patients treated with TKIs [46–48]. Interestingly, the observation that imatinib binds at the catalytic site of other TKs (for example KIT) has suggested its use for the treatment of solid cancers characterized by the upregulation of different but otherwise imatinib susceptible TK [50]. So far, imatinib has been approved for the treatment of CML, GIST (gastrointestinal stromal tumor) and hypereosinophilic syndrome associated with FIP1L1-PDGFRA gene fusion that results from an interstitial chromosome 4q12 deletion and leads to a constitutive activation of the platelet-derived growth factor receptor-alpha (PDGFRA) tyrosine kinase [50, 51].

In the last years, a new TKI has come out to significantly improve disease-progression survival and overall response rates among patients with CLL including elderly and p53 mutated patients who are classically associated with bad prognosis [31]. Ibrutinib is a BTK inhibitor that has been approved by the FDA for the

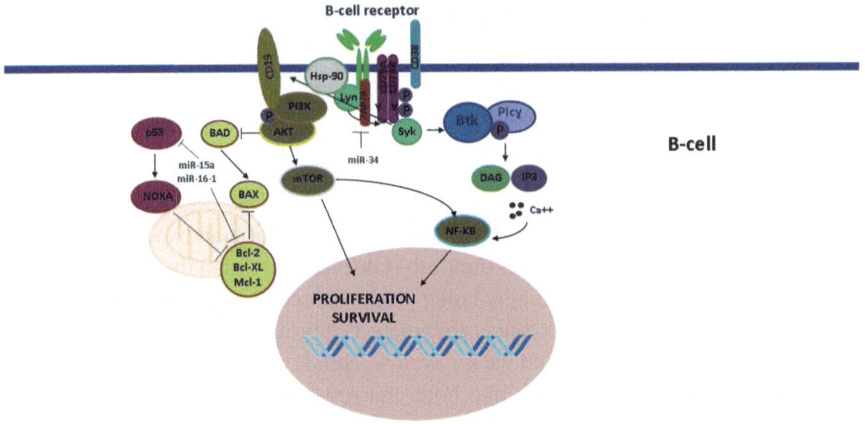

Fig. 2.1 BCR-mediated pathway in B-cell: the engagment of the B-cell receptor on the surface of B lymphocytes triggers intracellular signalling involving the Bruton's kinase and the PI3Kδ kinase that are able to modulate genome trascription and thus cell survival and proliferation. MicroRNAs may be part of this interplay through the repression and/or upregulation of crucial target genes

treatment of CLL as first or following lines of treatment, in monotherapy or in combination with monoclonal antibody [31, 52] (Fig. 2.1). However, ibrutinib received the approval also for the treatment of patients with mantle cell lymphoma (MCL) and Waldenstrom's Macroglobulinemia (WM) [53, 54].

Ibrutinib acts through an irreversible binding to the catalytic domain of the protein [55]. Ibrutinib has significantly improved the outcome of patients with CLL, MCL and WM. For example, in patients with 17p deleted CLL, ibrutinib as single agent has shown 97% and 80% of overall response rates in previously untreated and refractory/relapsed patients, respectively [8, 56]. In refractory/relapsed MCL, 68% and 21% of ORR and complete response (CR) has been observed after treatment with ibrutinib with a progression free survival of 18 months [54]. Finally, in previously treated patients with WM, ibrutinib favorably correlates with longer survival and clinical response [53].

The well known association between mutations at the binding site of BCR-ABL and imatinib resistance in CML, has prompted to search for mutations at the active site of ibrutinib in patients with B-cell malignancies who did not respond adequately to treatment or/and progressed after an initial response. In fact, in CLL as well as in CML the occurrence of mutations ad the drug active site may be responsible for failure or resistance of treatment [45, 57]. Interestingly, treatment failure may derive from mutations that activate pathways downstream the BTK so that the BTK inhibition is annealed or counterbalanced by the up regulation of independent pathways. In patients with CLL, BIRC3 mutations have been associated with activation of the non canonical NF-kB pathway [58]. Similarly, it has been speculated that in patients with WM who carry CXCR4 mutations the poor response to ibrutinib treatment could be related with the BTK-independent activation of the AKT signaling [59]. In other words, single target therapy may not be sufficient to control tumor growth if leukemic cells acquire the ability to up regulate collateral and not targetable pathways.

A further overlapping feature between BCR-ABL inhibitors and BTK inhibitors is the off target effect. After the results of the first clinical trials of ibrutinib were published, it has been observed an increased risk of atrial fibrillation and bleeding in patients treated with ibrutinib [60]. The off-target effect on the TEC kinase may explain the onset of atrial fibrillation [60]. The knowledge of the off-target effects and thus of the possible side effects has suggested how select patients to receive a specific treatment: the safety profile of a target therapy should guide patients who are good candidate to be treated, excluding the ones whose comorbidities may further increase the risk of drug toxicity. For example, a patient with a pre existing severe peripheral arteriopathy or ischemic heart disease should not receive treatment with nilotinib and ponatinib, two new generation BCR-ABL inhibitors.

More recently, encouraging results have been observed in patients with CLL and follicular lymphoma (FL) treated with idelalisib, an inhibitor of the tyrosine kinase PI3Kδ which is associated with the BCR [61] (Fig. 2.1). The promises and the pitfalls of idelalisib are the same encountered for other inhibitors: high clinical responses and survival improvement including heavily pre treated patients and unexpected toxicity, respectively [30, 61]. One may argue that for a potentially chronic treatment as it is for patients with CML or CLL treated with TK inhibitors, the safety profile should be carefully evaluated and minimized in order to avoid the cumulative toxicity over time that may discourage patients and affect his compliance. For this reason, in some cases a reduction of the dose has been successfully evaluated or the co-administration of medications to prevent or attenuate TKIs' toxicity [62, 63].

In conclusion, tyrosine kinases and tyrosine kinases inhibitors are an extraordinary model of how hematological malignancies develop and has opened a new way to treat cancer that underscores the possibility to select a specific treatment for each group of diseases.

Epigenetics in Hematological Malignancies

In addition to genomic aberrations determining gain of function of oncogenes or loss of function of tumor suppressor genes, numerous studies have shown that epigenetic phenomena may play a key role in the pathogenesis of hematological malignancies. Epigenetics include the complex cellular machinery that modulates the gene expression [64]. It is known that only the 3% of DNA corresponds with proteins, while the 97% of DNA has regulatory functions and that there is a significant amount of no codifying RNA. Therefore, either the direct damage of the oncogenes or tumor suppressor genes or the aberrations affecting genes which codify for epigenetic modulators partecipate with the process of tumor progression and clonal evolution.

Histone acetylation and DNA methylation represent two major epigenetic events involved in the pathogenesis of hematological malignancies [65–67]. The histones are proteins that act as spools around which DNA winds and depending on the compactness between DNA and histones, gene transcription may be promoted (if the chromatin is released from the histones) or inhibited (if the chromatin is tight with the histones). The acetylation of the histones lysine neutralizes the positive charge of the histones thus reducing the attraction between histones and chromatin. Therefore, highly acetylated histones are associated with transcription activation; on the contrary histone deacetylation represses gene transcription. Several studies have revealed the aberrant expression of histone deacytilation enzymes (HDAC) which remove the acetyl groups from lysine thus limiting the access of transcription factors upon the DNA [68]. Specifically, the hyperexpression of HDAC combines with the repression of genes involved in DNA repair, cell cycle, immune regulation, angiogenesis, and cell death. In DLBCL and in FL, CREBBP mutations have been observed that cause loss of function of the acetylation activity of the CREBBP protein. As a consequence, TP53 and BCL6 are less acetylated which determine their repression and activation, respectively [69–71].

A further epigenetic mechanism is methylation, i.e. the addition of methyl groups on histones lysine amino acids. The hypermethylation has been demonstrated to cause gene repression [72]. For example, EZH2 acts as methyl transferase and it has been found aberrantly expressed in DLBCL where it participates with the repression of genes involved in controlling B-cell proliferation within the germinal center [73].

As for the TKs, the observation that hematological malignancies are frequently characterized by hypermethylation or deacetylation phenomena has suggested the use of hypomethylating agents or inhibitors of HDAC to treat those malignancies where epigenetic aberration were more evident [10, 72]. Although HDAC inhibitors have provided *in vitro* activity against lymphoma cells, the *in vivo* efficacy is limited and new clinical trials are warranted to prove whether they should be included in treatment schedules for patients with hematological malignancies [72]. On the contrary, hypomethylating agents are routinely applied for the treatment of myelodsyplastic syndromes (MDS) and low blast count acute myeloid leukemia (AML) [10, 74]. Azacytidine and decitabine represent two hypomethylating agents which significantly improve survival in patients with MDS compared with best supportive care or low dose chemotherapy that were the only treatment options other than bone marrow transplantation till the introduction of hypomethylating agents [10]. Interestingly, it has not yet been clarified whether the efficacy of these agents correlates with the de-repression of specific genes or rather if the restoration of the global cell methyalation status is responsible of their clinical efficacy [75].

MicroRNAs and Hematological Malignancies

In several hematological malignancies, epigenetic modifications may have a critical role in clonal transformation and disease progression. MicroRNA deregulation represented the first epigenetic mechanism that has been identified in B-cells from patients with CLL [76]. MicroRNAs (miRNA) are 19–24 nucleotides noncoding RNAs (ncRNA) which regulate the expression of target messenger RNAs (mRNAs) [77]. Mature miRNA are processed from long, capped and polyadenylated precursors which are cleaved in the nucleus, exported to the cytoplasm where duplex miRNA are transformed into single strand miRNA and associated with RISC (RNA-induced silencing complex). By binding with the seed sequence, mainly at the 3′-untraslated region (3′-UTR) of the target messenger, miRNAs inhibit the translation or favor the destabilization of mRNAs depending on the degree of nucleotide pairing [77].

Recently, several studies have elucidated the physiological roles of miRNAs as key regulators of hematopoiesis [76, 78, 79]. The discovery that miRNAs are able to finely tune cell machinery at crucial points suggested the possibility that genetic aberrations at miRNAs gene sequences may contribute to cancer development [80]. Interestingly, miRNAs function, as oncogenes or tumor-suppressors respectively, is strictly dependent on the target genes but also may associate with the specific context of normal and tumor cells. The identification of specific miRNA expression profile among normal and tumor tissues has several diagnostic, prognostic and therapeutic implications. Although preliminary, recent findings support the possibility of using miRNAs expression to predict response to specific treatment or outcome among patients with hematological malignancies [81, 82] Finally, the inhibition of miRNAs by means of antagonist inhibitory molecules in pre-clinical models provided new insights in the understanding of the intricate miRNAs network in the pathogenesis of hematological malignancies and opened a new avenue in their treatment [83].

In patients with AML the aberrant expression of miRNAs, the association of miRNA expression profiles with cytogenetic characteristics and their impact on treatment response and outcome has been proven by different groups [82, 84, 85]. Nevertheless, the pathogenetic mechanisms, i.e. the pathways that miRNAs interfere with in leukemic cells, are in some cases lacking. Importantly, the inclusion of miRNA profile in the prognostication system for patients with AML, might contribute to better distinguish between low and high risk subtypes addressing clinicians in the choice of therapy strategy among several treatment options including standard chemotherapy, bone marrow transplantation and new biological agents. Furthermore, preclinical models and in vitro experiments suggest that targeting miRNAs can increase leukemia cells susceptibility to chemotherapy. For example, the CALGB group analyzed the gene expression signature of 72 patients aged ≥60 years with primary CN-AML harboring FLT3-ITD (FLT3-internal tandem duplications) and treated frontline with intensive chemotherapy [86]. Although the treatment included in some cases investigational drugs, patients with FLT3-ITD had significantly shorter DFS and OS compared with FLT3-WT (FLT3 wild type) (p 0.007 and <0.001, respectively). However, the difference disappeared among patients aged ≥70 years. As expected, FLT3-ITD samples correlated with increased expression of miR-155 and miR-125b-2. Additionally, miR-144, miR-451, miR-488, miR-486-5p were encountered among the most downregulated miRNAs in the same group of patients [86].

Notably, in patients with CLL the deletion of the long arm of chromosome 13 (del13q) occur in 48–50% of previously untreated patients with CLL [2, 87]. The analysis of the minimal deleted region in del(13q) in B-cells from patients with CLL led to the discovery that the microRNAs miR-15a and miR-16-1, encoded by an intron of DLEU2, have a role in the pathogenesis of CLL [76]. Subsequently other microRNAs have been demonstrated to be variably de-regulated in CLL and may interfere with multiple cellular pathways.

Although the prognostication work-up at diagnosis includes FISH (Fluorescense in situ hybridization) assessment, thus allowing for example the detection of del13q, microRNAs level measurements have not been validated prospectively in clinical trials. However, retrospective analysis of blood samples from patients with CLL have shown that microRNA down- or up-regulation affect response to treatment and outcome [81, 88].

In the last decade, several findings have substantially revolutionized the old concept that CLL is a disease originating from mature, not-dividing cells with indolent clinical course [89]. On the contrary, CLL is a disease arising from an early lymphoid progenitor, antigen naïve-B cell, harboring clonal passenger mutations, that would enter the lymph-node where secondary genetic events triggered by the BCR activation favor definitive clonal transformation and expansion of a CD5+ B-cell [14]. Secondly, mutated genes in CLL B-cells cluster in a few pathways (NOTCH1 signaling, mRNA splicing-processing and transport, DNA damage response and innate inflammatory response) that not only affect apoptosis, as it was always believed, but promote B-cell activation, division and clonal expansion [12]. Finally, it has been clearly showed that CLL is a deeply heterogeneous disease within tumor mass and among affected populations, meaning that as previously reported, multiple subpopulations are present at diagnosis and will differently grow along the disease, influencing clinical behaviour with the worst outcome in cases where more unstable and resistant clones will take advantage over the other cells. [12, 90].

In conclusion, several studies have been clearly proved that the levels of specific microRNAs are significantly different in B-cells from patients with CLL compared with B-cells from healthy individuals. Through epigenetic mechanism they can modulate gene expression and interfere with cellular pathways that are involved in cell cycle, apoptosis and BCR activation. Mouse models have confirmed the crucial role of microRNA abnormalities in the pathogenesis and progression of hematological malignancies and allowed the experimentation of small interfering molecules that inhibit miRNA, thus restoring the normal balance within miRNA cellular networks. Prospective clinical studies are desirable for the validation of microRNAs as prognostic and molecular monitoring tool and in order to verify the possible use of small interfering molecules in clinical practice.

Conclusion

Since the introduction of the first TKI to treat patients with CML, many others targeted molecules have been tested and spread in clinical practice (Table 2.1). Although chemotherapy remains the backbone of treatment in several hematological malignancies, the safety and efficacy of the new compounds has radically changed the way to diagnose, to treat and to monitor hematological malignancies. Furthermore in several cases the addition of target therapy has significantly improved the responses to treatment and the survivals. In future, the knowledge of the molecular mechanisms that cause leukemias would offer new hints for the construction and the experimentation of new compounds.

Table 2.1 Targeted therapy in hematology: in the last decades several molecules targeting epigenetic machinery and tyrosine kinases have been introduced in clinical practice

	Target	Disease	Phase of experimentation
Imatinib, nilotinib, dasatinib, ponatinib, bosutinib	ABL (tyrosine kinase)	CML and Ph positive ALL	Phase 2–3
Ibrutinib, acalabrutinib	Bruton's kinase (BTK)	CLL, mcl, Wm	Phase 2–3
Idelalisib	PI3Kδ (phosphoinositide 3-kinase)	CLL, FL	Phase 2–3
Azacitidine, decitabine	DNA methyltransferase	MDS, AML	Phase 2–3
Vorinostat, Romidepsina, Panobinostat	Histone deacethylase	T-NHL, mm	Phase 2–3
EPZ7438, EPZ5676	Histone demethylase	NHL, AML	Preclinical and phase 1–2
microRNA antagonists	microRNAs	AML, CLL	Preclinical

Abbreviations: *CML* chronic myeloid leukemia, *ALL* acute lymphoblastic leukemia, *CLL* chronic lymphocytic leukemia, *WM* Waldenstrom Macroglobulinemia, *FL* follicular lymphoma, *MDS* myelodysplastic syndrome, *AML* acute myeloid leukemia, *T-NHL* T-cell non-Hodgkin Lymphoma, *MM* multiple myeloma

References

1. Campo E, Swrdlow SH, Harris NL, et al. The 2008 WHO classification of lymphoid neoplasms and beyond: evolving concepts and practical applications. Blood. 2011;117:5019–32.
2. Döhner H, Stilgenbauer S, Benner A, et al. Genomic aberrations and survival in chronic lymphocytic leukemia. N Engl J Med. 2000;343:1910–6.
3. Döhner H, Estey EH, Amadori S, et al. Diagnosis and management of acute myeloid leukemia in adults: recommendations from an international expert panel, on behalf of the European LeukemiaNet. Blood. 2010;115:453–74.
4. Solal-Celigny P, Roy P, Colombat P, et al. Follicular lymphoma international prognostic index. Blood. 2004;104:1258–65.
5. Hoppe RT, Advani RH, Ai WZ, et al. Hodgkin lymphoma version 1.2017, NCCN clinical practice guidelines in oncology. J Natl Compr Cancer Netw. 2017;15:608–38.
6. O'Donnell MR, Tallman MS, Abboud CN, et al. Acute myeloid leukemia, version 3.2017, NCCN clinical practice guidelines in oncology. J Natl Compr Cancer Netw. 2017;15:926–57.
7. Wierda WG, Zelenetz AD, Gordon LI, et al. NCCN guidelines insights: chronic lymphocytic leukemia/small lymphocytic leukemia, version 1.2017. J Natl Compr Cancer Netw. 2017;15:293–311.
8. Byrd JC, Brown JR, O'Brien S, et al. Ibrutinib versus ofatumumab in previously treated chronic lymphoid leukemia. N Engl J Med. 2014;371:213–23.
9. Coiffier B, Lepage E, Briere J, et al. CHOP chemotherapy plus rituximab compared with CHOP alone in elderly patients with diffuse large-B-cell lymphoma. N Engl J Med. 2002;346:235–42.
10. Fenaux P, Mufti GJ, Hellstrom-Lindberg E, et al. Efficacy of azacitidine compared with that of conventional care regimens in the treatment of higher-risk myelodysplastic syndromes: a randomised, open-label, phase III study. Lancet Oncol. 2009;10:223–32.
11. Hughes TP, Kaeda J, Branford S, et al. Frequency of major molecular responses to imatinib or interferon alfa plus cytarabine in newly diagnosed chronic myeloid leukemia. N Engl J Med. 2003;349:1423–32.
12. Landau DA, Carter SL, Stojanov P, et al. Evolution and impact of subclonal mutations in chronic lymphocytic leukemia. Cell. 2013;152:714–26.
13. Puente XS, Lopez-Otin C. The evolutionary biography of chronic lymphocytic leukemia. Nat Genet. 2013;45:229–31.

14. Ciccone M, Ferrajoli A, Keating MJ, et al. Snapshot: chronic lymphocitic leukemia. Cancer Cell. 2014;26:770.
15. Gaidano G, Foa R, Dalla-Favera R. Molecular pathogenesis of chronic lymphocytic leukemia. J Clin Invest. 2012;122:3432–8.
16. Ritchie EK, Guerin A, Wolff J, et al. Role of cost-sharing subsidies on the initiation of and adherence to tyrosine kinase inhibitor therapy by Medicare beneficiaries with chronic myeloid leukemia. J Clin Oncol. 2017;35:1744–5.
17. Winn AN, Keating NL, Dusetzina SB. Factors associated with tyrosine kinase inhibitor initiation and adherence among Medicare beneficiaries with chronic myeloid leukemia. J Clin Oncol. 2016;34:4323–8.
18. Hehlmann R, Hochhaus A, Baccarani M, LeukemiaNet E. Chronic myeloid leukaemia. Lancet. 2007;370:342–50.
19. Rowley JD. Letter: a new consistent chromosomal abnormality in chronic myelogenous leukaemia identified by quinacrine fluorescence and Giemsa staining. Nature. 1973;243:290–3.
20. Druker BJ, O'Brien SG, Cortes J, et al. Chronic myelogenous leukemia. Hematology Am Soc Hematol Educ Program. 2002;1:111–35.
21. Rowley JD. The critical role of chromosome transactions in human leukemias. Annu Rev Genet. 1998;32:495–519.
22. Greaves MF, Wiemeis J. Origins of chromosome translocations in childhood leukemia. Nat Rev Cancer. 2003;3:639–49.
23. Soekarman D, van Denderen J, Hoefsloot L, et al. A novel variant of the bcr-abl fusion product in Philadelphia chromosome-positive acute lymphoblastic leukemia. Leukemia. 1990;4:397–403.
24. Mullighan CG, Goorha S, Radtke I, et al. Genome-wide analysis of genetic alterations in acute lymphoblastic leukaemia. Nature. 2007;446:758–76.
25. Baccarani M, Cortes J, Pane F, et al. Chronic myeloid leukemia: an update of concepts and management recommendations of European LeukemiaNet. J Clin Oncol. 2009;27:6041–51.
26. Diehl LF, Karnell LH, Menck HR. The American College of Surgeons Commission on cancer and the American Cancer Society. The National Cancer Data Base report on age, gender, treatment, and outcomes of patients with chronic lymphocytic leukemia. Cancer. 1999;86:2684–92.
27. Solomon BM, Rabe KG, Slager SL, et al. Overall and cancer-specific survival of patients with breast, colon, kidney, and lung cancers with and without chronic lymphocytic leukemia: a SEER population-based study. J Clin Oncol. 2013;31:930–7.
28. Stilgenbauer S. Prognostic markers and standard management of chronic lymphocytic leukemia. Hematology Am Soc Hematol Educ Program. 2015;2015:368–77.
29. Byrd JC, Harrington B, O'Brien S, et al. Acalabrutinib (ACP-196) in relapsed chronic lymphocytic leukemia. N Engl J Med. 2016;374:323–32.
30. Furman RR, Sharman JP, Coutre SE, et al. Idelalisib and rituximab in relapsed chronic lymphocytic leukemia. N Engl J Med. 2014;370:997–1007.
31. O'Brien S, Furman RR, Coutre SE, et al. Ibrutinib as initial therapy for elderly patients with chronic lymphocytic leukaemia or small lymphocytic lymphoma: an open-label, multicentre, phase 1b/2 trial. Lancet Oncol. 2014;15:48–58.
32. Aoki Y, Isselbacher KJ, Pillai S. Bruton tyrosine kinase is tyrosine phosphorylated and activated in pre-B lymphocytes and receptor-ligated B cells. Proc Natl Acad Sci U S A. 1994;91:10606–9.
33. Woyach JA, Johnson AJ, Byrd JC. The B-cell receptor signaling pathway as a thera- peutic target in CLL. Blood. 2012;120:1175–84.
34. Compagno M, Lim WK, Grunn A, et al. Mutations of multiple genes cause deregulation of NF-kappaB in diffuse large B-cell lymphoma. Nature. 2009;459:717–21.
35. Davis RE, Ngo VN, Lenz G, et al. Chronic active B-cell-receptor signalling in diffuse large B-cell lymphoma. Nature. 2010;463:88–92.
36. Lenz G, Davis RE, Ngo VN, et al. (2008) oncogenic CARD11 mutations in human diffuse large B cell lymphoma. Science. 2008;319:1676–9.
37. Treon SP, Xu L, Yang G, et al. MYD88 L265P somatic mutation in Waldenstrom's macroglobulinemia. N Engl J Med. 2012;367:826–33.

38. Chiorazzi N, Ferrarini M. Cellular origin(s) of chronic lymphocytic leukemia: cautionary notes and additional considerations and possibilities. Blood. 2011;117:1781–91.
39. Ciccone M, Agostinelli C, Rigolin GM, et al. Proliferation centers in chronic lymphocytic leukemia: correlation with cytogenetic and clinicobiological features in consecutive patients analyzed on tissue microarrays. Leukemia. 2012;26:499–508.
40. Hughes T, Deininger M, Hochhaus A, et al. Monitoring CML patients responding to treatment with tyrosine kinase inhibitors: review and recommendations for harmonizing current methodology for detecting BCR-ABL transcripts and kinase domain mutations and for expressing results. Blood. 2006;108:28–37.
41. Ravandi F, Jorgensen JL, Thomas DA, et al. Detection of MRD may predict the outcome of patients with Philadelphia chromosome-positive ALL treated with tyrosine kinase inhibitors plus chemotherapy. Blood. 2013;122:1214–21.
42. Jain P, Kantarjian H, Sasaki K, et al. Analysis of 2013 European LeukaemiaNet (ELN) responses in chronic phase CML across four frontline TKI modalities and impact on clinical outcomes. Br J Haematol. 2016;173:114–26.
43. O'Brien SG, Guilhot F, Larson RA, et al. Imatinib compared with interferon and low-dose cytarabine for newly diagnosed chronic-phase chronic myeloid leukemia. N Engl J Med. 2003;348:994–1004.
44. Druker BJ, Talpaz M, Resta DJ, et al. Efficacy and safety of a specific inhibitor of the BCR-ABL tyrosine kinase in chronic myeloid leukemia. N Engl J Med. 2001;344:1031–7.
45. Soverini S, Hochhaus A, Nicolini FE, et al. BCR- ABL kinase domain mutation analysis in chronic myeloid leukemia patients trea- ted with tyrosine kinase inhibitors: recommendations from an expert panel on behalf of European LeukemiaNet. Blood. 2011;118:1208–15.
46. Cortes JE, Kim DW, Kantarjian HM, et al. Bosutinib versus imatinib in newly diagnosed chronic-phase chronic myeloid leukemia: results from the BELA trial. J Clin Oncol. 2012;30:3486–92.
47. Kantarjian H, Shah NP, Hochhaus A, et al. Dasatinib versus imatinib in newly diagnosed chronic-phase chronic myeloid leukemia. N Engl J Med. 2010;362:2260–70.
48. Saglio G, Kim DW, Issaragrisil S, et al. Nilotinib versus imatinib for newly diagnosed chronic myeloid leukemia. N Engl J Med. 2010;362:2251–9.
49. Nicolini FE, Ibrahim AR, Soverini S, et al. The BCR-ABLT315I mutation compromises survival in chronic phase chronic myelogenous leukemia patients resistant to tyrosine kinase inhibitors, in a matched pair analysis. Haematologica. 2013;98:1510–6.
50. Demetri GD, von Mehren M, Blanke CD, et al. Efficacy and safety of imatinib mesylate in advanced gastrointestinal stromal tumors. N Engl J Med. 2002;347:472–80.
51. Baccarani M, Cilloni D, Rondoni M, et al. The efficacy of imatinib mesylate in patients with FIP1L1-PDGFRalpha-positive hypereosinophilic syndrome. Results of a multicenter prospective study. Haematologica. 2007;92:1173–9.
52. Chanan-Khan A, Cramer P, Demirkan F, et al. Ibrutinib combined with bendamustine and rituximab compared with placebo, bendamustine, and rituximab for previously treated chronic lymphocytic leukaemia or small lymphocytic lymphoma (HELIOS): a randomised, double-blind, phase 3 study. Lancet Oncol. 2016;17:200–11.
53. Dimopoulos MA, Trotman J, Tedeschi A, et al. Ibrutinib for patients with rituximab-refractory Waldenström's macroglobulinaemia (iNNOVATE): an open-label substudy of an international, multicentre, phase 3 trial. Lancet Oncol. 2017;18:241–50.
54. Wang ML, Rule S, Martin P, et al. Targeting BTK with ibrutinib in relapsed or refractory mantle-cell lymphoma. N Engl J Med. 2013;369:507–16.
55. Advani RH, Buggy JJ, Sharman JP, et al. Bruton tyrosine kinase inhibitor ibrutinib (PCI-32765) has significant activity in patients with relapsed/refractory B-cell malignancies. J Clin Oncol. 2013;31:88–94.
56. O'Brien S, Jones JA, Coutre SE, et al. Ibrutinib for patients with relapsed or refractory chronic lymphocytic leukaemia with 17p deletion (RESONATE-17): a phase 2, open-label, multicentre study. Lancet Oncol. 2016;17:1409–18.
57. Burger JA, Landau DA, Taylor-Weiner A, et al. Clonal evolution in patients with chronic lymphocytic leukaemia developing resistance to BTK inhibition. Nat Commun. 2016;20(7):11589.

58. Woyach JA. Patterns of resistance to B cell-receptor pathway antagonists in chronic lymphocytic leukemia and strategies for management. Hematology Am Soc Hematol Educ Program. 2015;2015:355–60.
59. Roccaro AM, Sacco A, Jimenez C, et al. C1013G/CXCR4 acts as a driver mutation of tumor progression and modulator of drug resistance in lymphoplasmacytic lymphoma. Blood. 2014;123:4120–31.
60. McMullen JR, Boey EJ, Ooi JY, et al. Ibrutinib increases the risk of atrial fibrillation, potentially through inhibition of cardiac PI3K-Akt signaling. Blood. 2014;124:3829–30.
61. Salles G, Schuster SJ, de Vos S, et al. Efficacy and safety of idelalisib in patients with relapsed, rituximab- and alkylating agent-refractory follicular lymphoma: a subgroup analysis of a phase 2 study. Haematologica. 2017;102:e156–9.
62. Cortes JE, Kim DW, Pinilla-Ibarz J, et al. A phase 2 trial of ponatinib in Philadelphia chromosome-positive leukemias. N Engl J Med. 2013;369:1783–96.
63. Jain P, Kantarjian H, Jabbour E, et al. Ponatinib as first-line treatment for patients with chronic myeloid leukaemia in chronic phase: a phase 2 study. Lancet Haematol. 2015;2:e376–83.
64. Tough DF, Lewis HD, Rioja I, et al. Epigenetic pathway targets for the treatment of disease: accelerating progress in the development of pharmacological tools: IUPHAR review 11. Br J Pharmacol. 2014;171:4981–5010.
65. Okano M, Bell DW, Haber DA, et al. DNA methyltransferases Dnmt3a and Dnmt3b are essential for de novo methylation and mammalian development. Cell. 1999;99:247–57.
66. Pradhan S, Bacolla A, Wells RD, et al. Recombinant human DNA (cytosine-5) methyltransferase. Expression, purification, and comparison of de novo and maintenance methylation. J Biol Chem. 1999;274:33002–10.
67. Taverna SD, Li H, Ruthenburg AJ, et al. How chromatin-binding modules interpret histone modifications: lessons from professional pocket pickers. Nat Struct Mol Biol. 2007;14:1025–40.
68. West AC, Johnstone RW. New and emerging HDAC inhibitors for cancer treatment. J Clin Invest. 2014;124:30–9.
69. Berger SL. The complex language of chromatin regulation during transcription. Nature. 2007;447:407–12.
70. Pasqualucci L, Dominguez-Sola D, Chiarenza A, et al. Inactivating mutations of acetyltransferase genes in B-cell lymphoma. Nature. 2011a;471:189–95.
71. Pasqualucci L, Trifonov V, Fabbri G, et al. Analysis of the coding genome of diffuse large B-cell lymphoma. Nat Genet. 2011b;43:830–7.
72. Miranda TB, Cortez CC, Yoo CB, et al. DZNep is a global histone methylation inhibitor that reactivates developmental genes not silenced by DNA methylation. Mol Cancer Ther. 2009;8:1579–88.
73. Knutson SK, Wigle TJ, Warholic NM, et al. A selective inhibitor of EZH2 blocks H3K27 methylation and kills mutant lymphoma cells. Nat Chem Biol. 2012;8:890–6.
74. Damaj G, Duhamel A, Robin M, et al. Impact of azacitidine before allogeneic stem-cell transplantation for myelodysplastic syndromes: a study by the Société Française de Greffe de Moelle et de Thérapie-Cellulaire and the Groupe-francophone des Myélodysplasies. J Clin Oncol. 2012;30:4533–40.
75. Tobiasson M, Abdulkadir H, Lennartsson A, et al. Comprehensive mapping of the effects of azacitidine on DNA methylation, repressive/permissive histone marks and gene expression in primary cells from patients with MDS and MDS-related disease. Oncotarget. 2017;8:28812–25.
76. Calin GA, Dumitru CD, Shimizu M, et al. Frequent deletions and down-regulation of micro-RNA genes miR15 and miR16 at 13q14 in chronic lymphocytic leukemia. Proc Natl Acad Sci U S A. 2002;99:15524–9.
77. Lund E, Güttinger S, Calado A, et al. Nuclear export of microRNA precursors. Science. 2004;303:95–8.
78. Fazi F, Racanicchi S, Zardo G, et al. Epigenetic silencing of the myelopoiesis regulator microRNA-223 by the AML1/ETO oncoprotein. Cancer Cell. 2007;12:457–66.
79. Larsen MT, Häger M, Glenthøj A, et al. miRNA-130a regulates C/EBP-ε expression during granulopoiesis. Blood. 2014;123:1079–89.

80. Berindan-Neagoe I, Monroig Pdel C, Pasculli B, et al. MicroRNAome genome: a treasure for cancer diagnosis and therapy. CA Cancer J Clin. 2014;64:311–36.
81. Ferrajoli A, Shanafelt TD, Ivan C, et al. Prognostic value of miR-155 in individuls with monoclonal B-cell lumphocytosis and patients with B chronic lymphocytic leukemia. Blood. 2013;122:1891–9.
82. Marcucci G, Maharry KS, Metzeler KH, et al. Clinical role of microRNAs in cytogenetically normal acute myeloid leukemia: miR-155 up regulation independently identifies high-risk patients. J Clin Oncol. 2013;31:2086–93.
83. Velu CS, Chaubey A, Phelan JD, et al. Therapeutic antagonists of microRNAs deplete leukemia initiating cell activity. J Clin Invest. 2014;124:222–36.
84. Garzon R, Volinia S, Liu CG, et al. MicroRNA signatures associated with cytogenetics and prognosis in acute myeloid leukemia. Blood. 2008;111:3183–9.
85. Marcucci G, Radmacher MD, Maharry K, et al. MicroRNA expression in cytogenetically normal acute myeloid leukemia. N Engl J Med. 2008;358:1919–28.
86. Whitman SP, Maharry K, Radmacher MD, et al. FLT3 internal tandem duplication associates with adverse outcome and gene- and microRNA-expression signatures in patients 60 years of age or older with primary cytogenetically normal acute myeloid leukemia: a cancer and leukemia group B study. Blood. 2010;116:3622–6.
87. Zenz T, Döhner H, Stilgenbauer S. Genetics and risk-stratified approach to therapy in chronic lymphocytic leukemia. Best Pract Res Clin Haematol. 2007;20:439–53.
88. Zenz T, Mohr J, Eldering E, et al. miR-34a as part of the resistance network in chronic lymphocytic leukemia. Blood. 2009;113:3801–8.
89. Zenz T, Mertens D, Stilgenbauer S. Biological diversity and risk-adapted treatment of chronic lymphocytic leukemia. Haematologica. 2010;95:1441–3.
90. Landau DA, Tausch E, Taylor-Weiner AN, et al. Mutations driving CLL and their evolution in progression and relapse. Nature. 2015;526:525–30.

Chapter 3
The Emerging Role of Exosomes in Cancer Progression and Their Potential as Therapy Targets

Carolina F. Ruivo and Sónia A. Melo

Abstract Exosomes are a specific population of extracellular vesicles (EVs) that originate from an endocytic process. Virtually every cell type secretes exosomes and their size ranges from 40 to 150 nm. Exosomes are surrounded by a lipid bilayer and contain functional cargo that comprises proteins, lipids and genetic material such as protein, RNA and DNA. In the recent years, several studies have reported the role of exosomes as mediators of intercellular communication. Exosomes serve as vehicles used by cancer cells and stromal cells to influence both local and distant metastatic sites, by reprogramming recipient cells. This chapter will focus on the mechanisms underlying the role of exosomes in tumor development, metastasis, immune escape, therapy resistance, microenvironment reprogramming and angiogenesis. Furthermore, we will also discuss the potential to target exosomes as a new therapeutic strategy in cancer.

Keywords Exosomes · Intercellular-communication · Cancer · Therapy

C. F. Ruivo
Instituto de Investigação e Inovação em Saúde, Universidade do Porto, Porto, Portugal

Institute of Pathology and Molecular Immunology of the University of Porto (IPATIMUP), Porto, Portugal

S. A. Melo (✉)
Instituto de Investigação e Inovação em Saúde, Universidade do Porto, Porto, Portugal

Institute of Pathology and Molecular Immunology of the University of Porto (IPATIMUP), Porto, Portugal

Medical Faculty of the University of Porto (FMUP), Porto, Portugal
e-mail: smelo@ipatimup.pt

S. Fayyaz, A. A. Farooqi (eds.), *Recent Trends in Cancer Biology: Spotlight on Signaling Cascades and microRNAs*, https://doi.org/10.1007/978-3-319-71553-7_3

Exosomes: Structure, Cargo and Origin

Normal and cancer cells release vesicles into the extracellular space. Typically, these vesicles have sizes in the range of hundreds to thousand nanometers and consist of a lipid bilayer that encloses proteins from the cytosol and from organelles, as well as nucleic acids such as RNA and DNA [1, 2]. Besides these common features, evidence shows that extracellular vesicles [3] are not homogeneous in their morphology and molecular content [4, 5]. This diversity is partially due to the fact that released vesicles derive from different subcellular locations. Larger vesicles are classically associated with a plasma-membrane origin [6], while smaller vesicles with a diameter inferior to 150 nm are associated with an endosomal origin [7]. Exosomes fall into this last category of EVs. With sizes ranging from 40 to 150 nm, they are produced within multivesicular bodies (MVB) that upon fusion with the plasma membrane release the exosomes to the extracellular space (Fig. 3.1). EVs heterogeneity is currently a highly discussed topic in the field and effort is being made to develop a systematic classification according to the already well-established features for each vesicle type [10]. In this chapter we discuss studies that respect the most accepted criteria to classify EVs, as it is size range, endosomal-origin associated markers and isolation methods.

Johnstone first observed exosomes in 1987; he was interested in understanding how the transferrin receptor (TfR) was secreted by reticulocytes [11]. By tracing this receptor with electron microscopy (EM), they found that TfR co-localized with nanosized round-shaped structures inside MVBs of endosomal origin [11]. The endocytic pathway comprises the processes of internalization of extracellular components, lipids and membrane proteins [12]. Upon endocytosis, endosomes are formed by the inward folding of the PM resulting in sac-like structures commonly found in the cell cytoplasm [13]. These initial endosomes, also known as early endosomes, collect the endocytosed cargo and are responsible for its sorting. Early endosomes mature into late endosomes and during this maturation process, they form vesicles that bud into their lumen, intraluminal vesicles (ILVs) that correspond to future exosomes (Fig. 3.1). Due to their vesicular anatomy, these endosomes are called MVBs. MVBs can then fuse with lysosomes leading to their content degradation or fuse with the PM in an exocytosis process [13]. While ILVs are produced, they capture proteins and nucleic acids present in the cytoplasm [4], particularly the protein loading is carried by the endosomal sorting complex (ESCRT), whose -0, -I and -II complexes recognize and sequestrate ubiquinated proteins in the endosomal membrane [16], while ESCRT-III is responsible for the inward budding of the plasma membrane [17]. Biogenesis of ILVs can also occur via an ESCRT-independent mechanism, for example through the action of sphingolipid ceramide [18]. Although these sorting mechanisms are not yet fully understood, it is believed that exosomes cargo is somewhat tailored by the cell. Therefore, the cargo of exosomes does not necessarily fully mimics the composition of the donor cell and can be enriched in certain components [19].

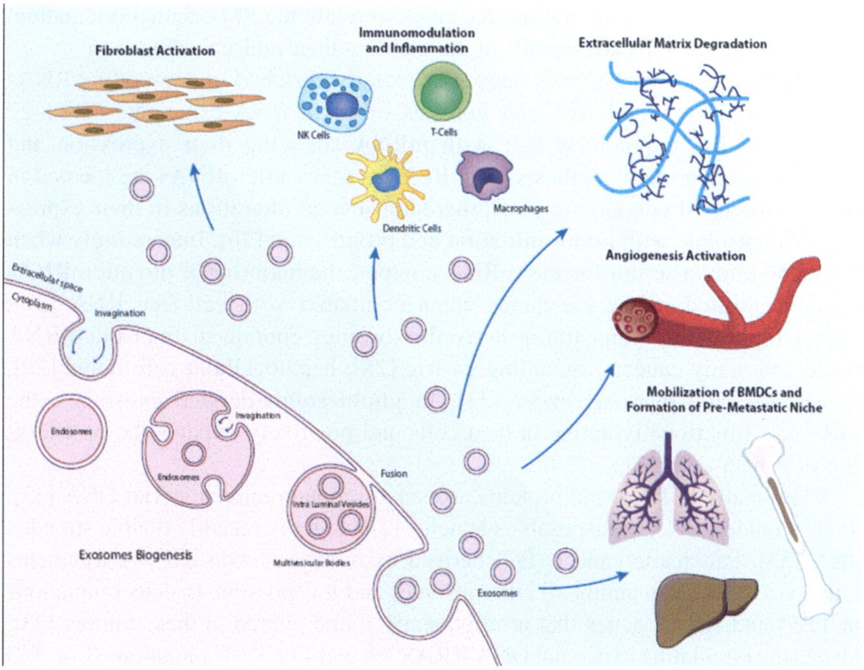

Fig. 3.1 Exosomes-mediated communication in cancer. The biogenesis of exosomes initiates through the invagination of the cellular membrane into an early endosome. After the inward budding of the membrane of the endosome the ILVs are formed. The mature endosome with the ILVs are called MVB then fuses with the plasma membrane and releases its content to the extracellular space, as exosomes. In the extracellular space exosomes interact with nearby and distant cells, playing several roles that ultimately contribute to cancer formation and fuels disease progression. Exosomes are capable of modulating the expression and differentiation of fibroblasts [8, 9], degrading the ECM, and prepare the tissue for the arrival of cancer cells, thus promoting the formation of the pre-metastatic niche. The angiogenic and immunomodulatory features of exosomes have also been recently addressed which furthers implicates exosomes and its effectors in the carcinogenic process

Rab proteins are GTPases involved in intracellular vesicular transport and play an important role in guiding and processing the MVBs. Distinct Rabs act during exosomes biogenesis. Rab11 and Rab35 carry their role in the early endosomes promoting the docking and fusion of MVBs [20]. On the other hand, RAB27A and RAB27B are involved in the release of exosomes to the extracellular space by locating the MVBs close to the PM and managing its docking in order for the fusion to occur [21, 22]. As a result of their endosomal origin, exosomes are characterized by the presence of proteins involved in membrane transport and fusion processes, such as the mentioned Rabs, annexins and flotilins, components of the ESCRT complex, tumor susceptibility gene 101 (TGS101), heat shock proteins (HSP60, HSP70 and HSP90) [23], integrins and tetraspanins including CD81, CD63 and CD9 [7, 24]. Likewise, the double inward membrane invagination that originates the MVBs, allows the membrane-domains and receptors to retain the PM original orientation, meaning these proteins are capable of maintaining their molecular function.

Regarding RNA cargo, exosomes are specially enriched in non-coding RNAs such as microRNAs. MicroRNAs consists of small RNA fragments of 20–22 nucleotides that imprecisely pair with mRNA silencing their expression and inhibiting their protein synthesis [25]. Because many microRNAs are located in cancer-associated genomic regions, there are several alterations in their expression that correlate with tumor initiation and progression [26]. Interestingly when analyzing human serum for microRNA content, the majority of the microRNAs are concentrated in the exosomes when compared with cell-free RNA [27]. Recent reports show that tumor-derived exosomes contain distinct microRNA profiles in many cancers, including gastric [28], hepatocellular carcinoma [29], breast [30], and ovarian cancer [31]. In glioblastoma-derived exosomes, the miR-21 is functionally active in host cells and positively impacts the proliferation of cancer cells [3].

Additionally to RNA and proteins, exosomes contain mitochondrial DNA [32], single stranded DNA, transposable elements [2] and more recently double stranded DNA [33]. Pancreatic cancer (PC) derived exosomes contain dsDNA fragments with a size >10-kb spanning all chromosomes and it is possible to detect mutations on TP53 and KRAS genes that are frequently found altered in these tumors [33]. Also using circulating exosomal DNA $KRAS^{G12D}$ and $TP53^{R273H}$ mutations (Fig. 3.2) are detectable in both PC and precursor lesions. Up until now, the majority of the available studies show that exosomal DNA has great potential to be used as a biomarker, however unlike proteins and RNA, no biological function has been attributed to this cargo. Could exosomal DNA be delivered to a recipient cell and be transcribed? Is exosomal DNA degraded once is delivered? These are a few examples of the many questions that remain to be elucidated.

Exosomes-Mediated Communication Promotes Tumor Development

Cell-to-cell communication is a critical process that mediates homeostasis in a multicellular organism [34, 35]. Many physiological processes rely on a coordinate cellular response achieved by intercellular communication. Synapsis and gap junctions are classical examples of short-distant cellular communication that allow signals to travel fast and efficiently enough to allow a nervous response [36]. When it comes to cancer, it has been established that cancer cells communicate between themselves and with other cells from the tumor microenvironment. Several studies show that cancer cells secrete soluble factors such as chemokines, cytokines and growth factors that modulate immune cells activity and activate fibroblasts [37]. For example, transforming growth factor beta TGF-β secreted by cancer cells mediates a paracrine signaling between breast

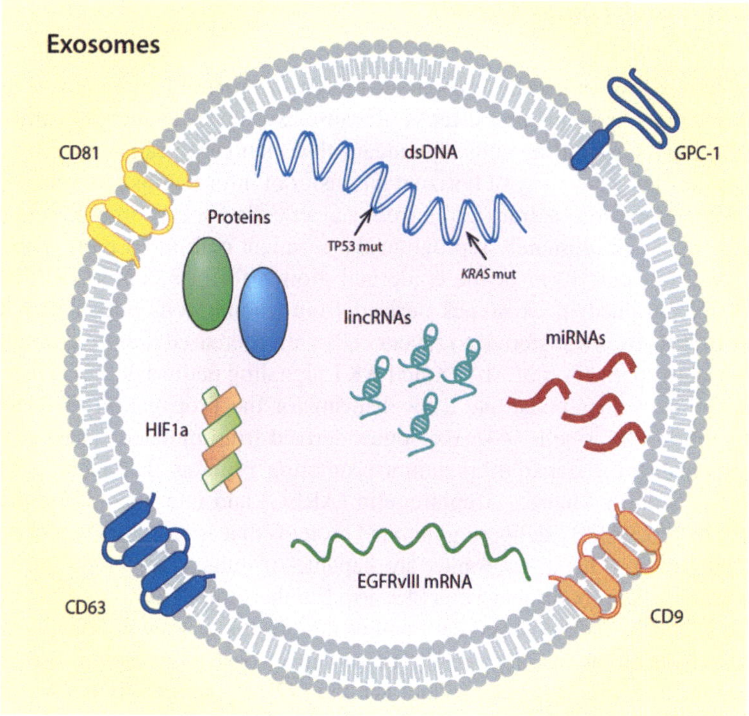

Fig. 3.2 Cargo of cancer exosomes. Reported exosomal content includes both proteins and nucleic acids such as mRNA, microRNA and dsDNA. The surface markers CD63, CD9 and CD81 are known to be broadly expressed in exosomes and therefore are commonly used as exosomal biomarkers. Protein cargo specific for cancer exosomes is being currently explored; HIF1a [14] and GPC-1 [15] are examples of intra and surface proteins respectively, that are being explored as biomarkers

cancer cells and fibroblasts that activates and differentiates fibroblasts into cancer associated-fibroblasts (CAFs), hence promoting tumor progression [38]. There are three main mechanisms proposed for exosomes-mediated intercellular communication: direct interaction of membrane proteins with receptors in the target cell activating intracellular signaling processes, through the cleavage of exosomal membrane proteins by proteases in the extracellular space, resulting in differently sized fragments that may act as ligands that can be internalized by the recipient cell or they can fuse with the target cell membrane and release their contents directly into the cytoplasm [39, 40]. Exosomes are found in every body fluid (blood, saliva, urine, cerebral fluid etc.) [41–43], meaning that these vesicles can travel long distances within the organism and they seem to be sufficiently stable to hold in circulation. This opens the possibility for exosomes to work as vehicles of inter-tissue/organ communication that can support cancer progression.

Exosomes and Tumor Growth

Cancer derived-exosomes have oncogenic proteins as part of their cargo, as well as mRNA and pro-oncogenic microRNAs. Because exosomes can interact with and be internalized by surrounding cells, this means that their cargo can be directly delivered, constituting a new way of horizontal transfer of information between cells [1]. When oncogenic cargo is transferred from cancer cells via exosomes to less active cancer cells, it can ultimately reprogram the recipient cell into a more aggressive state. The oncogenic form of the epidermal growth factor receptor, EGFRvIII is particularly enriched in exosomes derived from glioma cells. Upon their cargo release, EGFRvIII is transferred to cancer cells with decreased tumorigenic activity, resulting in the activation of MAPK and AKT signaling pathways [44]. These cancer derived-exosomes constitute a mechanism for the propagation of oncogenic activity within tumor cells [44]. Exosomes derived from mutated cancer cells are also responsible for transmitting tumor-promoting proteins that include KRAS, EGFR, SRC family kinases, Amphiregulin (AREG) and integrins to non-mutated cancer cells [45, 46]. By delivering mutated proteins that are responsible for driving tumor progression, these exosomes are capable of enhancing three-dimensional cancer growth and promote invasion [45, 46]. But the simple transport of oncogenic material is not the only process that justifies the pro-tumor growth effect of cancer derived-exosomes on recipient cells. Breast cancer derived-exosomes are able to maturate miRNA in a cell independent fashion [47]. By selectively loading RISC-Loading Complex associated proteins such as DICER along with pre-miRNA into cancer exosomes, this allows the maturation of miRNA outside the cell of origin hence transforming pre-miRNA into functional miRNA that upon exosomes delivery silences genes expression on non-transformed cells [47]. DICER is also found associated to CD63, a well described exosomal marker, in colorectal cancer and its presence on exosomes is independent of the cells KRAS status [48]. Such a process occurring in cancer-exosomes points to a poorly explored exosomal feature of cell-independent activity. Are exosomes able to conduct even more complex cellular processes outside the cytoplasm? Exploring this topic can certainly add new knowledge to the biological significance of exosomes in cancer and even in normal physiological processes.

Exosomes and Tumor Microenvironment

A tumor is comprised of many cells and biological components besides the cancer cells themselves. Classically, the malignant mass that we call a tumor has also many non-malignant cells such as fibroblasts, endothelial cells, immune cells and other non-cellular components [49]. Depending on the cancer type and stage, the frequency and type of normal cells will vary [50, 51]. There is an evident interaction between cancer and tumor-microenvironment (TMC) that is based on the exchange of secreted factors such as cytokines, chemokines, growth factors and enzymes [49]. This intercellular communication strongly correlates with the promotion of malignant features like tumor growth, invasion and metastasis [52, 53]. Exosomes have also been pointed as mediators of the communication between cancer cells and TMC. Fibroblasts are commonly the most frequent cellular component of the tumor microenvironment. When associated to cancer, fibroblasts are also known as cancer-associated fibroblasts CAFs and present an activated phenotype similar to what happens in a wound healing process [54]. CAFs are well described for promoting tumor proliferation, production of ECM and for modulating metabolism [54]. Many studies support that CAFs-derived exosomes also have a positive impact on cancer proliferation and migration [1, 8, 47]. Breast cancer-associated fibroblasts secrete CD81-positive exosomes that activate Wnt-PCP pathway in recipient cancer cells through production of Wnt11, resulting in the overexpression of Fzd, Vangl and Dvl, which stimulate protrusive activity and mobility [9]. Exosomes derived from CAFs also play a role in metabolic reprogramming of cancer cells. Prostate and pancreatic cancer patient CAFs-derived exosomes are able to reduce oxygen-dependent metabolism by downregulating mitochondrial oxidative phosphorylation and promoting glycolysis [55]. Additionally, CAFs-derived exosomes contain pro-glycolysis metabolites such lactate, acetate, aminoacids, lipids and intermediates of the Krebs cycle. The recipient cancer cells in nutrient/oxygen-deprived environment continue to proliferate by using these metabolites.

On the other hand, cancer cells are also known to promote the recruitment and activation of fibroblasts. TGF-β is a well-established factor responsible for the differentiation of fibroblasts into CAFs (also known as myofibroblasts) (Fig. 3.2). Cancer-derived exosomes containing TGF-β can bind to type II receptor and ALK5 receptor, forming a complex that results in the phosphorylation of SMAD2 and 3 that can then bind to SMAD4 resulting in its activation. Following SMAD cascade, this protein translocates to the nucleus and activates the expression of several genes including αSMA that is associated with the myofibroblast phenotype [8]. This resulting activated phenotype that is driven by cancer-exosomes cargo is different from the one generated in response to soluble TGF-β [8]. Unlike the soluble form of TGF-β, cancer exosomes that carry this factor also activate the secretion of angiogenic factors by CAFs. Interestingly, downregulation of RAB27A gene, involved in the late stages of exosomes biogenesis, in cancer cells leads to a failure in triggering fibroblasts activation *in vivo*. Altogether, this demonstrates that exosomes play a crucial role on both sides of tumor-microenvironment communication. On one side cancer-derived exosomes promote the activated state of fibroblasts, on the other CAFs-exosomes enhance migration and support proliferation of cancer cells.

Exosomes and Immune Modulation

The immune escape in cancer includes processes that involve antigen masking. Transformation of cells into cancer cells originates a plethora of new antigens, the so-called neoantigens. Since tumors principally arise due to mutations in key oncogenic genes, they will express non-mutated and mutated antigens that will be recognized by the immune system. In order to become "invisible" to the immune selection, cancer cells often present loss of the major histocompatibility MHC complex and silence antigen presenting mechanisms. On the other hand, if the tumor retains antigenicity, it can reduce its immunogenicity instead by expressing immune inhibitor molecules such as PD-L1 and FASL [56].

Exosomes from both tumor and immune cells show expression of MHC class I and class II [57]. Cancer-derived exosomes express functional MHC-Peptides complexes that work similarly to an antigen-presenting cell by directly present or cross present the complexes to CD8+ T-Cells [57]. This process of exosomes-dependent antigen presentation is considered to be anti-cancer because it primes an immune response against the tumor. However, cancer-derived exosomes can also perform an immune inhibitory action (Fig. 3.2). In fact, this dual role can be observed in the response of subsets of T-Cells that react differently to carcinoma-derived exosomes [58]. Resting TRegs treated with tumor derived exosomes increase the expression of CD39 and CD73, while in activated TRegs there is an overexpression of immune-suppressive genes. T-cell $CD39^{+}CD73^{+}$ phenotype is associated with ATP catalysation into AMP that has pro-inflammatory effects and promotes secretion of IL-17 [59]. Additionally, uptake of tumor-derived exosomes in activated $CD4^{+}$ T-cells leads to a decreased expression of immune suppressive genes like COX2, CTLA-4, Fas, Fas Ligant (FASL) and TGF-β. Exosomes from the same tumor-origin potentiate contrary responses when it comes to immune cells. Depending on the recipient cell type, tumor-derived exosomes can potentiate an immune response or suppress an anti-tumor reaction. The suppression of an immune reaction is also achieved by cancer-derived exosomes that express FASL. By binding to FAS, this ligand induces apoptosis in CD8+ T-Cells when they are treated with these exosomes [60]. However, this effect can be reversed when exosomes are treated with antibodies against FASL that block the interaction with T-Cells FAS. Besides lymphocytes, tumors are also classically associated with inflammatory pro-oncogenic environments modulated by myeloid cells like macrophages [61]. Several studies show that macrophages uptake exosomes *ex-vivo* and *in-vivo* [62, 63]. Particularly, cancer-derived exosomes stimulate NF-KB that results in the production of pro-inflammatory cytokines such as IL-6, TNFα, GCSF, and CCL2 [62]. When genetically abrogated in breast cancer cells Toll-like receptor 2 (TLR2), part of NF-KB signaling, cancer-derived exosomes do no longer show this effect. On the other hand, engineered PC-derived exosomes that express miRNA-155 and miR-125b2, are able to invert M2 polarized macrophages into M1 polarization [64]. M1 polarization state corresponds to the activated state that reacts to immune stimuli such as interferon γ, while M2 macrophages promote inflammation by secretion of angiogenesis and fibrosis. This means that cancer-exosomes can themselves be used to alter the immune composition of the tumor microenvironment when it comes to macrophages. Due to the complex role of immune cells in cancer, it is crucial to further address the immunomodulatory properties of cancer derived-exosomes in the context of tumor progression and explore how these apparent contradictory exosomes-mediated effects occur *in vivo*.

Exosomes and Extracellular Matrix (ECM)

ECM is composed by a network of macromolecules such as proteins, collagen and proteoglycans frequently modified with sugar chains that are secreted by stromal and cancer cells. It is well established that ECM not only provides support and mechanic cues but its composition is also an authentic reservoir of growth factors and signaling molecules that have great impact on the tumor growth [65]. Fibronectin (FN) is one of the most frequent ECM constituents and it is released via fibrosarcoma-derived exosomes that promote the adhesion assembly of cancer cells to the ECM. This FN exosomes-mediated release also directs cancer cell mobility, since cells attach their protrusions to FN that is deposited according to integrins orientation that directly interact with the exosomes [66]. This exosomal-mediated process comprehends an autocrine mechanism of directional mobility and it also influences the migration speed of cancer cells. Cancer cells with higher mobility are known to transmit its migratory phenotype to non-motile cells. Uptake of exosomes derived from hepatocellular carcinoma leads to the activation of PI3K/AKT and MAPK pathways that result in the expression of metalloproteinases, particularly MMP-2 and MMP-9 [67]. This group of enzymes is responsible for degrading proteins and collagen part of the ECM. Traditionally, degradation of the ECM promotes migration leading to invasion and ultimately to metastization. Evidence shows that due to their cargo, cancer-derived exosomes are able to directly interplay with ECM via protein-interaction or indirectly by activating the expression of ECM degrading proteins.

Exosomes and Angiogenesis

Tumor growth implies an increased supply of oxygen, nutrients and a constant replacement of extracellular fluid that allows waste excretion. To support cells proliferation and metastization, tumors promote a pro-angiogenic environment leading to the formation of new vessels. Based on chemical factors secreted by cancer cells, angiogenic factors, endothelial cells are recruited to regions were the basement membrane was disrupted, where they proliferate and stabilize [68]. Tumors often present regions of hypoxia and the exosomes released from these hypoxic cancer cells are able to reach endothelial cells [69, 70]. Exosomes derived from hypoxic cancer cells promote endothelial proliferation via cytokines and growth factors that also stimulate pericytes and lead to the activation of PI3K/AKT pathway [71]. Hypoxia seems to have a great influence on exosomes biogenesis and their composition [72]. Cells under hypoxia produce a significant higher number of exosomes [71, 72]. Moreover, under hypoxic conditions, lung cancer exosomes specifically express miR-23a, that is known to target PHD2 [73]. PHD protein family members control HIF1α action. When HIF1α is downregulated in endothelial cells, these present tight junction loss and are more easily recruited [73]. By turning back on HIF1α, it increases endothelial cells proliferation and tube formation. Additionally, in hypoxic conditions, miR-23a besides targeting PHD2 also downregulates ZO-1, a tight junction protein, allowing endothelial barrier to be disrupted hence promoting cancer cells extravasation [73]. Cancer-derived exosomes are considerably enriched in angiogenic friendly-cargo and this particular cargo results from the hypoxic state of their cell of origin.

Exosomes and Metastasis

Metastasis corresponds to the process by which cancer cells leave the primary tumor, reach a secondary location and proliferate originating a new tumor mass [74]. The mechanisms used by cancer cells to gain a migratory phenotype, such as epithelial-to-mesenchymal transition, to survive in circulation and create a metastatic niche are far from being totally explored. In fact, the probability of a tumor cell that enters circulation, a circulating tumor cell (CTC), to actually give rise to a mass is below 0.02% [75]. Therefore, cancer cells should probably have other tricks under their sleeve to help them seed to different organs. The concept of pre-metastatic niche arises from the observation that bone marrow-derived hematopoietic progenitor cells (BMDCs) are recruited previous to the cancer cells arrival to a distant organ [76, 77]. These BMDCs express vascular endothelial factor receptor-1 and VLA-4 while maintaining their progenitor phenotype (CD133+, CD34+ and c-Kit expression). The role of these cells is to create a permissive "soil" for cancer cells to seed [76]. Interestingly, melanoma-derived exosomes re-educate BMDCs and enhance their recruitment through MET signaling. When treating mice with melanoma-derived exosomes it is possible to simulate BMDCs recruitment, which after cancer cells inoculation show to significant metastasis enhancement [6]. The protein cargo of exosomes derived from highly metastatic cells compared with poor metastatic melanoma cells is actually enriched in MET cascade proteins that are responsible for the BMDCs recruitment [6]. Melanoma patient's exosomes are also known to contain a proteomic signature that includes TYRP2, VLA-4, HSP70, HSP90 and MET [6]. The presence of this cargo also correlates with the patients' metastatic disease and tumor burden [6].

Cancer derived-exosomes are responsible for the formation of the liver pre-metastatic niche that is a typical metastasis site for PC [6]. PC-derived exosomes are uptaken by the liver Kupffer cells (KC) and induce TGF-β signaling [78]. This activation leads to ECM remodeling by hepatic stromal cells and deposition of FN that in turn promotes bone marrow derived-macrophages migration to the liver [78]. All of these reprogramming effects construct a favorable environment in the liver that helps cancer cells to proliferate and form a metastasis. PC-exosomes contain MIF, an anti-fibrotic factor that most likely is the molecular effector behind this niche formation. In fact, MIF-positive exosomes bind more frequently to Kupffer cells in the liver, increasing TGF-β expression that then will increase FN production and deposition [78]. FN also works as an anchor for bone marrow-derived macrophages to settle and create an inflammatory reaction that is also advantageous for the cancer cells [78].

Exosomes can also be behind a very peculiar characteristic of tumors, the so called metastatic organotropism. This term refers to the predisposition of cancer cells to metastasize only to certain organs. PC-derived exosomes express integrins that target specific niche cells, ITGαvβ5 is associated with KC cells in the liver, while ITGα6β4 and ITGα6β1 are associated with the uptake of these exosomes by lung resident fibroblasts and epithelial cells [79]. When interacting with the target cells, these exosomes promote a pro-migratory and inflammatory reaction mediated by the S100 gene family overexpression [79]. By injecting specific integrin-positive exosomes it is possible to recapitulate and simulate PC-metastization organotropism, also by using antibodies to block these integrins exosomes uptake is significantly reduced [79]. Altogether, these observations support that exosomes derived from tumors have a predominant role in promoting metastasis, however these observations need a deeper evaluation using better animal cancer-models that recapitulate the human disease.

Exosomes and Therapy Resistance Mechanisms

In cancer, a failed response to therapy can derive from intrinsic or acquired resistance. The first one corresponds to pre-existing factors that unable the drug action, while acquired resistance results from cellular response to the therapy [80]. The cellular mechanisms behind acquired resistance are extremely complex and rely on varied signaling pathways that are not fully understood [81]. The dissemination of therapy resistance due to communication between resistant and sensitive cancer cells is one of the processes that can help explain how a tumor becomes rapidly resistant to a certain drug. Exosomes derived from a breast cancer cell line resistant to docexatel express P-glycoprotein [82, 83]. P-glycoprotein works as a drug efflux pump that allows a cell to reduce the drug intracellular level. Upon exosomes transfer to sensitive cells with low levels of P-glycoprotein, these become resistant to docexatel [83]. Exosomes can efficiently transfer molecular cargo from resistant to sensitive cells thereby transferring resistance to therapy.

Together with cancer cells, TMC particularly fibroblasts also play a role in the overall tumor response to therapy [84]. Paracrine signaling and fibroblasts-derived exosomes communication towards cancer cells constitutes important mechanisms that promote therapy resistance. In response to radiotherapy, breast cancer stroma derived-exosomes transfer large non-coding RNA and transposable elements that activate anti-viral response in cancer cells via RIG-1 receptor and STAT pathway [85]. This results in the expansion of tumor initiating cells (TICs) that are therapy resistant thereby leading to tumor growth. Likewise, CAFs exosomes isolated from colorectal cancer also promote an increased number of cancer stem cells (CSCs) as well as their clonogenicity [86]. Cancer stem cells are known to be intrinsically resistant to therapy and their presence promotes tumor recurrence after treatment [87]. Alongside with the mentioned indirect promotion of therapy resistance via TICs or CSCs proliferation, fibroblasts exosomes also seem to directly support anti-drug response. When PC associated fibroblasts are exposed to therapy (gemcitabine) there is a significant increase in their exosomes production that leads to increase proliferation and survival of PC cells [88]. Upon gemcitabine treatment PC-associated fibroblasts show increased levels of SNAIL1 and microRNA-146a, hence modulating the cancer cells response to therapy [88]. Exosomes mediated-transfer of therapy resistance can constitute an important cellular response mechanism that tumors rely on to rapidly bypass the aggressions provoked by conventional chemo and radiotherapy.

Oncogenic Transformation and Its Impact on Exosomes

Cancer derived-exosomes have oncogenic proteins and mutated genetic material as part of their cargo [3]. Additionally, many studies have shown that this cargo is effectively delivered to neighbor cells and reprogram them, most frequently in favor of the tumor progression. Could this mean that cancer cells have the ability to use exosomes in their advantage and that this was a result of their malignant transformation they suffered? Currently, there is not enough evidence to fully support this hypothesis, however some studies are showing that oncogenic drivers have direct implications on the cargo and biogenesis of exosomes. Upon DNA damage, p53 is activated and promotes or represses the transcription of certain genes. This gene is frequently mutated in many cancers. Non-small cell lung cancer cell lines that are p53 wild type increase exosomes secretion when submitted to γ radiation [89]. On the other hand, p53 mutant cell lines do not show increased exosomes production after radiation treatment [89]. Interestingly, expression of TSAP6 gene allows the cells to produce exosomes after stress, independently of p53 status [89]. By comparing microRNA expression of colorectal cancer cells that express wild type or mutant KRAS, it is observed that KRAS mutant cells show an enrichment of miR-100 and miR-10b [90]. Additionally, KRAS activating mutations of MEK-ERK signaling regulate AGO2 secretion into exosomes by promoting a phosphorylation process that prevents AGO2 interaction with MVB, also affecting the sorting of specific miRNAs such as let-7a, miR-100 and miR-320a into exosomes [48]. There is also increasing evidence that oncogenic transformation affects exosomes uptake. Stimulation of EGFR increases exosomes uptake via macropinocytosis [91]. Also, PC cell lines with activating KRAS mutations have increased macropinocytosis when compared with KRAS wild type cancer cells [91, 92]. If oncogenic transformation deeply alters exosomes processing it is relevant to consider which and how different biological functions can be carried by exosomes in cancer compared to the normal physiological role of these vesicles.

Therapeutic Potential of Exosomes: Targets and Vehicles

Due to their pro-oncogenic properties, cancer derived-exosomes could constitute an important target when considering anti-tumor therapy. At the moment, there are no genes described exclusively dedicated to the biogenesis of exosomes. Nonetheless, several proteins associated with MVB and exosomes release show promising results as targets to block exosomes secretion by cancer cells. Most importantly, when blocking exosomes production many studies show that it is possible to inhibit the pro-oncogenic effects caused by these vesicles. GW4869, a neutral sphingomyelinase inhibitor that acts on ceramide, can effectively block the production of exosomes. Treating PC associated CAFs with GW4869 exosomes production is reduced both in the presence and absence of gemcitabine [88]. Since exosomes derived from PC-CAFs enhance cancer cell proliferation, especially when the CAFs are treated with gemcitabine, cancer cells significantly reduce their survival once CAFs-derived exosomes are blocked [88]. Similarly, exosomes produced by metastatic breast cancer cells lines induce invasion by miR-10b regulation of non-metastatic cells. GW4869 treatment reduces exosomes production in metastatic cells, and minimizes the invasion potential of the non-metastatic recipient cells [93]. RAB proteins family are implicated in MVB processing. RAB27A and B are involved in the late stages of MVB fusion and exosomes release. Melanoma-derived exosomes are able to potentiate the recruitment of BMDCs to metastatic niches [6]. Moreover, these cells show an increased expression of RAB27A that points to an increased exosomes secretion [6]. Using RNA interference to shut down this gene, exosomes release is significantly reduced [6]. This also prevents the recruitment of BMDCs via exosomes [6].

On the other hand, if considering exosomes as specialized vehicles of intercellular communication, they can constitute a way of delivering therapy to cancer or other cells because they are well described for delivering genetic information capable of reprogramming recipient cells. Interference RNA can be electroporated into exosomes and efficiently shutdown gene expression on target cells [94, 95]. Moreover, is also possible to redirect exosomes to certain cell types [96]. Exosomes can be directed to neural cells by using dendritic cells that are engineered to express Lamp2b fused with an RVG peptide that is neuron specific [96]. In this study engineered exosomes were filled with siRNA against GADPH and BACE-1 genes and were delivered *in vivo* via intravenous injection and inhibited the target genes expression by 60%. Studies like this one illustrate how exosomes are biocompatible and stable *in* vivo, and open up the possibility to engineer exosomes to target them to specific cell populations. Epidermal growth factor receptor, EGFR, is frequently overexpressed in many cancers such breast, lung and kidney, and its signaling, which is activated by EGF, promotes cell division and therefore proliferation. A small peptide called GE11 specifically binds to EGFR, but because it is significantly less mitogenic than EGF, GE11 can work as its competitor [97]. Exosomes engineered to express GE11 at their surface is done using a platelet-derived growth factor receptor transmembrane domain fused with the GE11 sequence, hence promoting its expression on the plasma membrane and consequently on exosomes [97]. Using this strategy, GE11 positives exosomes are able to target EGFR expressing cancer cells, both *in vitro* and *in vivo*. Additionally, it is possible to deliver microRNA let-7a in a breast cancer model via GE11 exosomes [97]. Let-7a inside modified exosomes reduces also tumor growth [97].

Recently, it was demonstrate that fibroblasts-derived exosomes can be engineered to target one of the most common mutated genes in PC, KRAS [92]. The activated KRAS gene is a well-known driver of PC, and up until now no therapy system to target KRAS was successfully developed. In this study, exosomes were engineered to carry siRNA against KRASG12D, a mutation that is present in about 80% of PC cases [98]. Curiously, fibroblasts-derived exosomes express CD47 that is a ligand for the signal regulatory protein alpha SIRPα present in macrophages, that upon binding inhibits their destruction, working like a 'don't eat me signal' [99, 100]. Moreover, this activated form of KRAS promotes micropinocytosis that constitutes a process of exosomes uptake, hence enhancing the delivery of these exosomes to PC tumors. *In vivo* studies using genetically engineered mouse models (GEMMs) that spontaneously develop PC driven by KRASG12D mutation, show that the modified exosomes significantly decrease pancreatic tumors by suppressing cell proliferation, enhancing apoptosis and blocking RAS signaling [92]. Altogether, studies like the ones mentioned in this chapter illustrate the potential of exosomes to be used as tailored therapy deliver vehicles with very promising results *in vivo*. Due to their biocompatible and engineering properties, modified exosomes will soon be tested in a clinical setting.

Conclusion

Accumulating evidence points to the pro-oncogenic role of tumor and stroma-derived exosomes. As discussed, these vesicles seem to actively participate in very crucial processes that sustain tumor development such cell proliferation, activation of the microenvironment, immunosuppressive action and formation of the pre-metastatic niche. However, many of these effects are described based on *in vitro* observations, or based on *in vivo* models that do not fully recapitulate the natural behavior of exosomes in an organism. We hope in the future that the field will provide new demonstrations of the role of exosomes in cancer using models that allow to track exosomes secreted from the tumor and/or stroma in new animal models. Moreover, very few studies explore how blocking exosomes could hamper tumor progression; this is also a very crucial aspect that will define how important exosomes are for cancer development. On the other hand, the potential of exosomes as therapy delivery systems is being highly explored with great success. With the advent of more comprehensive and promising studies being published we predict that engineered exosomes can reach clinical studies in a near future.

Acknowledgements This work was supported by: FEDER - Fundo Europeu de Desenvolvimento Regional funds through the COMPETE 2020 - Operacional Programme for Competitiveness and Internationalisation (POCI), Portugal 2020, and by Portuguese funds through FCT - Fundação para a Ciência e a Tecnologia/Ministério da Ciência, Tecnologia e Inovação in the framework of the projects "Institute for Research and Innovation in Health Sciences" (POCI-01-0145-FEDER-007274), and "Papel dos Exosomas na Heterogeneidade Tumoral: Mais do Que Simples Vesículas" (PTDC/BIM-ONC/2754/2014); and Maratonas da Saúde; and by NORTE-01-0145-FEDER-000029, supported by Norte Portugal Regional Programme (NORTE 2020), under the PORTUGAL 2020 Partnership Agreement, through the European Regional Development Fund (ERDF). SAM is supported by FCT – Foundation for Science and Technology (IF/00543/2013). C. F. Ruivo is supported by FCT (PTDC/BIM-ONC/2754/2014). We thank Dr. Nuno Barros for the help with the design of the figures included in this chapter.

References

1. Valadi H, Ekstrom K, Bossios A, Sjostrand M, Lee JJ, Lotvall JO. Exosome-mediated transfer of mRNAs and microRNAs is a novel mechanism of genetic exchange between cells. Nat Cell Biol. 2007;9(6):654–9.
2. Balaj L, Lessard R, Dai L, Cho YJ, Pomeroy SL, Breakefield XO, et al. Tumour microvesicles contain retrotransposon elements and amplified oncogene sequences. Nat Commun. 2011;2:180.
3. Skog J, Wurdinger T, van Rijn S, Meijer DH, Gainche L, Sena-Esteves M, et al. Glioblastoma microvesicles transport RNA and proteins that promote tumour growth and provide diagnostic biomarkers. Nat Cell Biol. 2008;10(12):1470–6.
4. Théry C. Exosomes: secreted vesicles and intercellular communications. F1000 Biol Rep. 2011;3:15.
5. Yanez-Mo M, Siljander PR, Andreu Z, Zavec AB, Borras FE, Buzas EI, et al. Biological properties of extracellular vesicles and their physiological functions. J Extracell Vesicles. 2015;4:27066.
6. Peinado H, Aleckovic M, Lavotshkin S, Matei I, Costa-Silva B, Moreno-Bueno G, et al. Melanoma exosomes educate bone marrow progenitor cells toward a pro-metastatic phenotype through MET. Nat Med. 2012;18(6):883–91.
7. Raposo G, Stoorvogel W. Extracellular vesicles: exosomes, microvesicles, and friends. J Cell Biol. 2013;200(4):373–83.
8. Webber J, Steadman R, Mason MD, Tabi Z, Clayton A. Cancer exosomes trigger fibroblast to myofibroblast differentiation. Cancer Res. 2010;70(23):9621–30.
9. Luga V, Zhang L, Viloria-Petit AM, Ogunjimi AA, Inanlou MR, Chiu E, et al. Exosomes mediate stromal mobilization of autocrine Wnt-PCP signaling in breast cancer cell migration. Cell. 2012;151(7):1542–56.
10. Lotvall J, Hill AF, Hochberg F, Buzas EI, Di Vizio D, Gardiner C, et al. Minimal experimental requirements for definition of extracellular vesicles and their functions: a position statement from the International Society for Extracellular Vesicles. J Extracell Vesicles. 2014;3:26913.
11. Johnstone RM, Adam M, Hammond JR, Orr L, Turbide C. Vesicle formation during reticulocyte maturation. Association of plasma membrane activities with released vesicles (exosomes). J Biol Chem. 1987;262(19):9412–20.
12. Grant BD, Donaldson JG. Pathways and mechanisms of endocytic recycling. Nat Rev Mol Cell Biol. 2009;10(9):597–608.
13. Elkin SR, Lakoduk AM, Schmid SL. Endocytic pathways and endosomal trafficking: a primer. Wien Med Wochenschr. 2016;166(7–8):196–204.
14. Aga M, Bentz GL, Raffa S, Torrisi MR, Kondo S, Wakisaka N, et al. Exosomal HIF1alpha supports invasive potential of nasopharyngeal carcinoma-associated LMP1-positive exosomes. Oncogene. 2014;33(37):4613–22.

15. Melo SA, Luecke LB, Kahlert C, Fernandez AF, Gammon ST, Kaye J, et al. Glypican-1 identifies cancer exosomes and detects early pancreatic cancer. Nature. 2015;523:177.
16. Katzmann DJ, Babst M, Emr SD. Ubiquitin-dependent sorting into the multivesicular body pathway requires the function of a conserved endosomal protein sorting complex, ESCRT-I. Cell. 2001;106(2):145–55.
17. Schuh AL, Audhya A. The ESCRT machinery: from the plasma membrane to endosomes and back again. Crit Rev Biochem Mol Biol. 2014;49(3):242–61.
18. Trajkovic K, Hsu C, Chiantia S, Rajendran L, Wenzel D, Wieland F, et al. Ceramide triggers budding of exosome vesicles into multivesicular endosomes. Science. 2008;319(5867):1244–7.
19. Villarroya-Beltri C, Gutierrez-Vazquez C, Sanchez-Cabo F, Perez-Hernandez D, Vazquez J, Martin-Cofreces N, et al. Sumoylated hnRNPA2B1 controls the sorting of miRNAs into exosomes through binding to specific motifs. Nat Commun. 2013;4:2980.
20. Hsu C, Morohashi Y, Yoshimura S, Manrique-Hoyos N, Jung S, Lauterbach MA, et al. Regulation of exosome secretion by Rab35 and its GTPase-activating proteins TBC1D10A-C. J Cell Biol. 2010;189(2):223–32.
21. Ostrowski M, Carmo NB, Krumeich S, Fanget I, Raposo G, Savina A, et al. Rab27a and Rab27b control different steps of the exosome secretion pathway. Nat Cell Biol. 2010;12(1):19–30.
22. Pfeffer SR. Two Rabs for exosome release. Nat Cell Biol. 2010;12(1):3–4.
23. Lancaster GI, Febbraio MA. Exosome-dependent trafficking of HSP70: a novel secretory pathway for cellular stress proteins. J Biol Chem. 2005;280(24):23349–55.
24. Kowal J, Arras G, Colombo M, Jouve M, Morath JP, Primdal-Bengtson B, et al. Proteomic comparison defines novel markers to characterize heterogeneous populations of extracellular vesicle subtypes. Proc Natl Acad Sci U S A. 2016;113(8):E968–77.
25. Ambros V. The functions of animal microRNAs. Nature. 2004;431(7006):350–5.
26. Calin GA, Croce CM. MicroRNA signatures in human cancers. Nat Rev Cancer. 2006;6(11):857–66.
27. Gallo A, Tandon M, Alevizos I, Illei GG. The majority of MicroRNAs detectable in serum and saliva is concentrated in exosomes. PLoS One. 2012;7(3):e30679.
28. Ohshima K, Inoue K, Fujiwara A, Hatakeyama K, Kanto K, Watanabe Y, et al. Let-7 microRNA family is selectively secreted into the extracellular environment via exosomes in a metastatic gastric cancer cell line. PLoS One. 2010;5(10):e13247.
29. Kogure T, Lin WL, Yan IK, Braconi C, Patel T. Intercellular nanovesicle-mediated microRNA transfer: a mechanism of environmental modulation of hepatocellular cancer cell growth. Hepatology. 2011;54(4):1237–48.
30. Le MT, Hamar P, Guo C, Basar E, Perdigao-Henriques R, Balaj L, et al. miR-200-containing extracellular vesicles promote breast cancer cell metastasis. J Clin Invest. 2014;124(12):5109–28.
31. Taylor DD, Gercel-Taylor C. MicroRNA signatures of tumor-derived exosomes as diagnostic biomarkers of ovarian cancer. Gynecol Oncol. 2008;110(1):13–21.
32. Guescini M, Guidolin D, Vallorani L, Casadei L, Gioacchini AM, Tibollo P, et al. C2C12 myoblasts release micro-vesicles containing mtDNA and proteins involved in signal transduction. Exp Cell Res. 2010;316(12):1977–84.
33. Kahlert C, Melo SA, Protopopov A, Tang J, Seth S, Koch M, et al. Identification of double-stranded genomic DNA spanning all chromosomes with mutated KRAS and p53 DNA in the serum exosomes of patients with pancreatic cancer. J Biol Chem. 2014;289(7):3869–75.
34. Classen L, Tykocinski LO, Wiedmann F, Birr C, Schiller P, Tucher C, et al. Extracellular vesicles mediate intercellular communication: transfer of functionally active microRNAs by microvesicles into phagocytes. Eur J Immunol. 2017;47:1535.
35. Mittelbrunn M, Sánchez-Madrid F. Intercellular communication: diverse structures for exchange of genetic information. Nat Rev Mol Cell Biol. 2012;13(5):328–35.
36. Ahmed KA, Xiang J. Mechanisms of cellular communication through intercellular protein transfer. J Cell Mol Med. 2011;15(7):1458–73.
37. Calvo F, Sahai E. Cell communication networks in cancer invasion. Curr Opin Cell Biol. 2011;23(5):621–9.

38. Kojima Y, Acar A, Eaton EN, Mellody KT, Scheel C, Ben-Porath I, et al. Autocrine TGF-beta and stromal cell-derived factor-1 (SDF-1) signaling drives the evolution of tumor-promoting mammary stromal myofibroblasts. Proc Natl Acad Sci U S A. 2010;107(46):20009–14.
39. Mulcahy LA, Pink RC, Carter DRF. Routes and mechanisms of extracellular vesicle uptake. J Extracell Vesicles. 2014;3. https://doi.org/10.3402/jev.v3.24641.
40. Tian T, Zhu YL, Hu FH, Wang YY, Huang NP, Xiao ZD. Dynamics of exosome internalization and trafficking. J Cell Physiol. 2013;228(7):1487–95.
41. Pan J, Ding M, Xu K, Yang C, Mao LJ. Exosomes in diagnosis and therapy of prostate cancer. Oncotarget. 2017. 10.18632/oncotarget.18532.
42. Yan S, Han B, Gao S, Wang X, Wang Z, Wang F, et al. Exosome-encapsulated microRNAs as circulating biomarkers for colorectal cancer. Oncotarget. 2017;8:60149.
43. Giusti I, Di Francesco M, Dolo V. Extracellular vesicles in glioblastoma: role in biological processes and in therapeutic applications. Curr Cancer Drug Targets. 2017;17(3):221–35.
44. Al-Nedawi K, Meehan B, Micallef J, Lhotak V, May L, Guha A, et al. Intercellular transfer of the oncogenic receptor EGFRvIII by microvesicles derived from tumour cells. Nat Cell Biol. 2008;10(5):619–24.
45. Demory Beckler M, Higginbotham JN, Franklin JL, Ham AJ, Halvey PJ, Imasuen IE, et al. Proteomic analysis of exosomes from mutant KRAS colon cancer cells identifies intercellular transfer of mutant KRAS. Mol Cell Proteomics. 2013;12(2):343–55.
46. Higginbotham JN, Demory Beckler M, Gephart JD, Franklin JL, Bogatcheva G, Kremers GJ, et al. Amphiregulin exosomes increase cancer cell invasion. Curr Biol. 2011;21(9):779–86.
47. Melo Sonia A, Sugimoto H, O'Connell Joyce T, Kato N, Villanueva A, Vidal A, et al. Cancer exosomes perform cell-independent microrna biogenesis and promote tumorigenesis. Cancer Cell. 2014;26(5):707–21.
48. McKenzie AJ, Hoshino D, Hong NH, Cha DJ, Franklin JL, Coffey RJ, et al. KRAS-MEK signaling controls Ago2 sorting into exosomes. Cell Rep. 2016;15(5):978–87.
49. Balkwill FR, Capasso M, Hagemann T. The tumor microenvironment at a glance. J Cell Sci. 2012;125(23):5591–6.
50. Bindea G, Mlecnik B, Tosolini M, Kirilovsky A, Waldner M, Obenauf Anna C, et al. Spatiotemporal dynamics of intratumoral immune cells reveal the immune landscape in human cancer. Immunity. 2013;39(4):782–95.
51. Place AE, Jin Huh S, Polyak K. The microenvironment in breast cancer progression: biology and implications for treatment. Breast Cancer Res. 2011;13(6):227.
52. Condeelis J, Pollard JW. Macrophages: obligate partners for tumor cell migration, invasion, and metastasis. Cell. 2006;124(2):263–6.
53. Kessenbrock K, Plaks V, Werb Z. Matrix metalloproteinases: regulators of the tumor microenvironment. Cell. 2010;141(1):52–67.
54. Kalluri R. The biology and function of fibroblasts in cancer. Nat Rev Cancer. 2016;16(9):582–98.
55. Zhao H, Yang L, Baddour J, Achreja A, Bernard V, Moss T, et al. Tumor microenvironment derived exosomes pleiotropically modulate cancer cell metabolism. elife. 2016;5:e10250.
56. Kearney CJ, Lalaoui N, Freeman AJ, Ramsbottom KM, Silke J, Oliaro J. PD-L1 and IAPs co-operate to protect tumors from cytotoxic lymphocyte-derived TNF. Cell Death Differ. 2017;24:1705.
57. Raposo G, Nijman HW, Stoorvogel W, Liejendekker R, Harding CV, Melief CJ, et al. B lymphocytes secrete antigen-presenting vesicles. J Exp Med. 1996;183:1161–72.
58. Muller L, Mitsuhashi M, Simms P, Gooding WE, Whiteside TL. Tumor-derived exosomes regulate expression of immune function-related genes in human T cell subsets. Sci Rep. 2016;6:20254.
59. Wang LL, Tang HP, Shi GC, Wan HY, Tang W, Hou XX, et al. CD39/CD73 and the imbalance of Th17 cells and regulatory T cells in allergic asthma. Mol Med Rep. 2013;8(5):1432–8.
60. Abusamra AJ, Zhong Z, Zheng X, Li M, Ichim TE, Chin JL, et al. Tumor exosomes expressing Fas ligand mediate CD8+ T-cell apoptosis. Blood Cell Mol Dis. 2005;35(2):169–73.
61. Coussens LM, Werb Z. Inflammation and cancer. Nature. 2002;420(6917):860–7.

62. Chow A, Zhou W, Liu L, Fong MY, Champer J, Van Haute D, et al. Macrophage immunomodulation by breast cancer-derived exosomes requires Toll-like receptor 2-mediated activation of NF-κB. Sci Rep. 2014;4:5750.
63. Su MJ, Aldawsari H, Amiji M. Pancreatic cancer cell exosome-mediated macrophage reprogramming and the role of MicroRNAs 155 and 125b2 transfection using nanoparticle delivery systems. Sci Rep. 2016;6:30110.
64. Momen-Heravi F, Bala S, Bukong T, Szabo G. Exosome-mediated delivery of functionally active miRNA-155 inhibitor to macrophages. Nanomedicine. 2014;10(7):1517–27.
65. Venning FA, Wullkopf L, Erler JT. Targeting ECM disrupts cancer progression. Front Oncol. 2015;5:224.
66. Sung BH, Ketova T, Hoshino D, Zijlstra A, Weaver AM. Directional cell movement through tissues is controlled by exosome secretion. Nat Commun. 2015;6:7164.
67. He M, Qin H, Poon TCW, Sze S-C, Ding X, Co NN, et al. Hepatocellular carcinoma-derived exosomes promote motility of immortalized hepatocyte through transfer of oncogenic proteins and RNAs. Carcinogenesis. 2015;36(9):1008–18.
68. Nishida N, Yano H, Nishida T, Kamura T, Kojiro M. Angiogenesis in cancer. Vasc Health Risk Manag. 2006;2(3):213–9.
69. Huang Z, Feng Y. Exosomes derived from hypoxic colorectal cancer cells promote angiogenesis through Wnt4-induced beta-catenin signaling in endothelial cells. Oncol Res. 2017;25(5):651–61.
70. Hsu DH, Paz P, Villaflor G, Rivas A, Mehta-Damani A, Angevin E, et al. Exosomes as a tumor vaccine: enhancing potency through direct loading of antigenic peptides. J Immunother. 2003;26:440.
71. Kucharzewska P, Christianson HC, Welch JE, Svensson KJ, Fredlund E, Ringner M, et al. Exosomes reflect the hypoxic status of glioma cells and mediate hypoxia-dependent activation of vascular cells during tumor development. Proc Natl Acad Sci U S A. 2013;110(18):7312–7.
72. King HW, Michael MZ, Gleadle JM. Hypoxic enhancement of exosome release by breast cancer cells. BMC Cancer. 2012;12(1):421.
73. Hsu YL, Hung JY, Chang WA, Lin YS, Pan YC, Tsai PH, et al. Hypoxic lung cancer-secreted exosomal miR-23a increased angiogenesis and vascular permeability by targeting prolyl hydroxylase and tight junction protein ZO-1. Oncogene. 2017;36:4929.
74. Lambert AW, Pattabiraman DR, Weinberg RA. Emerging biological principles of metastasis. Cell. 2017;168(4):670–91.
75. Valastyan S, Weinberg RA. Tumor metastasis: molecular insights and evolving paradigms. Cell. 2011;147(2):275–92.
76. Peinado H, Zhang H, Matei IR, Costa-Silva B, Hoshino A, Rodrigues G, et al. Pre-metastatic niches: organ-specific homes for metastases. Nat Rev Cancer. 2017;17(5):302–17.
77. Paget S. The distribution of secondary growths in cancer of the breast. 1889. Cancer Metastasis Rev. 1989;8(2):98–101.
78. Costa-Silva B, Aiello NM, Ocean AJ, Singh S, Zhang H, Thakur BK, et al. Pancreatic cancer exosomes initiate pre-metastatic niche formation in the liver. Nat Cell Biol. 2015;17(6):816–26.
79. Hoshino A, Costa-Silva B, Shen T-L, Rodrigues G, Hashimoto A, Tesic Mark M, et al. Tumour exosome integrins determine organotropic metastasis. Nature. 2015;527(7578):329–35.
80. Lippert TH, Ruoff HJ, Volm M. Intrinsic and acquired drug resistance in malignant tumors. The main reason for therapeutic failure. Arzneimittelforschung. 2008;58(6):261–4.
81. Longley DB, Johnston PG. Molecular mechanisms of drug resistance. J Pathol. 2005;205(2):275–92.
82. Lv MM, Zhu XY, Chen WX, Zhong SL, Hu Q, Ma TF, et al. Exosomes mediate drug resistance transfer in MCF-7 breast cancer cells and a probable mechanism is delivery of P-glycoprotein. Tumour Biol. 2014;35(11):10773–9.
83. Bebawy M, Combes V, Lee E, Jaiswal R, Gong J, Bonhoure A, et al. Membrane microparticles mediate transfer of P-glycoprotein to drug sensitive cancer cells. Leukemia. 2009;23(9):1643–9.

84. McMillin DW, Negri JM, Mitsiades CS. The role of tumour-stromal interactions in modifying drug response: challenges and opportunities. Nat Rev Drug Discov. 2013;12(3):217–28.
85. Boelens MC, Wu TJ, Nabet BY, Xu B, Qiu Y, Yoon T, et al. Exosome transfer from stromal to breast cancer cells regulates therapy resistance pathways. Cell. 2014;159(3):499–513.
86. Hu Y, Yan C, Mu L, Huang K, Li X, Tao D, et al. Fibroblast-derived exosomes contribute to chemoresistance through priming cancer stem cells in colorectal cancer. PLoS One. 2015;10(5):e0125625.
87. Maitland NJ, Collins AT. Cancer stem cells – a therapeutic target? Curr Opin Mol Ther. 2010;12(6):662–73.
88. Richards KE, Zeleniak AE, Fishel ML, Wu J, Littlepage LE, Hill R. Cancer-associated fibroblast exosomes regulate survival and proliferation of pancreatic cancer cells. Oncogene. 2017;36(13):1770–8.
89. Yu X, Harris SL, Levine AJ. The regulation of exosome secretion: a novel function of the p53 protein. Cancer Res. 2006;66(9):4795–801.
90. Cha DJ, Franklin JL, Dou Y, Liu Q, Higginbotham JN, Demory Beckler M, et al. KRAS-dependent sorting of miRNA to exosomes. elife. 2015;4:e07197.
91. Nakase I, Kobayashi NB, Takatani-Nakase T, Yoshida T. Active macropinocytosis induction by stimulation of epidermal growth factor receptor and oncogenic Ras expression potentiates cellular uptake efficacy of exosomes. Sci Rep. 2015;5:10300.
92. Kamerkar S, LeBleu VS, Sugimoto H, Yang S, Ruivo CF, Melo SA, et al. Exosomes facilitate therapeutic targeting of oncogenic KRAS in pancreatic cancer. Nature. 2017;546(7659):498–503.
93. Singh R, Pochampally R, Watabe K, Lu Z, Mo Y-Y. Exosome-mediated transfer of miR-10b promotes cell invasion in breast cancer. Mol Cancer. 2014;13(1):256.
94. Wahlgren J, Statello L, Skogberg G, Telemo E, Valadi H. Delivery of small interfering RNAs to cells via exosomes. Methods Mol Biol. 2016;1364:105–25.
95. El-Andaloussi S, Lee Y, Lakhal-Littleton S, Li J, Seow Y, Gardiner C, et al. Exosome-mediated delivery of siRNA in vitro and in vivo. Nat Protoc. 2012;7(12):2112–26.
96. Alvarez-Erviti L, Seow Y, Yin H, Betts C, Lakhal S, Wood MJA. Delivery of siRNA to the mouse brain by systemic injection of targeted exosomes. Nat Biotech. 2011;29(4):341–5.
97. Ohno S, Takanashi M, Sudo K, Ueda S, Ishikawa A, Matsuyama N, et al. Systemically injected exosomes targeted to EGFR deliver antitumor microRNA to breast cancer cells. Mol Ther. 2013;21(1):185–91.
98. Miglio U, Oldani A, Mezzapelle R, Veggiani C, Paganotti A, Garavoglia M, et al. KRAS mutational analysis in ductal adenocarcinoma of the pancreas and its clinical significance. Pathol Res Pract. 2014;210(5):307–11.
99. Matlung HL, Szilagyi K, Barclay NA, van den Berg TK. The CD47-SIRPalpha signaling axis as an innate immune checkpoint in cancer. Immunol Rev. 2017;276(1):145–64.
100. Weiskopf K. Cancer immunotherapy targeting the CD47/SIRPalpha axis. Eur J Cancer. 2017;76:100–9.

Chapter 4
Tranquilizing and Awakening ATM to Promote Killing of Cancer Cells

Yi Chieh Lim, Shahzad Bhatti, and Ammad Ahmad Farooqi

Abstract Maintenance of genomic integrity is a major challenge, as DNA is exposed to incessantly ongoing nucleolytic attacks from both exogenous and endogenous sources. To overcome these stumbling blocks, cells have evolved a global DNA damage response (DDR), which is an intricate and hierarchically organized network of interweaved pathways that are "switched on" whenever genotoxic insults occurs. ATM, ATR, and DNA-PK are multitalented modulators of DNA damage signaling which play a significant role in synchronizing and orchestrating an array of proteins at the site of DNA damage. Overwhelmingly, genomic and proteomic data have helped us to map the landscape of ATM mediated regulation of myriad of proteins in normal and cancer cells. Complex information has shown to "diametrically opposed" roles of ATM kinase in different cancers. Scientists have investigated effects of ATM activation and inhibition in different cancers and it is now clear that context-dependent activation or inhibition can consequently improve apoptotic rate of cancer cells. In this chapter we will summarize the most recent advancements of ATM kinase in different cancers and critically evaluated the effects of ATM activation or inhibition on apoptosis and drug resistance of cancer cells.

Keywords ATM Kinase · DNA damage Signaling · Apoptosis, Cancer

Y. C. Lim
Translational Brain Cancer Research, QIMR Berghofer Medical Research Institute, Brisbane City, QLD, Australia

S. Bhatti (✉)
Department of Human Genetics, University of Health Sciences, Lahore, Pakistan
e-mail: drshahzadbhatti@yahoo.com

A. A. Farooqi
Institute of Biomedical and Genetic Engineering (IBGE), Islamabad, Pakistan

S. Fayyaz, A. A. Farooqi (eds.), *Recent Trends in Cancer Biology: Spotlight on Signaling Cascades and microRNAs*, https://doi.org/10.1007/978-3-319-71553-7_4

Introduction

The genome is a storage powerhouse where essential information required for the development, function and survival of all living organisms are kept. While genetic preservation is critical for cellular homeostasis and propagation, it is often under constant threats. Insults can originate exogenously from genotoxic agents, including different sources of radiation. Endogenously, it can also emerge from aberrant cellular metabolisms. To mitigate cellular risk posed by DNA lesions, a complex network of molecular circuitries is required to swiftly recognise and remove all deleterious insults. Cell–cycle progression, replication, repair and chromosomal modification are part of this orchestrated global signalling platform that is collectively known as the DNA damage response (DDR) cascade. In the event of a fatal damage where recovery is irreversible, execution of self–destruction via programmed death or cessation in growth termed cellular senescence must be initiated to avoid malignant development. To some extent, the fate of a cell following DNA damage is dependent on the ability to resolve harmful lesions where "self–sacrifice" is the last resort.

The Effect of DNA Double Strand Break: A Summarised View to the Hierarchical Signalling Cascade

Conceptually, surveillance of structural abnormalities is the first concerted action by DNA sensor proteins to safeguard the genome [1]. Transducer and mediator proteins are subsequently recruited to the site of damage to rely, amplify and sustain pathway signals while allowing effector proteins to initiate the desired biological response [2, 3]. Across the different forms of genotoxic lesions, double strand breaks (DSBs) are considered lethal and can occur spontaneously ranging from chromosome missegregation, replication fork collapse and ionizing radiation (IR). Early study indicates a single DSB lesion is sufficient to render cell death. Its efficacy is still regarded to date as a major cancer treatment modality. Mainly because one gray (1 Gy) of IR can deliver approximately 20–40 DSBs to a single cell with other accompanying forms of damage [4]. Apart from its efficiency in killing cells, DNA DSBs also have defined cellular functions. For instance, re–programming of the genome requires a well–controlled DSBs for V(D)J recombination and class–switch recombination [5]. The generation of a non–exhaustive repertoire of immunoglobulins and receptors is vital for developing B– and T–lymphocytes. Similarly, programmed DSB is essential for meiosis to trigger a genome–wide recombination between two adjoining chromatids follow by systematic segregation for the first cell division [6].

The primary core of DDR is regulated by specific members of the phosphatidylinositol 3–kinase–related kinase (PIKKs) family. Each member has a distinct role but all share considerable sequence homology which includes a PIKK domain near the carboxyl termini that is flanked by FRAP–ATM–TRRAP (FAT) and C–terminal FAT (FATC) domain [7]. At current, there are six members of the PIKKs family [8, 9]. Of which, only three have been extensively characterised with explicit roles in DNA damage [10]. Ataxia–telangiectasia mutated (ATM) is a master regulator of the DDR signalling cascade that involves in the overall resolution of DNA DSBs. The protein exists predominantly as dimers or oligomers in undamaged cells. Inactive ATM is converted to phosphorylated monomers by the presence of DNA damage. DNA–dependent protein kinase catalytic subunit (DNA–PK_{cs}) also responds predominately to lethal strand breakages but is tasked with the activation of non–homologous end joining (NHEJ) repair pathway. Ataxia–telangiectasia and rad3–related protein (ATR) is primarily recruited to the site of single strand (ss)DNA lesions via ATR–interacting protein (ATRIP) which occurs during the cell–cycle phase when DNA replicates. Unresolved damage poses a risk in fork stall, leading to the formation of DNA DSBs.

Sensing DNA Double Strand Break: The Detection Process

With the exception of some, sensor proteins often participate in the specific detection of selective lesions. Poly(ADP–ribose) (PAR) polymerase–1 (PARP–1) represents the former and can physically interact with a range of DNA structures including pyrimidine dimers, abasic (AP) sites, single strand breaks (SSBs) and DSBs [11]. PARP–1 binding affinity to the DNA is dependent on the action of zinc fingers (F1 and F2) [12]. These motifs are located at the N–terminus region of the gene and are critical for detecting DNA lesions. PARP–1 also has a paralog termed PARP–2 which has a binding preference towards defective DNA with gaps and flap structures [13]. On the other end of the spectrum, Ku heterodimer (Ku70–Ku80) is an abundant nuclear protein complex that also couples to duplex DNA ends [14]. The two subunits form an asymmetric ring with an expansive base and a narrow bridge to translocate along the DNA. Ku together with DNA–PK_{cs} recruit XRCC4 and DNA ligase IV (X4–L4) to initiate end–joining process [15]. In the context of DSBs, Ku and PARP are direct competitors [16]. Each has the intention of activating their own respective pathway through initial binding of damage DNA. PARP catalyzes the addition of PAR units onto the flanking DNA ends and acts cooperatively to facilitate Mre11, RAD50 and NBS1 (MRN) complex recruitment [17]. Knock–out study indicates PARP is dispensable for MRN recruitment and does not affect overall DDR activation. Instead, PARP merely acts as a "catalyst" to accelerate MRN to the site of damage. Collectively, the MRN complex is a heterotrimerically assembled nano-machinery that is vital for DDR signalling. Patient–derived cells with defect in Mre11, RAD50 or NBS1 genes show significant radiosensitivity with characteristic of chromosomal instability. Mre11 contains DNA–binding motifs which are needed to establish short resected 3′ single–strand tails. The globular–arranged ATPase (walker A and B motif) domain of RAD50 binds and partially unwinds dsDNA termini. Mre11–RAD50 interaction is dependent on NBS1 acting as an adaptor protein. The forkhead–associated (FHA) and BRCA1 C terminus (BRCT) domains of NBS1 are essential for proteins recruitment. Most importantly, NBS1 C–terminal motif is vital for the physical interaction with ATM, including its activation.

Signalling in DNA Double Strand Break: The Initiation Process of Signal Transduction and Amplification

The site of DNA damage in the nuclear termed ionizing radiation–induced foci (IRIF) is dependent on a systematic hierarchical recruitment of different DDR factors, including chromatin modifiers to restore genome integrity. MRN complex senses and tethers to the DNA ends which is followed by the autophosphorylation of ATM at serine 1981 [18]. The addition of DNA–damage induced chromatin response is also a key element to ensure a robust DDR. Tyrosine kinase C–Abl activates Tip60 acetyltransferase which in turn methylates adjacent histone H3 lysine–9 (H3k9me3) of the DSB, leading to the configuration of an open–relaxed chromatin structure [19]. The action of Tip60–H3K9me3 then promotes ATM acetylation and is facilitated by forkhead box O3 (FOXO3a) interaction with ATM [20, 21]. There is evidence to suggest transmembrane receptor NOTCH1 can act antagonistically to FOXO3a by binding the FATC domain of ATM [20]. In the presence of NOTCH1, ATM is still recruited to DSBs but lacks Tip60–mediated acetylation to initiate consequent activation. The major substrate of active ATM at the IRIF is histone H2AX and phosphorylation of serine 139 yields γH2AX. So far, the FHA and BRCT domains as mentioned are frequently found in proteins that are associated with DDR [22]. The FHA recognizes phosphorylated threonine residues in an amino acid sequence–specific context whereas the tandem BRCT domains create a sequence–specific binding of phosphorylated peptides. Several lines of evidence including crystallography data reveals BRCT repeats of MDC1 build the recognition module for γH2AX which is tailored to recognize its C–terminus region [23]. Interaction between these two molecules are generally accepted as the initiation process of DNA DSB where cascade signalling begins. Within the IRIF, casein kinase 2 (CK2) phosphorylation of MDC1 at the ser–asp–thr (SDT) repeats mediates NBS1 interaction via its FHA and BRCT domains. Analysis of MDC1–depleted cells reveal its FHA domain also binds and concentrates active ATM at the site of damage [24]. Most importantly, the interaction of MDC1 is vital to create a positive feedback loop for phosphorylated ATM accumulations which in turn activate distant H2AX molecules and create available domains of γH2AX that extend for kilobases along the chromatin away from the DSB.

With the continuity of γH2AX and MDC1 interaction, the latter also promotes the recruitment of E3 ubiquitin–protein ligase RNF8 [25, 26]. An important observation based on sequence analysis elutes RNF8 localisation to the

IRIF is dependent on its FHA domain interaction with ATM-phosphorylated Thr-Gln-Xaa-Phe (TQXF) motifs of MDC1 whereas its RING domain is important for the translocation of 53BP1 and RAP80 to initiate subsequent DNA DSB repair [25, 26]. Chromatin–bound RNF8 cooperates with E2 conjugating enzyme UBC13 by adding lysine (K)63–linked ubiquitin chains to histone H1. Unlike K48–linked ubiquitin chains that lead to protein degradation, formation of K63–linked chains act as a non–proteolytic platform to recruit downstream DDR factors. Similar to RNF8, RNF168 also works in concert to extend the K63–linked chains by ubiquitinating histone H2A on K13–K15 (H2AK13Ub and H2AK15Ub) [27]. 53BP1 binds specifically to H2AK15Ub through its ubiquitination–dependent recruitment motif (UDR). To retain its presence at the IRIF, an interaction between Tudor domain of 53BP1 and histone H4 lysine 20 dimethylation (H4K20me2) is required [28]. In addition to 53BP1 recruitment, RAP80 also preferentially translocates to the K63–linked ubiquitin chains. The binding interaction is dependent on RAP80 N–terminal ubiquitin–interacting motif (UIM) [29]. Currently, associated members (Abraxas, BRCC36, BRCC45 and MERIT40) are required for the overall stability of RAP80 complex at the IRIF and essential for the recruitment of BRCA1 [30]. The antagonism between 53BP1 and BRCA1 is important for the choice of DNA DSB during cell–cycle progression and consequent cell survival [31].

Assisting in the Response of DNA Double Strand Break: Halting the Cell–Cycle Process

The cell–cycle is separated into four distinct (G_1–, S–, G_2– and M–) and two gap (transit through G_1– into S– and G_2– into M–) phases where cells are screened for aberration (damage) in the genome prior to DNA synthesis (S–phase) and cell division (M–phase). To regulate cell–cycle arrest, two downstream effector kinase proteins namely checkpoint proteins 1 and 2 (CHK1 and CHK2) are required. Activated checkpoint proteins target and degrade the cyclin/cyclin–dependent kinases (CDKs) complexes that are essential for cell–cycle progression. As the slowing or arrest occurs between the phases of the cell–cycle, additional DDR–related proteins initiated by transducer proteins can establish the recovery of damaged DNA.

The majority of cell population resides in G_1–phase (resting) of the cell–cycle. In preparation of cellular division, cells will begin to transit from G_1– to S–phase for DNA replication. The process is dependent on CDK4/6–Cyclin D and CDK2–Cyclin E complexes to promote S–phase entry. There are two distinct mechanisms that can initiate arrest at G_1/S phase if DNA damage is present. Phosphorylation of ATM–CHK2 axis triggers P53 activation by degrading its negative regulators (MDM2 and MDMX). P53 accumulation transcriptionally upregulates P21 to interact with CDK2–Cyclin E and CDK4/6–Cyclin D complexes. Due to the need for the translation of nascent protein and subsequent post–translational modifications (PTM), complete activation requires several hours. The second option also involves an ATM–dependent phosphorylation of CHK2 except the process is rapid. Activated ATM–CHK2 axis initiates the ubiquitination of CDC25A, which prevents the interaction of CDK2 and Cyclin–E. This mechanistic of action halts replicon firing. It is believed extensive DNA damage during the early phases of G_1 can promote cells to undergo permanent arrest or G_0 through P53 activation. If DNA damage occurs during late G_1–phase, it is likely ATM–CHK2 activation would not prevent cell–cycle arrest, thus allowing cells to bypass the G_1/S checkpoint entirely.

Protecting the genome integrity during S–phase is vital to ensure an unaltered DNA is inherited by daughter cells. It is not surprising that S–phase of the cell–cycle contains multiple checkpoints. Phosphorylation of CDC25A requires ATR–CHK1 activation. The continuity of CDC25A ubiquitination limits its accumulation and is necessary for cells to progress through S–phase. In return, DNA replication can continue at multiple origins which are distributed across several chromosomes. Presence of DNA damage causes replication fork stall, resulting in non–duplicated regions of the chromosomes. For this reason, the intra–S checkpoint activation of ATM–CHK2 axis or CHK1 phosphorylation through ATR acts in a similar fashion as G_1/S arrest. Modulation by ubiquitination of CDC25A phosphatase leads to the inhibition of CDK2–Cyclin E/A. The process also prevents the association of CDC45 factor in establishing origin of replication, leading to a halt in S–phase progression. With the delay in cell–cycle progress chromosomal maintenance by means of DNA repair can be initiated to ensure integrity of replication forks do not cause further damage to the chromosome. Regardless, PARP preferentially binds to the gaps of stalled–replication fork structure and assist in the recruitment of MRN complex and ATM to initiate DDR [32]. The process is a step–wise assembly and continues to recruit mediators (MDC1, BRCA1, FancD2 and etc.). There is still discussion at length in regards to the different DDR factors associated with S–phase progression. Currently, only one signalling pathway has been well described, ATM–MDC1–NBS1 phosphorylation of structural maintenance of chromosomes–1 (SMC1) pathway. SMC1 is a cohesin protein required for sister chromatid cohesion. During early S–phase, SMC1 activation prohibits DNA synthesis to allow repair of damaged DNA.

As cells progress from S– to G_2–phase, the CDK1–Cyclin B complex drives the final mitotic entry before cellular division. Held in a negative feedback loop at G_2–phase, the complex is kept in a dormant state throughout phosphorylation by WEE1. When dephosphorylated by CDC25A, WEE1 frees CDK1–Cyclin B complex, thereby allowing cells to enter mitosis. In the event of DNA damage, ATM–CHK2 or ATM–ATR–CHK1 activation phosphorylates CDC25A, rendering CDK1–Cyclin B complex inactive. It is also caution to note that checkpoint activation does not necessary represent an absolute safety mechanism. Presence of DNA damage can still escape the G_2/M checkpoint cell–cycle arrest. In lower mammalian systems, it has been shown that unresolved DNA damage can induce G_2 arrest but abrogates checkpoint arrest while proceeding with cell division.

Resolving DNA Double Strand Break: Selecting the Right Choice of DNA Repair

It is known that components of the DDR initiates two main forms of repair, the non–homologous end–joining (NHEJ) and homologous recombination (HR). NHEJ is highly efficient in resolving DSBs and does not require excessive processing to adhere the broken DNA ends [33]. The repair pathway is functionally present in almost all phases of the cell–cycle but is highly prevalent during G_1–phase. In contrast, HR is predominately at work during S– and G_2–phase of the cell–cycle [34]. Its activity in DSB repair is dependent on the presence of an intact homologous duplex sequence that serves as a tool for template copying. Processing of the DNA ends during early DDR is a key determinant for the choice of DSB repair and is correlated to CDKs activities during cell–cycle progression. In G_1–phase, low CDK level does not stimulate end resection. Instead, presence of DSBs mediates an ATM response that drives 53BP1 translocation to the IRIF via RNF8 and RNF168 ubiquitination. Additional factors such as RIF1, PTIP and REV7 are also recruited by 53BP1 to work in concert as a complex to stimulate the blocking of DNA end resection [31, 35]. Displacement of BRCA1 from chromatin favours Ku (70/80) heterodimer to form a synaptic complex with DNA DSB termini and acts as a scaffold to recruit DNA–PK_{cs}. Depending on the nature of the DSBs, different factors are required. Artemis, WRN, and APLF are essential tools for DNA resection whereas filling of DNA gaps can be performed by members of the family X polymerases which include polymerases μ and λ [36]. The final step of NHEJ repair is the ligation of the broken ends by the X4–L4 complex.

Rescue study of BRCA1–deficient cells showed 53BP1–deletion can revert HR, indicating both proteins are at constant play to restore genome integrity. The decision to switch from 53BP1–dependent NHEJ during G_1–phase to the preferential BRCA1–mediated HR from S– to G_2–phase can be predominately explained by the function of CtIP [37]. Cells increases CDK activity to transit from G_1– to S–phase. Increment of CDK activity also turns on resection by multisite phosphorylation of CtIP to promote BRCA1 binding [38–40]. Prior to DNA processing, BRCA1 requires chromatin remodeller SMARCAD1 to redirect 53BP1 protein away from the IRIF [41]. MRN complex together with CtIP–BRCA1 can then initiate resection at the site of DSB. The "two–step" model suggests MRN–CtIP–BRCA1 fulfils the initial role of short range resection which is followed by the recruitment of EXO1 and BLM to generate longer stretches of ssDNA substrates for HR repair [42, 43]. To prevent excessive resection, HELB helicase translocates to the site of IRIF and is critical to attenuate EXO1 and BLM activity [44]. Once regions of ssDNA are formed, replication protein A (RPA) proteins effectively bind to the naked strand and potentially unwind secondary structure to facilitate further resection. The RPA–ssDNA complex configuration also acts as a precautionary measure to prevent premature filament assembly and homology invasion. Consequently, recombination mediators facilitate the displacement of RPA with RAD51 proteins at the 3′ end of the ssDNA overhang. RAD51 loading onto ssDNA is partly assisted through MMS22L–TONSL complex

[45]. The newly formed RAD51–ssDNA nucleoprotein filament then catalyses an ATP–dependent homology search for compatible DNA sequence. Strand invasion at the sister chromatid generates a displacement–loop (D–loop) intermediate where RAD51 disengages to allow 3′–ssDNA priming and initiation of DNA synthesis. In conjunction with ligation of the repaired–strands, two Holliday junctions are established. For the final part of HR repair, the double Holliday junctions (dHj) are then cleaved to restore the original double strand DNA.

Therapeutic Intervention of DNA Damage Response: The Use of Natural Products to Activate ATM and Initiate Tumourgenic Death

Substantial information in regards to the role play of ATM kinase to improve clinical outcome with different drugs have been added to the existing pool of knowledge that is related to the strategies of targeting cancer cells. We will start with a summary of research findings which highlights the role of "active ATM" in causing tumourgenic death.

Ellipticine is an alkaloid with known properties of binding to DNA strands. Treatment of this agent has shown to upregulate ATM [46]. Phosphorylation of Chk1 is achieved through ATM dependent activation and has been noted at different concentration of this agent. The overall effect causes downstream inactivation of CDC25C, Cdk1 and Cyclin B1. It is clear from this study that phytochemical with alkylating property can promote ATM-Chk1-CDC25C-Cdk1 signaling response. In bladder cancer (BCa) cells, ellipticine treatment causes G_2/M arrest and subsequent activation of cell-death. Similar findings include the use of oridonin to target lung cancer [47] and anthocyanins (delphinidin and cyaniding) for leukemia [48]. Regardless of P53 status, these agents are capable of modulating ATM-Chk1-CDC25C or ATM-Chk2-CDC25C pathway to eliminate cancer cell population. Apart from using checkpoint response to elicit tumour death, several phytochemicals are also capable of achieving ATM activity and initiate checkpoint arrest while preventing DNA repair. Strigolactones [49] and Kaempferol treatment in osteosarcoma and leukemia cells [50] respectively have been reported to upregulate DDR while preventing the restoration of genomic insults. The accumulation of DNA damage promotes tumour death.

Eliminating DNA Damage Response: ATM Pathway Inhibition as a Strategy to Kill Cancer Cells

ATM activation to elicit DNA damage dependent apoptosis is a potential clinical strategy. Alternatively, its removal as a transducer protein to prevent functional DDR is also efficacy in sensitizing tumour cells to genomic insults. Human disabled homolog 2 interaction protein (DAB2IP), a tumor suppressor was noted to be frequently downregulated and is associated with ATM upregulation in BCa cells [51]. Loss of ATM can reverse tumourigenicity by compromising survival components such as NF-κB and mitogen-activated protein kinase (MAPK) pathways. Additionally, the lack of transducer protein also sensitizes DAB2IP-deficient BCa cells to IR due to a lack of DNA DSB repair [51]. In colorectal cancer, P53 status plays a critical role in survival. Tumour cells are highly resistant to single strand lesions that are potentially fatal when DSBs arise from replication fork collapse. Loss of ATM in colorectal cancer cells are sensitive to clinically approved low dose Olaparib, a PARP inhibitor. It is likely through the reduction of homologous recombination repair, ATM deficient cancer cells are unable to compensate for DNA DSB during replication [52]. Recent evidence suggests ATM is also involved in the progression of cancer metastasis. Using genetic knockdown model from colorectal cancer cells, loss of ATM showed inverse correlation with epithelial E-cadherin upregulation. Mesenchymal N-cadherin/vimentin and CD44 protein levels were noted to have low expression. ATM deficient tumour cells are non-migratory but such phenomenon can be reversed by restoring CD44 expression level. It is likely ATM regulates cancer cell migration through CD44 [53].

In a recent study, it is found that histone H2AX regulates hypoxia-inducible factor 1 alpha (HIF1) signaling which is primarily needed as a transcription factor. Under oxygen deprived condition, TNF receptor associated factor 6 (TRAF6) triggers mono-ubiquitylation of H2AX (mUb-H2AX) and is associated with active ATM which then phosphorylates H2AX. It is important to note that during hypoxia condition, recruitment of ATM is critical for H2AX formation and subsequent transcriptional activation. The downstream effect allows the binding and recruitment of HIF1 protein to initiate transcription activation of target genes. H2AX also physically interacts with HIF1 in the nucleus to stabilize and prevent proteasome degradation. Most importantly, therapeutically targeting of TRAF6–ATM–H2AX signalling axis such as ATM itself is an effective strategy to reverse hypoxia effect and reduce tumourigenicity [54]. In acute myeloid leukemia (AML), the use of FMS-like tyrosine kinase 3 (FLT3) inhibitors have demonstrated great efficacy in clinical trials but failed to achieve cancer remission. Synthetic lethal screen using genome-wide short hairpin (Sh)RNA approach identifies ATM as a critical factor for FLT3 inhibition in AML [55]. While ATM is well characterized as a transducer protein for DDR, in the context of AML this particular protein plays a role in promoting antioxidant response through activation of glucose-6-phosphate dehydrogenase (G6PD). Loss of ATM suppresses G6PD activity, leading to an increase in reactive oxygen species. Combination targeting of FLT3 and ATM/G6PD can improve treatment response for AML.

Conclusion

Cancer cells are highly susceptible to fatal DNA damage owing to the changes in both intrinsic and external factors. These lesions are detrimental to overall survival and can promote a range of complications. However, all cancer cells are equipped with a global 'self-defense' mechanism which can confer protection and preserving genome integrity. Over the years, the study of DNA DSB response has been an extensive journey with new interacting partners and functions being identified. It is interesting to see if further development in the elucidation of molecular mechanisms and inhibitors will translate potential clinical benefits, particularly in the area of ATM and how pathway targeting of specific cancer cells can prevent tumour recurrence.

References

1. Uziel T, et al. Requirement of the MRN complex for ATM activation by DNA damage. EMBO J. 2003;22(20):5612–21.
2. Lee J-H, Paull TT. ATM activation by DNA double-strand breaks through the Mre11-Rad50-Nbs1 complex. Science. 2005;308(5721):551–4.
3. Stewart GS, et al. MDC1 is a mediator of the mammalian DNA damage checkpoint. Nature. 2003;421:961–6.
4. Asaithamby A, Chen DJ. Cellular responses to DNA double-strand breaks after low-dose y-irradiation. Nucleic Acids Res. 2009;37(12):3912–23.
5. Schlissel M, et al. Double-strand signal sequence breaks in V(D)J recombination are blunt, 5'-phosphorylated, RAG-dependent, and cellcycle regulated. Genes Dev. 1993;7:2520–32.
6. Lange J, et al. ATM controls meiotic double-strand-break formation. Nature. 2011;479:237–40.
7. Bosottia R, et al. FAT: a novel domain in PIK-related kinases. Trends Biochem Sci. 2000;25(5):225–7.
8. Abraham RT. PI 3-kinase related kinases: 'big' players in stress-induced signaling pathways. DNA Repair (Amst). 2004;3:883–7.
9. Murr R, et al. Histone acetylation by Trrap–Tip60 modulates loading of repair proteins and repair of DNA double-strand breaks. Nat Cell Biol. 2006;8(1):91–9.
10. Blackford AN, Jackson SP. ATM, ATR, and DNA-PK: the trinity at the heart of the DNA damage response. Mol Cell. 2017;66(6):801–17.
11. Sukhanova MV, et al. Single molecule detection of PARP1 and PARP2 interaction with DNA strand breaks and their poly(ADP-ribosyl)ation using high-resolution AFM imaging. Nucleic Acids Res. 2016;44(6):e60.
12. Ali AAE, et al. The zinc-finger domains of PARP1 cooperate to recognize DNA strand breaks. Nat Struct Mol Biol. 2012;19:685–92.
13. Riccio AA, et al. PARP-2 domain requirements for DNA damage-dependent activation and localization to sites of DNA damage. Nucleic Acids Res. 2016;44(4):1691–702.
14. Walker JR, et al. Structure of the Ku heterodimer bound to DNA and its implications for double-strand break repair. Nature. 2001;412:607–14.
15. McElhinny SAN, et al. Ku recruits the XRCC4-Ligase IV complex to DNA ends. Mol Cell Biol. 2000;20(9):2996–3003.
16. Cheng Q, et al. Ku counteracts mobilization of PARP1 and MRN in chromatin damaged with DNA double-strand breaks. Nucleic Acids Res. 2011;39(22):9605–19.
17. Haince J-F, et al. PARP1-dependent kinetics of recruitment of MRE11 and NBS1 proteins to multiple DNA damage sites. J Biol Chem. 2008;283(2):1197–207.
18. Czornak K, et al. Mystery of DNA repair: the role of the MRN complex and ATM kinase in DNA damage repair. J Appl Genet. 2008;49(4):383–96.
19. Kaidi A, Jackson SP. KAT5 tyrosine phosphorylation couples chromatin sensing to ATM signalling. Nature. 2012;498:70–4.
20. Adamowicz M, et al. NOTCH1 inhibits activation of ATM by impairing the formation of an ATM-FOXO3a-KAT5/Tip60 complex. Cell Rep. 2016;16(8):2068–76.

21. Sun Y, et al. A role for the Tip60 histone acetyltransferase in the acetylation and activation of ATM. Proc Natl Acad Sci U S A. 2005;102(37):13182–7.
22. Gerloff DL, et al. BRCT domains: a little more than kin, and less than kind. FBBS Lett. 2013;586(17):2711–22716.
23. Stucki M, et al. MDC1 directly binds phosphorylated histone H2AX to regulate cellular responses to DNA double-strand breaks. Cell. 2005;123(7):1213–26.
24. Lou Z, et al. MDC1 maintains genomic stability by participating in the amplification of ATM-dependent DNA damage signals. Mol Cell. 2006;21(2):187–200.
25. Huen MSY, et al. RNF8 transduces the DNA-damage signal via histone ubiquitylation and checkpoint protein assembly. Cell. 2007;131(5):901–14.
26. Mailand N, et al. RNF8 ubiquitylates histones at DNA Double-strand breaks and promotes assembly of repair proteins. Cell. 2007;131(5):887–900.
27. Mattiroli F, et al. RNF168 ubiquitinates K13-15 on H2A/H2AX to drive DNA damage signaling. Cell. 2012;150(6):1182–95.
28. Botuyan MV, et al. Structural basis for the methylation state-specific recognition of histone H4-K20 by 53BP1 and Crb2 in DNA repair. Cell. 2006;127(7):1361–73.
29. Sims JJ, Cohen RE. Linkage-specific avidity defines the lysine 63-linked polyubiquitin binding preference of RAP80. Mol Cell. 2010;33(6):775–83.
30. Coleman KA, Greenberg RA. The BRCA1-RAP80 complex regulates DNA repair mechanism utilization by restricting end resection. J Biol Chem. 2011;286:13669–80.
31. Feng L, et al. Cell cycle-dependent inhibition of 53BP1 signaling by BRCA1. Cell Discov. 2015;1:15019.
32. Bryant HE, et al. PARP is activated as stalled foprks to mediate Mre11-dependent replication restart and recombination. EMBO J. 2009;27(17):2601–15.
33. Karanjawalaa ZE, et al. The embryonic lethality in DNA ligase IV-deficient mice is rescued by deletion of Ku: implications for unifying the heterogeneous phenotypes of NHEJ mutants. DNA Repair. 2002;1(12):1017–26.
34. Mao Z, et al. DNA repair by nonhomologous end joining and homologous recombination during cell cycle in human cell. Cell Cycle. 2008;**7**(18):2902–6.
35. Boersma V, et al. MAD2L2 controls DNA repair at telomeres and DNA breaks by inhibiting 5′ end resection. Nature. 2015;521:534–40.
36. Davis AJ, Chen D. DNA double strand break repair via non-homologous end-joining. Transl Cancer Res. 2013;2(3):130–43.
37. Wang H, et al. The interaction of CtIP and Nbs1 connects CDK and ATM to regulate HR–mediated double-strand break repair. PLoS Genet. 2013;9(2):e1003277.
38. Cruz-Garcia A, et al. BRCA1 accelerates CtIP-mediated DNA-end resection. Cell Rep. 2014;9:451–9.
39. Huertas P, Jackson SP. Human CtIP mediates cell cycle control of DNA end resection and double strand break repair. J Biol Chem. 2009;284(14):9558–65.
40. Yun MH, Hiom K. CtIP-BRCA1 modulates the choice of DNA double-strand-break repair pathway throughout the cell cycle. Nature. 2009;459:460–3.
41. Densham RM, et al. Human BRCA1–BARD1 ubiquitin ligase activity counteracts chromatin barriers to DNA resection. Nat Struct Mol Biol. 2016;23:647–55.
42. Nimonkar AV, et al. Human exonuclease 1 and BLM helicase interact to resect DNA and initiate DNA repair. Proc Natl Acad Sci U S A. 2008;105(44):16906–11.
43. Sartori AA, et al. Human CtIP promotes DNA end resection. Nature. 2007;450(7169):509–14.
44. Tkáč J, et al. HELB is a feedback inhibitor of DNA end resection. Mol Cell. 2016;61(3):405–18.
45. Piwko W, et al. The MMS22L–TONSL heterodimer directly promotes RAD51-dependent recombination upon replication stress. EMBO J. 2016;35(23):2584–601.
46. Tao S, Meng S, Zheng X, Xie L. ATM participates in the regulation of viability and cell cycle via ellipticine in bladder cancer. Mol Med Rep. 2017;15(3):1143–8. https://doi.org/10.3892/mmr.2017.6141.
47. Zheng M, Zhu Z, Zhao Y, Yao D, Wu M, Sun G. Oridonin promotes G2/M arrest in A549 cells by facilitating ATM activation. Mol Med Rep. 2017;15(1):375–9. https://doi.org/10.3892/mmr.2016.6008.

48. Tsai TC, Huang HP, Chang KT, Wang CJ, Chang YC. Anthocyanins from roselle extract arrest cell cycle G2/M phase transition via ATM/Chk pathway in p53-deficient leukemia HL-60 cells. Environ Toxicol. 2017;32(4):1290–304. https://doi.org/10.1002/tox.22324.
49. Croglio MP, Haake JM, Ryan CP, Wang VS, Lapier J, Schlarbaum JP, Dayani Y, Artuso E, Prandi C, Koltai H, Agama K, Pommier Y, Chen Y, Tricoli L, LaRocque JR, Albanese C, Yarden RI. Analogs of the novel phytohormone, strigolactone, trigger apoptosis and synergize with PARP inhibitors by inducing DNA damage and inhibiting DNA repair. Oncotarget. 2016;7(12):13984–4001. 10.18632/oncotarget.7414.
50. Wu L-Y. Kaempferol induces DNA damage and inhibits DNA repair associated protein expressions in human promyelocytic leukemia HL-60 cells. Am J Chin Med. 2015;43:365–82.
51. He H, Chang R, Zhang T, Yang C, Kong Z. ATM mediates DAB2IP-deficient bladder cancer cell resistance to ionizing radiation through the p38MAPK and NF-κB signaling pathway. Mol Med Rep. 2017;16(2):1216–22. https://doi.org/10.3892/mmr.2017.6689.
52. Wang C, Jette N, Moussienko D, Bebb DG, Lees-Miller SP. ATM-Deficient colorectal cancer cells are sensitive to the PARP inhibitor olaparib. Transl Oncol. 2017;10(2):190–6. https://doi.org/10.1016/j.tranon.2017.01.007.
53. Liu R, Tang J, Ding C, Liang W, Zhang L, Chen T, Xiong Y, Dai X, Li W, Xu Y, Hu J, Lu L, Liao W, Lu X. The depletion of ATM inhibits colon cancer proliferation and migration via B56γ2-mediated Chk1/p53/CD44 cascades. Cancer Lett. 2017;390:48 57. https://doi.org/10.1016/j.canlet.2016.12.040.
54. Rezaeian AH, Li CF, Wu CY, Zhang X, Delacerda J, You MJ, Han F, Cai Z, Jeong YS, Jin G, Phan L, Chou PC, Lee MH, Hung MC, Sarbassov D, Lin HK. A hypoxia-responsive TRAF6-ATM-H2AX signalling axis promotes HIF1α activation, tumorigenesis and metastasis. Nat Cell Biol. 2017;19(1):38–51. https://doi.org/10.1038/ncb3445.
55. Gregory MA, D'Alessandro A, Alvarez-Calderon F, Kim J, Nemkov T, Adane B, Rozhok AI, Kumar A, Kumar V, Pollyea DA, Wempe MF, Jordan CT, Serkova NJ, Tan AC, Hansen KC, DeGregori J. ATM/G6PD-driven redox metabolism promotes FLT3 inhibitor resistance in acute myeloid leukemia. Proc Natl Acad Sci U S A. 2016;113(43):E6669–78. Epub 2016 Oct 10

Chapter 5
The Role of GPER Signaling in Carcinogenesis: A Focus on Prostate Cancer

Marília I. Figueira, Henrique J. Cardoso, and Sílvia Socorro

Abstract The G protein-coupled estrogen receptor, GPER, also known as GPR30, belongs to the seven transmembrane receptor superfamily and is involved in the rapid non-genomic estrogenic responses. Nevertheless, GPER regulation of transcriptional activity also has been reported. GPER downstream signaling includes the activation of the phosphatidylinositol 3-kinase (PI3K)/Akt and mitogen-activated protein kinase (MAPK) pathways, the stimulation of adenylyl cyclase, and the mobilization of intracellular calcium. Over the last decade, the discovery of GPER specific agonists and antagonists has been crucial to understand its physiological functions, mechanisms of action and its putative usefulness as a therapeutic target. The GPER seems to have an important role in endocrine, reproductive, immune, nervous, and cardiovascular systems, and alterations in its expression or activity have been associated with several pathological conditions such as cardiovascular diseases, obesity, diabetes, Parkinson, stroke, and cancer. GPER has been linked with the carcinogenic process, though some ambiguity exists concerning its protective or causative role in different tissues, or even in the same tissue. This chapter summarizes the existent knowledge concerning the structural and molecular aspects of GPER, its known ligands and activated pathways, as well as its role over the known hallmarks of cancer: exacerbated proliferation, resistance to apoptosis, stimulated migration and invasion, induction of angiogenesis, and the metabolic reprogramming. A special focus will be given to prostate cancer.

Keywords GPER · GPR30 · Prostate cancer

M. I. Figueira · H. J. Cardoso · S. Socorro (✉)
CICS-UBI, Health Sciences Research Centre, Faculty of Health Sciences, University of Beira Interior, Covilhã, Portugal
e-mail: ssocorro@fcsaude.ubi.pt

S. Fayyaz, A. A. Farooqi (eds.), *Recent Trends in Cancer Biology: Spotlight on Signaling Cascades and microRNAs*, https://doi.org/10.1007/978-3-319-71553-7_5

Introduction

Estrogens are very important hormones playing a determinant role in mammalian biology by its wide action regulating several processes in a great variety of tissues. As all hormones, estrogens act through specific receptors that after recognizing and binding their ligands transmit the information to downstream effectors. The estrogen receptor α (ERα) was the first ER to be known in 1973 [1]; and was more than two decades after in 1996, that a second ER, the ERβ, was described [2]. Since then, several isoforms of ERα and ERβ have been identified known to have tissue specific functions or acting as negative regulators of full-length ERs, namely, the ERα-A-E, the ERα36 and the ERβ1–5 [3–9]. The ERα and ERβ are known as the classical ERs exerting their actions as transcription factors and regulating the gene expression network in target tissues [10]. Generally, after ligand binding in the cytoplasm, hormone–receptor complexes are translocated to the nucleus, where they dimerize and bind DNA modulating gene expression. Other alternative mechanisms of estrogens actions have been described, such as, the interaction of ERs with other transcription factors, and the hormone independent receptor phosphorylation and activation [11, 12]. In addition to the genomic actions, classical ERs also can trigger non-genomic actions. It has been described that ERs can associate with the cell membrane and lead to the activation of protein kinase signaling pathways, or cross-talk with the intracellular secondary messengers produced in response to the activation of G protein-coupled receptors (GPCRs) [11, 12].

The landscape of estrogens signaling has gained increasing interest and complexity with the report of the G protein-coupled estrogen receptor (GPER), originally known as GPR30 [13]. GPER is a member of the GPCRs superfamily, and thus, displays the typical seven transmembrane helices and predominant location at the cell membrane (Fig. 5.1) [14].

GPER expression has been found in multiple tissues and systems, including nervous, cardiovascular, gastrointestinal, immune, urinary and reproductive systems, and bone among others [15–19]. GPER is responsible for mediating the non-genomic actions of estrogens and is involved in the control of many biological processes. Its important role in several diseases including stroke, Parkinson, diabetes, cardiac diseases, bone diseases, epilepsy, autoimmune diseases, infection, and cancer has been suggested [20–32]. Concerning cancer, GPER has been detected in many types of tumors, including breast, ovarian, endometrial, prostate, testis, lung, thyroid, and others [33–38]. Also, its role in cancer development as a causative agent and associated with tumors aggressiveness, or as a tumor suppressor has been debated in the available literature [39, 40].

In this chapter, the molecular and structural aspects of GPER, as well as, its ligands, signaling activated pathways, and downstream effectors will be presented. Moreover, the GPER expression in cancer cases, and its role modulating the typical features of cancer cells will be discussed, with a special focus on prostate cancer (PCa).

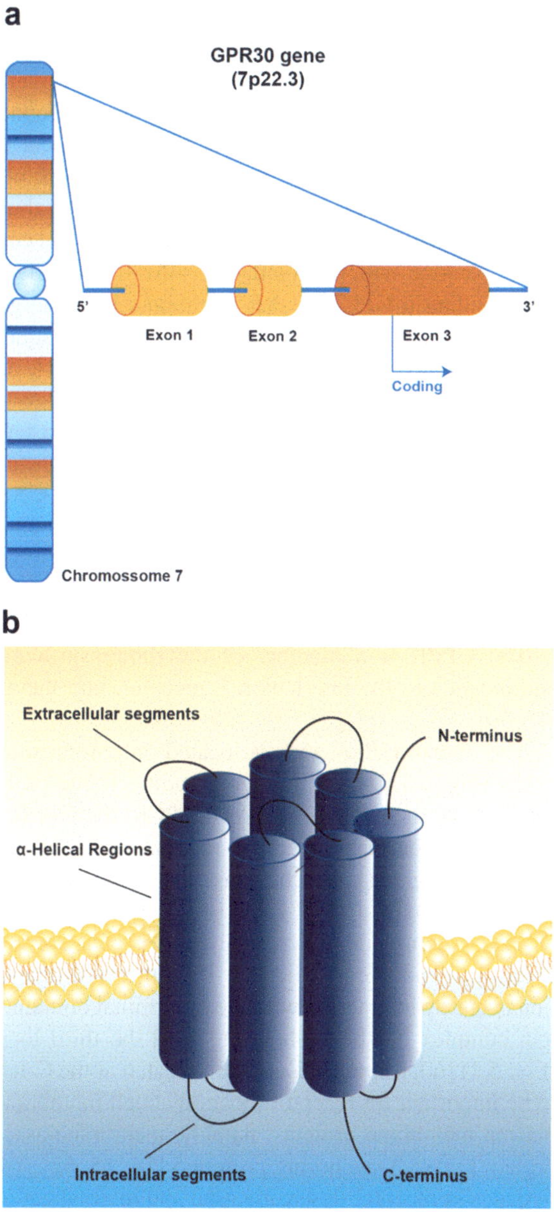

Fig. 5.1 Schematic representation of the human GPER: structural and molecular aspects. (**a**) Localization of *GPER* gene in the chromosome 7 and its structure consisting of three exons, with the coding region encoded by a single exon, the exon 3. (**b**) Structure of the GPER protein with the typical seven transmembrane α-helical regions, and the four extracellular and four cytosolic segments. The N-terminus is extracellular whereas the C-terminus is cytosolic

Structural, Molecular and Cellular Aspects of GPER

The "life history" of GPER has begun some time ago, when it aroused from the shadows, like a ghost. Before the biochemical and structural characterization of GPER, researchers just have known about the possibility of alternative estrogens' actions via non-genomic signaling mechanisms producing rapid cell effects. Over the years, as an attempt to explain the estrogenic effects in a time-frame not compatible with gene transcription and *de novo* protein synthesis, many theories were proposed concerning the existence of a membrane-bound ER [41, 42]. Amongst other hypotheses, it was shown that the classical nuclear ERs, the ERα and ERβ, were able to be translocated to the cell membrane through several mechanisms that involve post-translational modifications, such as palmitoylation or phosphorylation, interaction with caveolin-1, binding to adaptor and co-regulator molecules, and also the association with G proteins [43–45]. It was only later on that the idea of estrogens binding a GPCR at cell membrane has emerged [46]. At the end of 1990s several independent groups aimed at identifying a new GPCR characterized by its reduced homology to other GPCRs [14, 47–51], and for which no ligand was known. This orphan receptor was named GPR30, even before 17β-estradiol (E_2) has been identified as its ligand. Indeed, it was only a few years later that the "fate" of this orphan receptor has changed after the demonstration that it binds E_2 with the activation of downstream signaling pathways [13, 52–54]. This lead the scientific community to rename the GPR30 that started to be known as GPER, a designation formally established by the International Union of Basic and Clinical Pharmacology in 2007 [55]. Even so, GPER's "life" has been anything but consensual, and still today some groups question its role as an ER [56, 57].

The GPER is a member of the rhodopsin-like receptor superfamily highly homologous to the interleukin 8 receptor, and the angiotensin II receptor type 1 [47–50].

The human *GPER* gene is located on chromosome 7p22.3, and contains three exons (Fig. 5.1a), with the open reading frame of 1126 base pairs encoded by a single exon, the exon 3 [58]. As a GPCR, the GPER has a structure encompassing seven transmembrane α-helical regions, and four extracellular and four cytosolic segments (Fig. 5.1), which spans 375 amino acids and has a theoretical molecular mass of approximately 41 kDa [14, 15, 59]. The extracellular N-terminus has aspartic acid residues in its terminal region, which are targets for glycosylation; the C-terminal end is cytosolic [14, 15, 59]. Thus, ligand binding and receptor activation occur by interaction with the N-terminal domain, and the heterotrimeric G protein complex has been shown to bind the third loop of the intracellular domain (Fig. 5.1) [60, 61]. A PDZ domain located at the C-terminal region of GPER seems to be important for receptor location at cell membrane by its interaction with other plasma membrane proteins, namely, membrane-associated guanylate kinases and protein kinase A-anchoring protein 5 [14, 15, 62]. The C-terminal region also participates in receptor desensitization and internalization depending on phosphorylation by G protein-coupled receptor kinases [14, 15, 59]. Recently, a biosensor-based approach identified four distinct calmodulin-binding domains in the GPER protein [63, 64], though the effect of calmodulin regulating GPER-dependent signaling is at the moment unknown.

Although GPER was firstly identified as a membrane receptor, the question of its subcellular localization has deserved intense discussion. Several references exist indicating the presence of GPER in the endoplasmic reticulum, the Golgi apparatus,

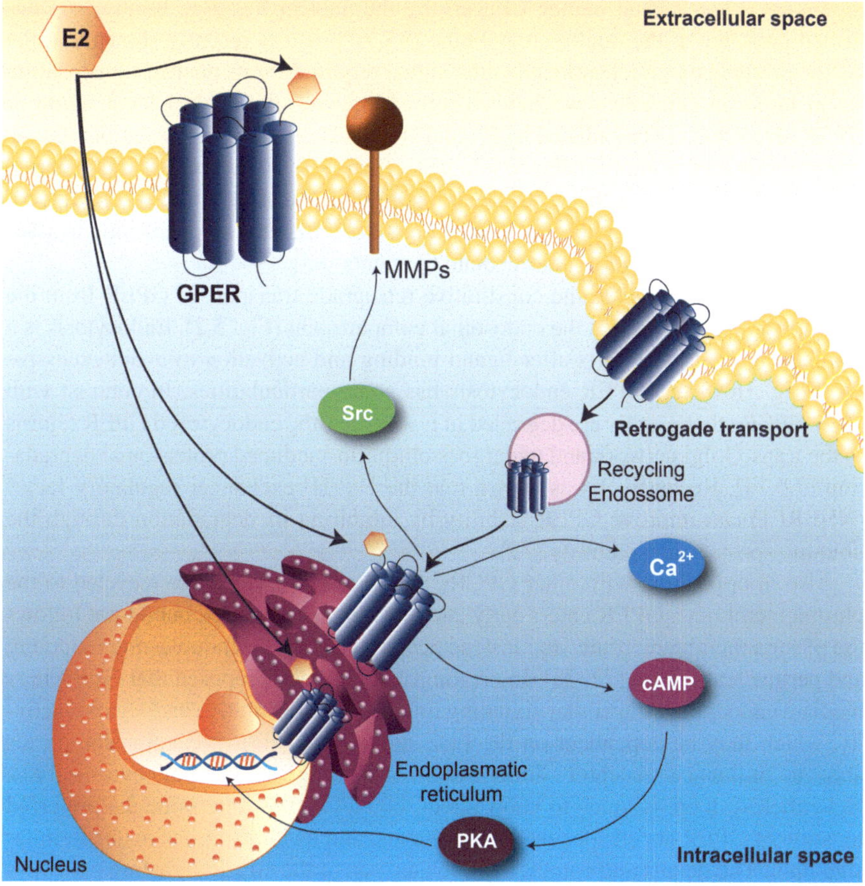

Fig. 5.2 GPER subcellular localization. The receptor has been shown to be localized at the cell membrane and/or also in the endoplasmic reticulum and nucleus. Although not a consensual matter, there are reports indicating that E_2 can interact with a signaling-active GPER at endoplasmic reticulum promoting cell responses [54, 65–71]. The Src, metalloproteinases (MMPs), calcium (Ca^{2+}), and cAMP/protyein kinase A (PKA) mediated signaling pathways are detailed throughout the text and in Fig. 5.3. The GPER, as other GPCRs, enters the constitutive endocytic pathway but it is not recycled to the plasma membrane being instead accumulated in the endoplasmic reticulum and perinuclear region [72, 73]

and also in the nucleus (Fig. 5.2) [54, 65–71]. The GPER expression in the endoplasmic reticulum and Golgi complex could be attributed at a first glance to the process of protein receptor synthesis and intracellular traffic before its translocation to the cell membrane. In fact, the GPER detected at the plasma membrane seems to be translocated from the cytoplasm 1 h after E_2 stimulation, which also suggests that GPER activation is triggered at the cell membrane [54, 65]. However, some research groups defend that the GPER at the endoplasmic reticulum is signaling-active mediating cell responses by the mobilization of intracellular calcium (Ca^{2+}) [52], though this is not a consensual matter. Concerning the nuclear location, besides its confirmed presence at the nuclear membrane [35, 67], it was recently shown that the GPER protein has a nuclear localization signal required for its nuclear translocation by an importin-dependent mechanism [69]. Moreover, the nuclear localization of GPER also seems to be induced by E_2, and is associated with the regulation of gene transcription and promotion of physiological effects [69, 70]. Another study reported that GPER activation at the plasma membrane with the stimulation of cAMP production is followed by receptor internalization and intracellular localization in association with the cytokeratin intermediate filaments [68].

It is also of worth note the constitutive retrograde transport of GPER from the plasma membrane towards the endosomal compartment (Fig. 5.2). Endocytosis is a common feature of GPCRs after ligand binding and activation to avoid excessive signaling. However, GPER endocytosis has some particularities. In contrast with other GPCRs that usually are degraded in lysosomes, the endocytosed GPER returns to the trans-Golgi network and undergoes ubiquitin-mediated proteasomal degradation [72–74]. Recently, it was shown that the Na^+/H^+ exchanger regulatory factor (NHERF1) can improve GPER stability by inhibiting its degradation through the ubiquitin-proteasome pathway [75].

Also in opposition with other GPCRs that after endocytosis are recycled to the plasma membrane, GPER enters early and recycling endosomes, but do not follows the plasma membrane route, and instead accumulates in the endoplasmic reticulum and perinuclear region [72, 73]. Interestingly, it has been proposed that GPER may mediate endocytic intracellular signaling linked to its constitutive endocytosis activity, which may be dependent on the PDZ domain that forms receptor complexes able to influence receptor dimerization, signaling and/or endocytosis [73]. Nevertheless, there is much to know about the factors that determine the preferred location of GPER at cell specific compartments, and its implications in intracellular signaling. The details and comprehensive analysis of the GPER signaling pathways will be explored in the following section of this chapter.

Overview of GPER Signaling Pathways

GPER is widely associated with the non-genomic effects of estrogens and estrogen-like substances, which are generally characterized as being: (1) very rapid, in minutes or even seconds; (2) insensitive to DNA transcription and protein synthesis inhibitors; and (3) activated by steroids coupled to high-molecular-weight molecules, such as bovine serum albumin (BSA), and thus incapable of cross the plasma membrane. GPER-mediated responses have been described in various cell types and involve several molecular targets and signaling pathways (Fig. 5.3), including, the mobilization of second messengers, such as Ca^{2+} and nitric oxide (NO), the

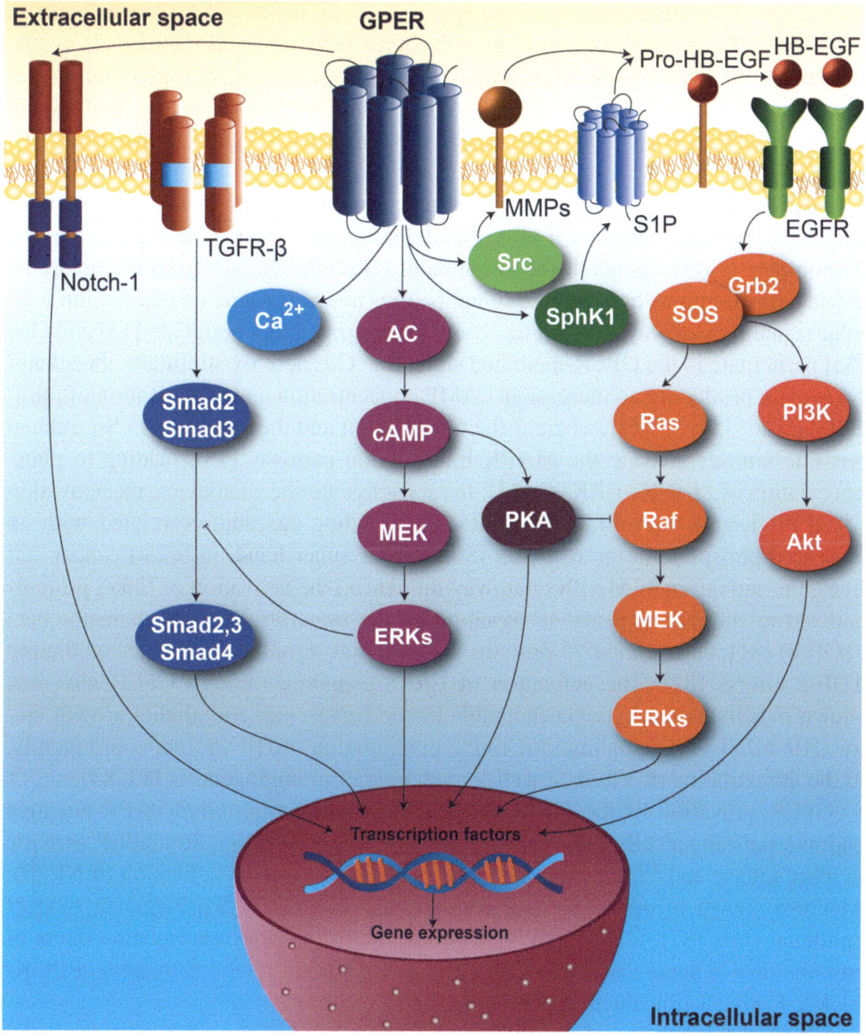

Fig. 5.3 GPER signaling pathways. GPER activates several downstream signaling cascades including the PI3K/AKT pathway, the Src pathway, the MAPKs/ERK1/2 pathway and the Notch signaling. The activation of MAPKs pathway influences the activation of Smad proteins and leads to the inhibition of the TGF-β signaling. GPER activation also leads to stimulation of adenylyl cyclase (AC) with an increase of cAMP concentrations and mobilization of intracellular Ca^{2+}. GPER activation of intracellular signaling pathways also has been coupled with EGFR transactivation. GPER activation of Src-kinase stimulates MMPs that cleave the pro-HB-EGF leading to the release of free HB-EGF and EGFR activation. Alternatively, EGFR activation can be mediated by sphingosine 1 phosphate (S1P) a product of sphingosine kinase 1 (SphK1) activity

interaction with other membrane receptors, such as insulin-like growth factor-1-receptor (IGF-1R) and epidermal growth factor receptor (EGFR), as well as, the activation of effector molecules, namely, the Src kinase and phosphatidylinositol 3-kinase (PI3K), the serine/threonine protein kinase Akt, mitogen-activated protein kinase (MAPK) family members, and the protein kinases A and C (PKA and PKC, respectively) [76].

After ligand binding, GPER initiates intracellular signaling by activating heterotrimeric G proteins, composed of α, β and γ subunits, with the latter known as the βγ-complex [77]. In general, the Gα-subunit (Gαs) dissociates from the heterotrimeric Gαβγ complex and the free subunits (Gαs and Gβγ) activate distinct intracellular signaling pathways [78]. It has been demonstrated that both Gβγ [13] and Gαs [53] participate in the GPER-mediated signaling. Gαs activity stimulates the adenylyl cyclase producing an increase in cAMP concentration and mobilization of intracellular Ca^{2+} [52, 53, 79], whereas the Gβγ subunit and the downstream Src-related tyrosine kinases activate the MAPK transduction pathway [13], leading to phosphorylation of MAPKs/ERK1/2 [13]. In rat pachytene spermatocytes, the activation of GPER led to the initiation of ERK1/2 signaling cascade, correlated with an increased phosphorylation of c-Jun [80]. On the other hand, in breast cancer cell lines, the activation of MAPKs pathway influenced the activation of Smad proteins and led to the down-regulated signaling of the transforming growth factor beta (TGF-β) [81], since Smad2/3 proteins are essential signaling molecules mediating TGF-β effects [82]. The activation of ERKs signaling through GPER also was shown to activate the hypoxia-inducible factor 1α/vascular endothelial growth factor (HIF1α/VEGF) signaling, due to the upregulation of HIF1α and, consequently, to the upregulation of VEGF, a well-known marker of angiogenesis [83, 84].

GPER activation of the PI3K/Akt pathway has been reported in ER-negative endometrial cancer cells and seems to stimulate cell growth [85, 86]. GPER activity, in a Src kinase and PI3K/Akt dependent mechanism with the activation of NF-kB, was also shown to up-regulate cyclooxygenase 2 (COX2) expression in oviduct epithelial cells [87]. GPER may also contribute to the pro-tumorigenic effects of endothelin-1 in hepatocarcinoma cells and breast cancer via the activation of PI3K/ERK/c-Fos/AP1 transduction pathway [88].

GPER actions were also implicated in Notch signaling, inducing both the γ-secretase-dependent activation of Notch-1 and the expression of the Notch target gene Hes-1 [85, 89]. It was reported that GPER activation leads to the association of the intracellular domain of Notch-1 with the Hes-1 promoter inducing transcription of a Hes-1-reporter gene, as well as, to the down-regulation of VE-Cadherin and increased expression of Snail, a Notch target gene acting as a repressor of cadherin expression [89].

Interestingly, GPER activation of intracellular signaling pathways has been coupled with EGFR transactivation, through the Src-dependent stimulation of metalloproteinases (MMPs) (Fig. 5.3). MMPs have been shown to cleave the pro-HB-EGF leading to the release of free HB-EGF with resultant EGFR activation [13, 64, 90, 91]. Moreover, sphingosine 1-phosphate (S1P), a product of sphingosine kinase 1 (Sphk1), seems to be another intermediate in the GPER-mediated transactivation of the EGFR [92]. SphK1 activation stimulated by E_2 leads to the release of S1P, activating the S1P receptor Edg-3 and resulting in pro-HB-EGF cleavage [92]. A cross-talk of GPER with the IGF signaling pathway and GPER interaction with the IGFR and the ErbB receptors also has been described and associated with the progression of diverse types of cancer [93, 94].

Despite the transcriptional responses induced by estrogens are mainly associated to ERα- and ERβ-mediated actions, also the rapid signaling pathways activated by GPER have been shown to regulate gene expression in broad range of mammalian tissues and cells. These includes the upregulation of expression of several genes involved in relevant biological responses, for example, the transcription factor c-fos, the connective-tissue growth factor (CTGF), the early growth response 1 (Egr-1), pS2, sirtuins, cyclins A, D1, D2 and E, and the apoptotic regulators Bcl-2 and Bax; cyclins A1 and B1 were identified as down-regulated genes by GPER [80, 95–101].

Other important aspect to consider in GPER signaling is its interaction with the nuclear steroid receptors, namely, with the classical ERs, the glucocorticoid receptors, the mineralocorticoid receptors, and the vitamin D receptor. This cross-talk was elegantly reviewed by Prossnitz et al. [102] and will not be further explored in this chapter.

In the end, it is important to mention the receptor activity-modifying proteins (RAMPs) that interact with the GPCRs modifying their function, and acting as pharmacological switches and chaperones regulating signaling transduction [103]. Although GPER interaction with RAMP3 has been reported and linked to the receptor presentation at the cell membrane [104], the role of this regulatory proteins in the modulation of GPER function/activity still is in its infancy and more research is needed to clarify the liaison between RAMPs and GPER.

GPER Agonists and Antagonists: A Therapeutic Approach

From the time that was considered an orphan receptor to nowadays, many compounds have been identified with the ability to bind GPER, which substantially enriched its "molecular relationships". Among the panoply of compounds that bind GPER, some of them are activators (agonists) whereas others inhibit receptor function/activity (antagonists). The identification of molecules able to modulate GPER activity (Table 5.1) and the disclosure of the underlying signaling pathways is an exciting field of research considering the exploitation of GPER as a therapeutic target.

Also, many studies have been performed to understand better the GPER structure and its binding characteristics. The specificity of GPER cavities and its several structural changes endows this receptor of a great ability to accept diverse ligands, some of them with large volume, and also that the same ligand can recognize different binding sites dependently on the structural conformation of GPER [105].

Table 5.1 List of compounds able to modulate GPER activity, agonists and antagonists[a]

Class	Compound	Effect
Hormone ligands	17α-Estradiol	No
	17β-Estradiol-17-D-glucuronide	+
	2-Hydroxyestradiol	−
	2-Methoxy-estradiol	+
	7β-OH-EpiA	−
	ACTH	?
	Aldosterone	+
	Cortisol	?
	E_2	+
	Estriol	−
	Estrone	?
	Progesterone	No
	Testosterone	No
Synthetic ligands	4-Hydroxytamoxifen	+
	DES	No
	DPN	No
	ICI 182,780	+
	K-1	−
	MIBE	−
	PPT	+
	STX	?
	Raloxifene	+
Phytoestrogens	(−)-Epicatechin	+
	Baicalein	−
	Coumestrol	No
	Daidzein	?
	Equol	?
	Genistein	+
	Icariin	+
	Icaritin	+
	Oleuropein	+
	Prunetin	+
	Puerarine	+
	Quercetin	+
	Resveratrol	+
	SDG/ENL	?
	Tectoridin	+

(continued)

Table 5.1 (continued)

Class	Compound	Effect
Xenoestrogens	2,29,59-PCB-4-OH	?
	Atrazine	+
	BDE-47	+
	Bis(4-hydroxyphenyl)[2-(phenoxysulfonyl)phenyl] methane	?
	BPA	+
	DDT	?
	DDE	?
	Kepone	+
	Methoxychlor	+
	Nonylphenol	+
	TDP	+
Other compounds	Arsenite	+
	Butylparaben	+
	C4PY	–
	Cadmium	+
	$CuSO_4$	?
	Methylparaben	+
	Nicotinamide	+
	Nicotinic acid/niacin/vitamin B3	+
	Propylparaben	+
	$ZnCl_2$	+
Specific ligands	G1	+
	G15	–
	G36	–
	GPER-L1	+
	GPER-L2	+
	PBX1	–
	PBX2	–

+, agonist; – antagonist; ?, direct action on GPER needs confirmation; No, no effect reported on GPER activity

[a]Corresponding references are cited and discussed throughout the text

Natural Hormone Ligands

The first identified ligands for GPER, which gave the receptor's name, were estrogens, particularly the E_2. Competitive binding assays in GPER-transfected COS7 cells using E_2 conjugated to the fluorophore Alexa 633 found a high-affinity for E_2 (Kd 6 nM) contrarily to the non-physiological estrogen stereoisomer 17α-estradiol that fails to displace the fluorophore [52]. Similarly, studies with tritiated-hormone on membrane fractions also showed high-affinity of GPER for E_2 (Kd 3 nM) but low-affinity for 17α-estradiol [53].

Nonetheless, the involvement of some E_2 metabolites in GPER activation has been reported. For example, the oxidised metabolite 2-methoxy-estradiol has been reported to act as an agonist of GPER [106] whereas the 2-hydroxyestradiol has been shown to antagonize GPER action [106, 107]. Glucuronidation and sulfonation of E_2 generate other metabolites like 17β-estradiol-17-D-glucuronide, which also seems to be an agonist of GPER at high concentrations; 50 mM 17β-estradiol-17-D-glucuronide increased the levels of cAMP and PKA activity [108].

Other estrogens like estrone and estriol exhibited very low affinity for GPER, whereas progesterone and testosterone were unable to bind GPER [53]. In the case of estriol, it seems to act as an antagonist of GPER [109, 110].

Considering the weak androgens dehydroepiandrosterone (DHEA) and epiandrosterone, no studies exist reporting GPER binding or activity. However, the endogenous metabolite 7β-hydroxy-epiandrosterone (7β-OH-EpiA) was shown to be capable of antagonizing estrogenic actions mediated by GPER [111].

Curiously, aldosterone, a mineralocorticoid steroid hormone, also seems to have affinity for GPER, with agonistic activity increasing the ERK phosphorylation in rat aortic endothelial cells and the migration of renal cancer cells [112–114]. Moreover, it was shown that the non-genomic effect of aldosterone causing rapid sodium intake involves GPER [115]. However, other studies were unable to detect aldosterone binding to cell membrane fractions [116, 117], suggesting that aldosterone could act through GPER, but without direct receptor binding [102].

Stress related hormones such as cortisol or adrenocorticotropic hormone (ACTH) seem to modulate GPER expression, although their ability to bind the receptor remains to be evaluated [53, 118].

Synthetic Ligands

Tamoxifen, a synthetic non-steroidal anti-estrogen is a recognized selective estrogen receptor modulator (SERM) with demonstrated efficacy as an adjuvant therapy for ER-positive and metastatic breast cancer [119]. It is a substituted triphenylethylene with low binding affinity for ERs compared with E_2, but with a similar binding affinity for both ER subtypes displaying tissue-dependent ER antagonist or partial agonist activity. The 4-hydroxy metabolite (4-hydroxytamoxifen) is a more potent ER ligand and its active metabolite [120, 121].

Another synthetic compound, the estrogen derivative ICI 182,780 binds ERs with high affinity and has anti-estrogenic activity, blocking the ER transactivation, impairing its dimerization and inducing its degradation [120]. Interestingly, both ERs antagonists, 4-hydroxytamoxifen and ICI 182,780, can bind GPER, acting as agonists [122, 123]. Moreover, the tamoxifen analog STX (a diphenylacrylamide) was shown to have rapid neurological responses through a mechanism unrelated to the classical ERs and dependent on a Gq protein, which suggests the involvement of GPER [124]. More recently, another SERM, raloxifene, also was shown to activate GPER in cells devoid of ERα, mediating Akt activation [35, 125].

The knowledge that the classical ER antagonists induce agonistic effects via GPER has stirred their application in cancer treatment. On the light of this information, extreme caution has to be placed on the application of anti-estrogens therapies when GPER is present in tumors, which can be overcome by the introduction of GPER analysis in clinical diagnosis. This would be particularly relevant in estrogen-related cancers, such as breast, endometrium and ovary.

The stilbene estrogen diethylstilbestrol (DES) a potent estrogenic compound binding both ER subtypes with an affinity similar to that of E_2, do not display affinity for GPER [13, 52, 53, 122, 123, 126], though weak binding capability able to induce activation of CREB, and of the downstream targets PKA, Src, and ERK1/2 has been reported [53, 127]. Also, DES seems to inhibit GPER expression [128].

Interestingly, the widely used ERα-selective agonist propylpyrazole triol (PPT) has shown agonistic activity for GPER in a range of 10–100 nM concentrations, whereas diarylpropionitrile (DPN) the ERβ-selective agonist had no activity through GPER [35]. Therefore, the use of PPT to ascertain ERα specific actions in different tissues and physiological conditions may have confounding effects triggered by GPER.

Also, there are reports of compounds disrupting the GPER action transactivating EGFR (Fig. 5.3). The anti-estrogenic benzopyran derivative 2-[piperidinoethoxyphenyl]-3-[4-hydroxyphenyl]-2H-benzo(b)pyran (K-1) showed to interfere with GPER mediated-EGFR activation, decreasing phosphorylation of MEK, ERK and of its downstream effectors such as c-jun and c-fos. Moreover, K-1 decreased the expression of β-catenin, proliferating cell nuclear antigen (PCNA), Cdk4, cyclin D1, c-myc, p-CREB and Bcl-xl and increased the expression of apoptotic markers like NOXA, PUMAα, p21, p27 and Bax [129]. The newly synthesized compound MIBE (ethyl 3-[5-(2-ethoxycarbonyl-1-methylvinyloxy)-1-methyl-1Hindol-3-yl] but-2-enoate) prevented the GPER-mediated EGFR and ERK activation, with consequent effects on gene transcription and cell proliferation, and was, thus, indicated as a GPER antagonist [130].

Membrane-Impermeable Ligands

The development of ligands with membrane impermeable properties, through steroid association with large proteins (e.g. BSA), oligomeric/polymeric dendrimers (PAMAM) or cyclodextrins, as well as the ionic estrogen derivatives (e.g. NMe^{3+}), have allowed studying membrane-associated ERs [131, 132], and to ascertain the existence of GPER effects triggered at cell membrane.

Revankar et al. [133] evaluated the properties of a set of 17α-substituted estrogen derivatives with differential cell permeability. For example, the cationic E_2-NMe^{3+} with a high-affinity for GPER, but membrane impermeable, at least for short periods of incubation and at low concentrations, was incapable of generating Ca^{2+} rise responses, whereas the related neutral carboxamide E_2-NB with membrane permeability similar to those of E_2 rapidly activated both the ERα and GPER [133]. These observations were quite relevant to support the intracellular location of GPER and its functionality driven by the endoplasmic reticulum. Nevertheless, this matter is far away from being solved and widely accepted among the scientific community. Moreover, these novel ligands are important to evaluate the subcellular location of GPER in healthy and pathological tissues and cells, and to decipher whether it varies with different environmental conditions, tissue properties, and/or developmental or disease stage. Further research is needed to clarify all these issues.

Phytoestrogens and Xenoestrogens

In the last years, with the increasing relevance of compounds mimicking estrogens and the concerns about endocrine disruption, many studies have identified some phytoestrogens and xenoestrogens as possible GPER ligands.

Phytoestrogens mainly encompass phenolic compounds, which are structurally divided into flavones, isoflavones, lignans, coumestans, and stilbenes [134]. The potential of phenolic compounds as preventive or therapeutic agents in several human diseases has been under intense investigation, and a large variety of these compounds also has shown to be able to modulate GEPR activity.

(−)-Epicatechin, a flavonoid present in cacao and many other food products, increases the activity of ERK1/2 through GPER activation [135]. Icariin is a prenylated flavonol glycoside isolated from plants of the genus *Epimedium*, which is used as a food supplement; icaritin is its hydroxylated derivative. Both icariin and icaritin seem to induce cellular effects through GPER, also inducing c-fos expression and EGFR-MAPK signaling activation [136, 137]. Baicalein, the main flavonoid derived from the root of *Scutellaria baicalensis* Georgi, a plant widely used in traditional Chinese medicine, also was indicated to interfere with GPER signaling [138]. In breast cancer cells, baicalein does not directly affect GPER expression, but it can inhibit GPER signal transduction pathways, namely, by suppressing the phosphorylation of ERK1/2, Akt, and Src, as well as the transcription of GPER-regulated genes, like c-fos, CTGF, CYR61, and Egr-1 [139].

Genistein, an isoflavone found in soy products, has shown high-affinity for GPER, with agonistic activity leading to MAPKs activation via a Gβγ-associated pathway and requiring Src-related and EGFR kinase activity. Moreover, genistein seems to upregulate c-fos expression [95, 140–142]. Quercetin is a genistein related flavonoid that naturally occurs as the aglycone form of other flavonoid glycosides. At 1 mM concentration, quercetin showed the same behavior of genistein, acting as a GPER agonist with stimulation of c-fos expression [95]. The isoflavone daidzein present in many plants and fruits exhibited both estrogenic and anti-estrogenic actions, despite no GPER binding affinity has been reported. However, it has been shown that daidzein mediates glutamate-induced effects via GPER [143]. Intestinal flora can metabolize daidzein to equol, which also do not have GPER binding ability described, but it has been suggested that it is involved in the generation of mitochondrial reactive oxygen species via GPER-mediated EGFR transactivation [144].

Nevertheless, puerarine an 8-C-glucoside of daidzein, was able to activate and modulate GPER activity and expression [145]. Similar findings were observed with prunetin, an isoflavone found in several agricultural foods, which seems to bind GPER with activation of ERK/MAPK through adenylyl cyclase and cAMP [146]. Tectoridin, the 7-glucoside of the isoflavone tectorigenin, also exerted estrogenic effects mediated through GPER, whereas it binds ERα poorly [147].

Coumestrol is a phenolic phytoestrogen of the class of coumestans known to activate ER-mediated signaling pathways, but its ability to bind and activate GPER was not described yet [140].

Resveratrol, a stilbene derivative with important biological actions and indicated as an anti-cancer compound [148] had its activity inhibited in the presence of GPER antagonists, which suggests that its role depends on GPER, likely by promoting PKC activation [149].

Oleuropein, the glycosylated conjugate of elenolic acid present in olives, also has been included in the group of GPER agonists [150].

The flaxseed derivative secoisolariciresinol diglucoside (SDG), upon metabolization by mammalian cells, originates the metabolite enterolactone (ENL), that shares the similar binding site of GPER specific agonist G1 as indicated by docking simulations, suggesting that it can bind GPER [151].

Xenoestrogens include a wide variety of nonsteroidal chemicals including nonylphenols, polybrominated diphenyl ethers, organophosphates, chlorinated hydrocarbons, biphenyls and phthalates that can be found in daily-life products such as detergents, surfactants, resins, lubricants, plastics, fire retardants, and pesticides [152].

The nonylphenols used in plastic industry, despite having the low affinity for classical ERs, showed a higher binding affinity for GPER [140]. Both bisphenol A (BPA) and nonylphenol, are strong competitors of E_2 displacing the tritiated-hormone from GPER-containing membrane fractions [140]. BPA induced a rapid activation of ERK1/2 and the transcription c-fos likely through an AP1-mediated pathway [153–157]. BPA also stimulates the upregulation of X-linked inhibitor of apoptosis protein (XIAP), with accompanying cell migration and angiogenesis [154, 158–160]. Moreover, BPA at 100 μmol/L seems to be able to increase GPER protein levels in the nucleus, but had no influence in the cytoplasmic fractions [161]. On the other hand, low doses of BPA did not seem to have stimulatory effects via GPER [162]. The 4,40-thiodiphenol (TDP), a BPA derivative that exhibits more potent estrogenic activity than BPA does, can increase the GPER expression and activate the GPER-PI3K/Akt and ERK1/2 pathways [163].

Commercial dichlorodiphenyltrichloroethane (DDT) products consist of a mixture of compounds with the p,p9- and o,p9-isomers, both of which showed binding affinity for GPER, although it was higher in the case of p,p9-isomer [164, 165]. The degradation of DDT produces isomers of dichlorodiphenyldichloroethylene (DDE), p,p9-DDE and o,p9-DDE, which also exhibit binding affinity for GPER [140]. The 2,20,4,40-tetrabromodiphenyl ether (BDE-47), a flame retardant frequently used in furniture, infant products, and electronics, also seems to have effects as a GPER agonist [166].

Several pesticides also have been indicated as activators of GPER. Methoxychor showed ability to bind GPER and activate the GPER/EGFR signaling transduction pathway [140], whereas atrazine, induced ERK1/2 activation and the expression of c-fos and other E_2-target genes. Atrazine and others chlorotriazines, including cyanazine and simazine were also shown to regulate GPER expression in breast cancer cells [167].

Polychlorinated biphenyls (PCBs) are another group of environmental xenoestrogens, including the phenolic derivative 2,29,59-PCB-4-OH, which exhibited low binding affinity for GPER [140]. Contrastingly, the high-affinity binding of GPER to another chlorinated endocrine disrupting chemicals, such as the insecticide kepone was indicated [140].

Also, the lipophilic bis(4-hydroxyphenyl)[2-(phenoxysulfonyl)phenyl] methane, present in cell culture medium as an impurity of phenol red (phenolsulfonphthalein) used as a pH indicator, exhibits estrogenic properties, however, its interaction with GPER remains unclear [164, 168].

In an exhaustive study of Peter Thomas' group [140] comparing the estrogenic activity of a panoply of phytoestrogens and environmental contaminants, it was possible to hierarchize several of the compounds mentioned above concerning its ability to bind GPER. The phytoestrogen genistein was the most effective competitor for GPER, and BPA, and nonylphenol also displayed relatively high binding affinity. Kepone, DDT, 2,2′,5′,-PCB-4-OH and o,p′-DDE had lower affinity for GPER whereas o,p′-DDT, p,p′-DDE, methoxychlor and atrazine caused less than 50% displacement of tritiated-estrogen from GPER-containing membrane fractions [140]. Moreover, it was shown that genistein, BPA, nonylphenol and kepone act as GPER agonists sharing the mechanistic of adenylyl cyclase activation [140].

Other Compounds

Calixpyrroles belong to the category of macrocyclic compounds and are made up of pyrrole units assembled by quaternary carbon atoms at their 2,5-positions [169, 170]. A derivative of these compounds, the calix[4]pyrrole derivative [meso-octamethylcalix[4]pyrrole (C4PY)], acts as an antagonist of GPER preventing the ERK activation by E_2 although without trigger ERK phosphorylation. Moreover, C4PY inhibited the E_2 and GPER agonist-induced expression of c-fos and Egr-1 and Akt phosphorylation [171].

The parabens, methyl-, propyl- and butylparaben, also seem to activate GPER, inducing the activation of the ERK1/2 pathway, an effect mediated by the G protein $\beta\gamma$ dimer. No activation was perceived concerning the cAMP/PKA pathway, and propylparaben was the only paraben able to activate Akt [172].

Reports also exist indicating that metallic compounds actions are mediated by GPER. Copper, an essential trace element naturally occurring in soil, water and air, has been implicated in tumor initiation and progression [173]. A recent study showed that GPER (together with HIF-1α) is required for the copper sulfate

($CuSO_4$)-induced VEGF expression and proliferation of breast cancer cells [174]. Nevertheless, the ability of copper to directly activate GPER signaling requires confirmation. Arsenite and cadmium, environmental contaminants acting as endocrine disruptors, were shown to activate GPER leading to MAPK activation [175], with cadmium directly activating the receptor [176]. Cadmium-induced effects mediated by GPER included the rapid activation of ERK1/2 and Akt and stimulation of breast cancer cells proliferation [177–179]. Zinc is essential in the regulation of several cellular functions, and a recent study also demonstrated that the stimulatory effects of zinc chloride ($ZnCl_2$) in breast cancer cells and cancer associated fibroblasts (CAFs) with the activation of ERK and Akt pathways are dependent on GPER [180].

Nicotinic acid, also designated niacin or vitamin B3, and nicotinamide were reported as agonists of GPER, inducing GPER-mediated ERK activation and gene expression, stimulating cell proliferation and migration [181].

It is important to refer that, despite the high homology of GPER amino acid sequence with the chemokine receptor subfamily of GPCRs, chemokines did not seem to activate GPER [47].

GPER Specific Ligands

The majority of endogenous and synthetic compounds with affinity for GPER also has the ability to bind the classical nuclear ERs. For this reason, it was necessary to find out specific ligands for GPER that would allow separating the contribution of each receptor subtype to the physiological effects of estrogens.

Virtual and biomolecular screening for chemicals with similarity to estrogens in a library with thousands of compounds, retrieved 100 strong candidates to be tested for GPER binding activity. One of these compounds presented agonist activity for GPER and was inactive for the classical ERs. It was called G1 (GPR30-specific compound 1, a substituted dihydroquinoline, with a tetrahydro-3H-cyclopenta[c] quinoline core structure) [182] that still today is the true selective GPER agonist being widely used in *in vitro* and *in vivo* studies.

Binding affinity studies demonstrated a high-binding affinity of G1 for GPER yielding a K_i of 11 nM whereas for E_2 the K_i was 5.7 nM. No significant binding of G1 was perceived for ERα and ERβ at concentrations up to 1 μM [182]. Moreover, functional studies showed that the G1 effects inducing Ca^{2+} mobilization and PI3K activation were exclusive of cells harboring GPER, being absent in cells expressing either ERα or ERβ [182].

More recently, an antagonist of GEPR closely resembling G1 but lacking the ethanone moiety was identified, the so-called G15 [183]. Competitive binding assays demonstrated that G15 binds GPER with an affinity of, approximately, 20 nM. Similarly to G1, G15 poorly binds ERα or ERβ at concentrations up to 10 μM [183]. G15 seems to be able to reduce Akt phosphorylation, and NF-kβ signaling [184].

However, some low-affinity cross-reactivity of G15 with the classical ERα was described. For this reason, efforts were made to identify an antagonist with improved selectivity for GPER. In this way, an isosteric G1 derivative, the G36 was synthesized. This new GPER antagonist showed diminished capability of activating ERα, while maintaining its GPER antagonist activity. Moreover, G36 was shown to selectively inhibit GPER activation of PI3K but not of ERα. Also, it inhibited estrogen and G1 effects on Ca^{2+} mobilization andERK1/2 activation, with no effect on EGF-mediated ERK1/2 activation [185]. Thus, the identification of G36 as a GPER antagonist with improved ER counter selectivity represented a significant step towards the development of new highly GPER selective therapeutics for cancer and other diseases. Indeed, both G1 and G36 have been envisaged as therapeutic tools [35, 186–188].

Other two original compounds, GPER-L1 and GPER-L2, were developed as GPER selective agonists [189], representing also valuable tools to disclose the pharmacologic actions of GPER and to discriminate more accurately the actions driven by each particular ER subtype. These compounds showed high affinity for GPER and were also capable of upregulating the expression of GPER target genes with higher efficacy than G1 or E_2 [189]. More recently, as a strategy to find out a new GPER antagonist, two novel compounds with a structure based on pyrroloquinoxaline were synthesized; PBX1 and PBX2, which were described as selective antagonists having the ability to suppress GPER-dependent signaling [190].

The development of new and selective agonists and antagonists of GPER significantly advanced the current knowledge about the physiological roles of this receptor, controlling biological processes in healthy and pathological conditions. Also, this allowed and reinforced the consistency of targeting GPER as a potential therapeutic strategy, which could be achieved by activating or inhibiting its activity, depending on the deregulation of GPER identified in each particular disease.

Expression Pattern of GPER in Cancer Tissues

GPER is expressed practically in all organs and tissues of the human body and has several important physiological functions in healthy conditions. However, GPER deregulated expression has been associated with several diseases, such as cancer, which may include either upregulated or downregulated expression levels.

Cancer-related expression of GPER has been described both in male and female cancers, with a great representation in hormone-dependent cancers, but reports in other oncological conditions also exist (Table 5.2). However, GPER has been mainly associated with female cancers, particularly breast cancer, but also with the endometrium and ovary cancers. In man, testicular and prostate cancers are important targets of this estrogen receptor.

GPER was firstly identified in human breast cancer cases, and cell lines and, thus, not surprisingly, its function in carcinogenesis is intimately related to breast cancer. Several studies revealed that GPER is strongly expressed in normal breast tissue, with a variable expression among tumor specimens. However, studies show that GPER is detected in the majority of invasive breast tumors whereas only 42% of intraductal tumors were GPER-positive [191, 192]. Indeed, GPER association with breast cancer aggressiveness has been reported [39].

Concerning subcellular location (Table 5.2), GPER seems to be found in the cytoplasm [193], around the nucleus and specifically at the plasma membrane in

Table 5.2 Summary of GPER expression and subcellular localization in different human cancers[a]

Tissue	Tumor subtype	Expression	Localization
Breast	Invasive tumors	++	Membrane Cytoplasm Nucleus
	Intraductal tumors	+	
	Inflammatory breast cancer	+	
	ER-positive	+	
	ER-negative	+	
	Triple-negative	+	
Ovary	–	+	?
Endometrium	ER-negative endometrial carcinoma	+	Cytoplasm
	ER-positive endometrial carcinoma	+	
	Carcinosarcoma	++	
Myometrium	Leiomyomas	++	Cytoplasm Nucleus
Testis	Sertoli cell tumor	+	Membrane Cytoplasm
	Leydig cell tumor	+	
	Seminomas	++	
	Embryonal carcinoma	+	
	Teratomas	+	
Lung	Adenocarcinomas	++	Cytoplasm Nucleus
	Squamous cell carcinomas	++	
	Non-small cell carcinoma	++	
Thyroid	Papillary carcinoma	++	Membrane Cytoplasm
	Follicular carcinoma	++	
Blood	T-cell leukemia	+	?
Bladder	Urothelial carcinoma	+	?
Pancreas	Adenocarcinoma	+	?
	Mucinous neoplasm	+	
Cervix	Carcinoma	+	?
Kidney	Adenocarcinoma	+	?
	Carcinoma	+	?
Pituitary	–	+	Cytoplasm
Prostate	Androgen sensitive	+/++	Membrane Cytoplasm Nucleus
	Castration resistant	++/+	

+, GPER expression; ++, high GPER expression; ?, unknown
[a]Corresponding references are cited and discussed throughout the text

breast cancer cases [194]. Moreover, the cytoplasmic location of GPER in breast cancer seems to be correlated with the less advanced tumor stages, better histologic differentiation, non-ductal histologic subtypes, and enhanced overall patient survival whereas GPER nuclear expression is associated with poorly differentiated carcinomas [195]. Also, GPER overexpression and plasma membrane localization were indicated as key events in breast cancer progression [194]. Oppositely, the absence of GPER at the plasma membrane is indicative of a better long-term prognosis of primary breast cancer [194], suggesting that membrane-triggered GPER signaling may favor tumor progression.

A significant association between GPER and ER expression, as well as a positive correlation of GPER with HER-2/neu, EGFR, C-erbB2 and CD133 expression, tumor size, metastasis, peri-nodal invasion, and lymph node status, were also described [39, 178, 192, 196–198]. Not consistently, other study found no correlation between GPER expression and ERα [193]. In the case of triple-negative breast cancers (TNBCs), which are defined by the absence of classical ERα, progesterone receptor (PR) and HER-2/neu, GPER expression is maintained [199–201] showing a nuclear location [195]. However, GPER expression seems to be restricted to a group of younger women and possibly associated with recurrence of disease since it happened in 22.2% of patients in the GPER-positive group against only 9.5% in the GPER-negative group [199]. This is further supported by findings showing that the *in vitro* inactivation of GPER reduces growth of TNBC cells, preventing the activation of the kinase Src [202]. Moreover, the phosphorylation of the EGFR was almost abolished, and *c-fos* induction was inhibited with GPER inactivation [202].

Other evidence exists implicating GPER in recurrence of breast cancer [198]. A study evaluating GPER expression in local and distant metastatic lesions found that almost 70% of specimens express GPER, and also identified a relationship between GPER positivity and resistance to tamoxifen because recurrent tumors displayed increased GPER expression [196, 203]. Accordingly, a more recent study evaluated 77 breast cancer tissues with 53 recurrent breast cancer specimens after tamoxifen treatment and found that GPER expression was increased in 73.58% of cases [204].

On the other hand, studies are proving that GPER expression decreases with the progression of breast cancer, in consequence of a reversible promoter inactivation by methylation [205, 206]. In addition, GPER down-regulated levels were suggested contributing to breast carcinogenesis and development of lymph node metastasis [34, 207]. Thus, GPER also has been proposed as a tumor suppressor gene [34, 207]. In line with these reports, the protective role of GPER in TNBC cells was described since its activation lead to inhibition of cancer cells growth [208]. Also, it was found that GPER expression in TNBC cases is negatively associated with high-grade tumors and lymph node metastasis, whereas displaying a significant and positive correlation with overall survival [209].

Concerning uterine tissues, GPER was described to be overexpressed in patients with endometriosis comparatively to the normal endometrium, suggesting the involvement of this receptor in disease onset [210, 211]. Endometriosis has been considered a risk factor for endometrial cancer, and in this way, it would be expectable a role for GPER in the endometrium cancer. Expectation confirmed by a study demonstrating that GPER modulates endometrial tumor growth by activation of its characteristic signaling pathways [35]. Besides that, GPER expression is augmented in endometrial carcinoma comparatively with non-neoplastic tissues, with 70% of endometrial carcinomas expressing GPER, against only 26.67% of normal endometrium [212]. However, contradictory reports also exist indicating a decreased expression of GPER in endometrial carcinoma comparatively to the normal endometrium [213, 214]. Moreover, GPER was found to be expressed in ER-negative endometrial carcinoma and in ER-positive and ER-negative endometrial cancer cell lines [212].

Endometrial cancers display GPER expression in the stroma and glandular tissue with protein expression being mainly cytoplasmic [215]. GPER expression was correlated with the grade of endometrial carcinoma but failed any association with FIGO stage, myometrial invasion, or peritoneal cytology [212]. On the other hand, others described that loss of GPER expression was significantly associated with more aggressive features of endometrial tumors and poor prognosis. This included high FIGO stage, non-endometrioid histology, presence of lymph node metastasis, aneuploidy, and ERα loss, which suggests a link between the absence of GPER and the development of an aggressive tumor phenotype [215].

In uterine carcinosarcoma, namely, in the glandular component, GPER expression is higher than that in the normal endometrium [216]. Moreover, GPER expression was higher in advanced stages of carcinosarcoma and seemed to be correlated with the expression of ERβ [216]. On the other hand, GPER is also highly expressed in uterine leiomyomas compared with their matched myometrium and, in this case, ERα also was overexpressed. Furthermore, GPER expression in uterine leiomyoma is detected in the nucleus, but in the myometrium, GPER localization is cytosolic [71].

GPER is expressed in ovarian cancer cell lines [217] and ovarian granulosa cell tumors [218, 219], and seems to be preferentially expressed in "high risk" epithelial ovarian cancer than in less aggressive ovarian tumors [40, 220]. Others also defended that GPER expression increases with the development and progression of ovarian cancer [40]. Accordingly, the high frequency of GPER expression in malignant ovarian endometriotic cysts comparatively with the benign forms was reported [221]. Moreover, GPER expression seems to be positively correlated with clinic pathological parameters, such as tumor stage, size, and lymph node metastasis, and also correlates with matrix MMP-9 expression, mainly in higher stage tumors [221]. In general, increased GPER expression is associated with aggressiveness of ovarian tumors and lower survival rates [40, 220, 221].

Testicular germ cell tumors (TGCTs) are the most common testicular solid cancer accounting for 90% of primary tumors and mainly affecting young men [222]. TGCTs are in majority seminomas, which were shown to overexpress GPER [37, 223–225]. The overexpression of GPER in seminomas comparatively to non-seminomas could be due to two polymorphisms frequently found in the promoter region of GPER in seminomas, rs3808350 and rs3808351 that cause a switch of the homozygous ancestral genotype GG to the homozygous AA genotype [223]. Thus, GPER has been considered a possible biomarker predictive of TGCTs [226]. It was also shown that GPER location in seminoma-derived cells was found at the cell membrane and cytoplasm, with a ratio of approximately 20%:80% [223]. The augmented expression of GPER was described for TGCTs, but GPER was also detected in Sertoli and Leydig cells tumors, embryonal carcinomas, and teratomas [224, 225].

The expression pattern of GPER and its biological roles in PCa will be discussed in a separate topic of the chapter.

Although GPER has been highly related to cancers of reproductive tissues, likely by their hormonal dependency, there are other cancer types displaying altered expression patterns of GPER.

In lung cancer, namely, adenocarcinomas, squamous cell carcinomas, and non-small cell lung carcinomas, GPER expression was significantly increased relatively to the surrounding non-tumor tissue, both at mRNA and protein level [36]. Also, non-small cell lung cancer cell lines display consistently higher expression of GPER relatively to immortalized normal lung bronchial epithelial cells [36]. Moreover, GPER expression in non-small cell lung cancer cells was shown to be nuclear and cytoplasmic, with cytoplasmic location following tumor stages IIIA–IV, lymph node metastasis, and poorly differentiated non-small cell lung cancer cells [227].

Although the expression pattern of GPER in human oral cancers needs to be fully characterized, this receptor was detected in non-neoplastic oral tissues, such as the submandibular and parotid glands, and tongue, and its functional role in oral squamous carcinoma cells also was suggested [184, 228].

In the case of thyroid cancer, a high expression of GPER comparatively to the nodular hyperplasia tissues was found, and associated with lymph node metastasis in papillary thyroid carcinoma [229]. Moreover, GPER, in cooperation with EGFR and CXCR1, seems to have an important role in the diagnosis screening to differentiate between malignant follicular thyroid carcinoma and benign follicular thyroid adenoma because it is highly expressed in malignancy [230].

GPER expression was also reported in other human cancers (Table 5.2), namely, leukemia [231], urothelial carcinoma [16, 232], pancreatic adenocarcinoma and mucinous neoplasm [233], cervical carcinoma [234], kidney carcinoma and adenocarcinoma [235], and pituitary tumors [236], which indicates a broad action of GPER in the control of cell fate and tissue homeostasis contributing to malignant transformation. Also, it is highly indicative that GPER expression is probably altered in other cancers that have not been evaluated yet.

The Role of GPER Controlling Cell Fate and Metabolism

GPER Actions in Cell Proliferation and Apoptosis

Abnormalities in the control of cell-fate decision commonly lead to the unbalance between proliferation and apoptosis being major drivers of carcinogenesis. GPER, the new actor in the estrogens' "drama" that is expressed in a broad range of

Table 5.3 Summary of GPER actions controlling cell fate in different human cancers

Tissue	Tumor subtype	Proliferation	Apoptosis	Invasion and migration
Adrenal cortex	Carcinoma	↓	↑	?
Bladder	–	↓	?	?
Breast	ER-negative	↑↓	↑	↑
	ER-positive	↓	↑↓	↑
	Triple-negative	?	↑	↑↓
Cervix	–	↓	?	?
Endometrium	–	↑	?	↑
Miometrium	Leiomyoma	↑	?	?
Blood	T-cell leukemia	?	↑	?
Kidney	Renal cell carcinoma	?	?	↑
Lung	–	↑	↑	↑
Lymphatic ganglions	Mantle cell lymphoma	↑	?	?
Nervous tissue	Neuroblastoma	?	?	↑
Oral mucosa	Squamous carcinoma	?	↓	?
Ovary	–	↑↓	↑↓	↑
	Granulosa cell tumors	?	?	↓
Prostate	Androgen sensitive	↑↓	↑↓	?
	Castration resistant	↑↓	↑↓	?
Testis	Leydig cell tumor	↓	↑	?
	Seminoma	↑	?	?
Thyroid	Papillary carcinoma	↑	?	?

↑, increase; ↓, decrease; ?, unknown action

neoplastic and non-neoplastic tissues, seems to be one of the molecular protagonists in the regulation of cell fate, influencing cell proliferation, apoptosis and malignant features, such as, cell migration and invasion (Table 5.3).

Some studies have reported that GPER is able to stimulate cell growth, increasing the proliferative activity. Estrogens and the GPER selective agonist G1 were shown to increase the mitotic index in the epithelial non-neoplastic MCF10A breast cells, and also the proportion of dividing cells in normal and malignant human breast and breast cancer explants [237]. The dependency of GPER to promote breast cells growth was further confirmed by the estrogenic effects observed on ER-negative breast cancer cells [178]. Moreover, GPER activation in MCF10A cells led to the activation of ERK in a process requiring EGFR transactivation [237]. Interestingly, in breast CAFs, E_2 seems to cause an interaction between GPER and phosphorylated EGFR, recruiting them to the cyclin D1 gene promoter [70]. In this CAFs, G1 activation of GPER resulted in transient increases in cell mitotic index, intracellular Ca^{2+}, and ERK1/2 phosphorylation, linked with the promotion of proliferation and cell-cycle progression [238]. This supports the idea that GPER is involved in the functional connection between breast tumor cells and CAFs [239]. GPER activation in breast and endometrial cancer cells also was associated with the upregulation of Egr-1, a transcription factor involved in the promotion of cell proliferation. It was reported that G1 induces transcription of Egr-1 through GPER/EGFR/ERK signaling and induces the recruitment of Egr-1 to the CTGF and cyclin D1 promoter sequences [98].

In endometrial cancer cells and leiomyoma cells GPER increased cell proliferation also through the MEK/ERK MAPK pathway [126, 212, 240–242]. Pretreatment of endometriotic cells with G1 stimulated cell proliferation accompanied by rapid Akt phosphorylation, which was reversed by the GPER antagonist G15 [243]. Accordingly, a recent study showed that targeting GPER by the overexpression of the regulatory miR-424, decreased E_2-induced cell proliferation by inactivation of the PI3K/Akt signaling [244].

In the case of ovarian cancer cells, an augmented proliferation was observed upon G1-stimulation and GPER activation, which was underpinned by an increased number of cells in S-phase, and upregulated levels of c-fos, cyclin D1, cyclin E, and cyclin A proteins [96, 245].

In primordial germ cells, precursors of both sperm and eggs, GPER activation by estrogens induced cell proliferation through Gβγ-subunit protein- and matrix metalloproteinase-dependent transactivation of the EGFR [17]. EGFR signaling activated the PI3K/Akt/β-catenin pathway, increasing c-fos, c-myc, and cyclin D1/E expression [17]. The proliferative effects of GPER over germ cells were confirmed at post-natal stages. G1 induced proliferation of the spermatogonial GC-1 cell line, with the upregulated expression of cyclin D1 and through a cross-talk between GPER and ERα, and activation of the EGFR/ERK/fos pathway [246].

Also in thyroid cancer cells, GPER activation caused transcriptional activity of c-fos promoter with increased expression of c-fos, cyclin A, and cyclin D1 [247].

The GPER also contributed to the proliferation and survival of mantle cell lymphoma cells, and its expression levels were correlated with Akt and MAPK phosphorylation, as well as, with cyclin D1 expression [248]. The activation of MAPK pathway by GPER also seems to be involved in the induction of proliferation in lung cancer cells [227].

There are also reports of GPER stimulation of cell proliferation through its activation by other compounds than estrogens or G1. It is the case of the estrogen-mimicking compound used in plastic production BPA that activates GPER with enhanced proliferation of breast cancer cells and augmented levels of c-fos, Egr-1, and CTGF proteins [98, 249]. BPA also was shown to increase proliferation of spermatogonial cells and the proliferative activity in testicular seminomas, effects mediated by the GPER [155, 156, 250, 251].

Also, the ER antagonist 4-hydroxytamoxifen, known to act as an agonist of GPER [126], stimulated breast [97, 98] and thyroid [247] cancer cells growth. Regarding thyroid cancer cells, similar effects were seen with the phytoestrogen genistein [247].

Overall, the facts described above argue for the “villain role” of GPER in tissue homeostasis by its ability to promote cell growth and proliferation. However, the other face of GPER as the “good guy” is sustained by many other studies that demonstrate it suppressive effects on cell proliferation, including in tissues where the pro-proliferative activity was described.

The activation of GPER inhibited the growth of ER-negative breast cancer cells with the cell cycle arrested at G2/M [252]. These effects were underpinned by the downregulated expression of G2-checkpoint regulator cyclin B and increased expression of p53, as well as, by enhanced p53 phosphorylation at serine 15, which promotes its nuclear translocation and inhibits ubiquitylation. The augmented levels of p53 were also accompanied by increased expression of cell cycle inhibitor p21 [252]. GPER also inhibited proliferation of ER-positive breast cancer cells, blocking cell cycle at G1, and decreasing the population of cells in S-phase, since both p53 and p21 proteins were upregulated by G1 administration [178]. G1 suppression of cell growth and arrest of cell cycle lead to caspase activation and apoptosis. Moreover, it was shown that G1 could bind the colchicine binding site of tubulin, inhibiting tubulin polymerization and the subsequent assembly of mitotic spindle apparatus in breast cancer cell mitosis [253].

In human cervical cancer cells, GPER activation induced G2/M cell cycle arrest and down regulated cyclin B expression, inhibiting cell proliferation through ERK1/2 and EGFR signaling [234]. Also, in human bladder-derived T24 carcinoma cells, E_2 inhibited cell growth via GPER mediated phosphorylation of ERK [232].

In ovarian cancer cells, it was shown that G1 blocks tubulin polymerization hampering the assembly of microtubules, which led to the cell cycle arrest and suppression of proliferation [254]. Similar G1 effects were observed in endothelial cells, with inhibition of DNA synthesis and accumulation of cells in S and G2 phases of cell cycle [255], and adult female rats hippocampus that displayed significantly diminished cell proliferation [256].

The activation of GPER was also associated with the decreased proliferation of adrenocortical carcinoma cells [186], and GPER-positive endometrial adenocarcinoma cell lines HEC-1A and RL95-2, but not in GPER-negative endometrial adenocarcinoma HEC-1B cells [213].

In the other plate of the balance of tissue homeostasis, there is apoptosis, the fundamental biological process of programmed cell death that also seems to be regulated by the GPER. Curiously, and as discussed for cell proliferation, GPER actions in the control of apoptosis are the two faces of the same coin. Studies exist reporting that GPER activation diminishes apoptosis, while others defend the opposite.

The apoptotic process may be triggered by two distinct and interrelated pathways, the intrinsic (or mitochondrial) and the extrinsic (or death receptor) pathway that converge at the activation of apoptosis effector caspase-3 [257]. Nevertheless, the GPER effects have been mainly related to the mitochondrial pathway and altered expression or activity of the Bcl-2 protein family of apoptosis regulators [258]. The antiapoptotic effect of GPER was stated, for example, in the ERα-negative ovarian cancer cells OVCAR5, by diminishing the expression of cleaved-caspase-3 [245]. In rat Sertoli cells, GPER seems to decrease apoptosis, by increasing the expression of Bcl-2 and decreasing Bax levels [259], respectively, anti- and pro-apoptotic members of the Bcl-2 family [258]. The same effect of GPER activation elevating Bcl-2 levels and reducing Bax with the consequent inhibition of apoptosis was seen in myocardial cells following ischemia/reperfusion injury [260]. The GPER activation through the ERK pathway also was seen to be cardioprotective by inhibiting the opening of mitochondria permeability transition pore mPTP [261]. Furthermore, GPER stimulation enhanced mitochondrial function and decreased oxidative stress in cardiac muscle [262]. Moreover, decreased TNF-α levels were reported following ischemia-reperfusion and GEPR activation [260]. Accordingly, E_2 enhanced Bcl-2 expression and CREB phosphorylation, preventing oxidative stress-induced apoptosis in keratinocytes by phosphorylating cAMP response element-binding protein via cAMP/PKA pathway, an effect mediated via membrane GPER [99]. The GPER also showed antiapoptotic and protective effects in spinal motor neurons after injury through the activation of PI3K/Akt pathway [263].

Other mechanism associated with the GPER inhibition of apoptosis was reported in the ER-positive MCF7 breast cancer cell line, and involves FOXO3a inactivation [264]. FOXO3 is a transcription factor that specifically induces the transcription of proapoptotic genes, such as Bim, p21, and p27 [265]. The activation of GPER leads to rapid FOXO3a translocation to the cytoplasm, and this process seems to be achieved by the p110α catalytic subunit of PI3K as a result of EGFR transactivation. Additionally, G1 stimulation of MCF7 cells resulted in decreased caspase activation under proapoptotic conditions [264]. In this report, the SERMs tamoxifen and raloxifene, as well as ICI 182,780, also were able to mediate FOXO3a inactivation in a GPER-dependent mechanism [264].

The antiapoptotic action of GPER was also evidenced by the use of GPER antagonist G15. Administration of this compound induced apoptosis of human oral squamous carcinoma cells, indicating that GPER activation is needed to sustain cell survival and diminish apoptosis [184].

Nevertheless, a substantial number of studies report the proapoptotic nature of GPER. In ER-positive MCF7 breast cancer cells, and contrastingly with the stated above, the proapoptotic role of GPER also was described. Knockdown of receptor was linked to decreased basal expression of tumor suppressor protein p53, and increased apoptosis and decreased cell-cycle progression [266]. This proapoptotic action of GPER was also found in the ER-negative SK-BR-3 breast cancer cells, concomitantly with enhanced expression of Bax, Bim, and cleaved-caspase-3, and diminished expression of Bcl-2 [205, 252]. Similar effects were seen in the triple-negative MDA-MB-231 and MDA-MB-468 breast cancer cells [208] that, despite devoid of classical ERs, are known to express GPER [208], which demonstrates the importance of GPER inducing apoptosis.

Also in other tumor types, the GPER effects inducing apoptosis were indicated. It is the case of tumor Leydig cells, in which GPER activation led to apoptosis, with decreased Bcl-2 and increased Bax expression, diminished cytochrome c release, and decreased activation of caspase-3 and poly (ADP-ribose) polymerase 1 (PARP-1) [267]. In ovarian cancer cells, the GPER agonist G1 caused DNA fragmentation, increased the expression of cell-cycle inhibitor p21, and decreased the expression of Bcl-2 and cleavage of PARP and fodrin, two important markers of apoptosis [254]. Enhanced caspase-3 activation and apoptosis in response to G1 stimulation and on the dependency of GPER actions were also reported in human ovarian endometriotic stromal cells [268], lung cancer cells [269] and adult T-cell leukemia cells [231].

Moreover, also in the adrenocortical H295R carcinoma cells, G1 treatment was shown to cause morphological changes in cell nuclei, DNA damage and apoptosis by the activation of the intrinsic pathway [186]. GPER agonist treatment increased Bax expression whereas decreasing Bcl-2, and increased cytosolic cytochrome c levels decreasing its content in the mitochondrial compartment. Accordingly, activation of the initiator caspase-9, as well as, the executioners caspase-3/7 was detected [186].

There is also the example of rat pachytene spermatocytes, in which GPER activation by E_2 or G1 up-regulates the expression of proapoptotic factor Bax whereas downregulating the cell cycle regulators cyclin A1 and B1 [80]. Moreover, it was demonstrated that the rapid EGFR/ERK/c-Jun pathway modulates gene expression towards the balance between cellular proliferation and apoptosis [80]. Similarly, in the mouse spermatocyte-derived cell line GC-2, GPER activation caused rapid ERK, c-Jun and p38 phosphorylation, Bax upregulation, Bcl-2 downregulation, cytochrome c release, caspase-3 and PARP activation, DNA damage and increased expression of cell cycle inhibitor p21 [270].

GPER regulation of cell proliferation and apoptosis has also been linked with the regulation of other associated processes, namely, oxidative stress and Ca^{2+} homeostasis. In lung cancer cells, it was shown that GPER regulates the NO levels, and superoxide dismutase (SOD), catalase and glutathione peroxidase (GPx) activity, contributing to decreased cell proliferation and increased apoptosis [269, 271]. GPER was able to increase T-type Ca^{2+} channels currents in trigeminal ganglion [272], and the activity of L-type Ca^{2+} channel α 1D subunit in endometrial carcinoma cells and breast cancer cells, which was required for the E_2-stimulated Ca^{2+} influx associated with the promotion of cell proliferation, and development of endometrial cancer [273, 274].

In the scenario of cell-fate decision and maintenance of tissue homeostasis, it became evident the opposite effects of GPER in the control of cell proliferation and apoptosis. This duality of action most likely depends on a panoply of factors that may include the type of tissue and physiological context, the dose of activators and time of exposure, as well as other unforeseeable variables. The different effects of GPER also would be explained by a regulatory role of GPER shaped by the tissue specificity in particular environmental and biological conditions.

GPER Influence Over Cell Migration and Invasiveness

Cell migration and invasiveness are coordinated biological processes that play a major role in cancer progression being closely related to metastization. The loss of cell-cell adhesion is a driven event for malignant cells evasion from their primary sites, which also depends on the degradation of the extracellular matrix, acquisition of an invasion phenotype, and finally, invading and metastasizing to other organs/ tissues [275]. Indeed, the great majority of cancer deaths (>90%) are caused by metastasis rather than by the primary tumors [276], which renders cell migration a therapeutically relevant target point. Both migration and invasion seem to be promoted by GPER in ER-negative breast cancer cells through the activation of ERK and Akt pathways, nuclear translocation of NF-κB and increased expression of interleukin-8 (IL-8) [277]. The role of this cytokine in the progression and metastasis of a variety of human cancers has been discussed [278]. GPER activation also leads to migration and invasiveness of TNBC cells and inflammatory breast cancer cells, responses mediated by the activation of the ERK1/2 pathway [279, 280]. Furthermore, GPER seems to upregulate β1-integrin expression, through EGFR/ERK signaling pathway, which promotes migration of CAFs and epithelial-mesenchymal transition of tamoxifen-resistant breast cancer cells [204], contributing to tamoxifen resistance via interaction with the tumor microenvironment [204, 281]. Moreover, GPER activation can lead to the increased expression of IL1β in CAFs and IL1 receptor 1 (IL1R1) in breast cancer cells, promoting the interaction between these two cell types, the upregulation of inflammatory target genes, inducing migration and invasion of breast cancer cells [282]. Accordingly, GPER inhibition with G15 seems to increase the sensitivity of epithelial breast cancer cells to doxorubicin by preventing epithelial-mesenchymal transition [283].

GPER actions prompting cell migration and invasion were also described in ER-positive breast cancer cells, through the activation of ErbB2-ERK signaling transduction pathway [284]. In this breast cancer cell type, GPER enhanced migration also by mediating the dramatic proteolysis of cyclin E, with the involvement of EGFR signaling [281]. Other results showed that E_2 and ICI 182,780 enhanced adhesion of MCF7 breast cancer cells to matrigel, with increased autolysis of calpain 1 and proteolysis of focal adhesion kinase (FAK), with calpain activation through the ERK1/2pathway [285]. Also, the xenoestrogen BPA stimulated migration of breast cancer cells and CAFs via GPER [249] by FAK, Src and ERK2-dependent pathways [286].

Interestingly, antagonizing GPER activity by natural compounds, like baicalein, diminished the expression of GPER target genes, including cysteine-rich 61 (CYR61) and CTGF, and suppressed E_2-stimulation of migration and invasion of breast cancer cells [138].

The pro-invasiveness effects of GPER were also identified in ovarian cancer cells OVCAR-5 and SKOV3 with modulation of expression and activity of MMP-2 and MMP-9, extracellular matrix proteins with a determinant role in cancer cell invasion and metastasis [287, 288]. Similar effects dependent on increased MMP-2 and MMP-9 production were seen in the endometrial Ishikawa and KLE cancer cells [241]. In the endometrial cancer cell line RL95-2, GPER mediated carcinogenesis and invasion via the activation of MEK/ERK MAPK pathway [240]. Moreover, the phosphorylation of FAK, a tyrosine kinase with a key role in tumor cells invasiveness [289], also was involved in the migration of Ishikawa and RL95-2 cells induced by GPER activation [290]. The association of MAPK pathway with cell migration and invasion in response to GPER activation also was described in lung cancer cells [199]. Also in lung cancer and mesothelioma, GPER contributes to the chemotaxis and migration, through IGF-I actions [291].

In renal cell carcinoma, GPER activation mediated cell invasiveness via PI3K/Akt/MMP-9 signals [235].

The environmental pollutant BDE-47 increased migration and invasion of neuroblastoma SH-SY5Y cells in a mechanism dependent on GPER [166]. The associated molecular events included the downregulated expression of E-cadherin and zona occludin-1, and the upregulated expression of MMP-9, via activation of PI3K/Akt signaling pathway [166].

Nevertheless, contrary evidence exists depicting the role of GPER suppressing cell migration and invasion (Table 5.3). A recent study has shown that the activation of PI3K/Akt and ERK1/2 pathways by the GPER specific agonist G1 suppress migration and invasion of TNBC cells. Other responses included the inhibition of the epithelial mesenchymal transition by reducing the phosphorylation, nuclear localization, and transcriptional activity of NF-kB, [209]. There is also evidence that E_2-effects mediated by the membrane GPER in human metastatic breast cancers lead to decreased cell adhesion, through a PKA-dependent mechanism requiring the activity of voltage-gated sodium channels (VGSCs) [292].

In granulosa cell tumors, E_2 decreased cell migration and matrix invasion, effects accompanied by GPER inhibition of ERK1/2 signaling through a non-genomic mechanism [219]. Other studies with ovarian cancer cells suggested that E_2 inhibits tumor invasion, by inhibiting EGF-induced cell migration and the expression of urokinase plasminogen activator receptor (uPAR) expression, effects mediated by GPER [293]. Also in lung, and contrarily to the studies mentioned above, the activation of GPER showed to inhibit the migration of non-small cell lung cancer cells, suppressing the activity of IKK-β and NF-κB [294].

Angiogenesis, Inflammation and the GPER

Angiogenesis and inflammation are other important hallmarks in tumor progression fueling cancer cell growth within the "hostile" tumor microenvironment. GPER activation increased VEGF levels in ER-negative breast cancer cells and CAFs via upregulation of the HIF1α in consequence of activation of the EGFR/ERK/c-fos signaling pathway [83]. Moreover, the receptor seems to be involved in the formation of human endothelial tube, enhancing angiogenesis and tumor progression [83].

The GPER knockout mice display a pro-inflammatory phenotype with augmented levels of the pro-inflammatory and immunomodulatory cytokines IL-1β, IL-6, IL-12, TNFα, monocyte chemotactic protein-1 (MCP-1), interferon γ-induced protein 10 (IP-10) and monokine induced by γ interferon (MIG), concomitantly with the decreased expression of the adipose tissue-specific cytokine adiponectin [295]. Contrastingly, the anti-inflammatory role of GPER was shown in endothelial cells, in which, receptor activation counteracted the TNF effects inducing upregulation of pro-inflammatory proteins, namely, intercellular cell adhesion molecule-1 (ICAM-1) and vascular cell adhesion molecule-1 (VCAM-1) [296]. Moreover, airway inflammation was suppressed by the GPER agonist G1 in a mouse model of asthma, apparently via IL-10 [297]. In human breast cancer cells, GPER inhibited the TNFα-induced IL-6 expression, probably through blockage of NF-κB promoter activity [298]. Consistent with that findings, a recent study in a GPER knockout mouse model, showed accelerated development of liver tumors, with immune cell infiltration, fibrosis, and production of inflammatory factors, including IL-6. More studies are warranted to establish the role of GPER in angiogenesis and inflammation.

GPER as a Regulator of Cancer Cell Metabolism?

In the last years, the ability of tumor cells to reprogram metabolism, sustaining their high energy requirements, proliferation, and survival, has been indicated as a characteristic of malignant transformation and recognized as a hallmark of cancer [299]. However, the most well-known adaptation of cancer cell metabolism was described in the 1930s by Otto Warburg and is known as the "Warburg effect" that consists in the increased glucose consumption followed by the augmented production of lactate instead of the oxidation of pyruvate in mitochondria, even under aerobic conditions [300]. In the light of available literature, GPER has "something to say" in the control of the metabolic process. GPER knockout female mice showed increased plasma glucose levels, leading to hyperglycemia and glucose intolerance, associated with the decreased expression of insulin in isolated pancreatic islets and diminished insulin release both *in vitro* and *in vivo* [301]. Likewise, glucose intolerance and insulin resistance were reported in the GPER knockout male mice [295]. In agreement, GPER seems to induce insulin secretion in pancreatic β-cells under low- and high-glucose conditions, through the activation of EGFR, ERK and PI3K/Akt pathways [302]. Curiously, insulin resistance is closely related to endometrial oncogenesis, and it seems to up-regulate Ten-Eleven Translocation 1 (TET1) that can up-regulate the GPER expression, activating PI3K/AKT signaling pathway and promoting cell proliferation [303]. Available data revealed an important role of GPER in glucose metabolism though in the particular case of cancer cells, information relating GPER with glycolytic metabolism is scarce, being an issue that deserves attention of direct studies. Nevertheless, Yu et al. [304] first reported the interplay between breast cancer cells and CAFs concerning glucose metabolism. This study showed that the activation of the PI3K/AKT signaling pathway in breast cancer cells induces the cytoplasmic translocation of GPER in CAFs, leading to the activation of GPER/cAMP/PKA/CREB signaling and to the aerobic glycolysis switch in CAFs. In turn, the glycolytic CAFs feed breast tumor cells with extra pyruvate and lactate augmenting mitochondrial activity. Moreover, cytoplasmic GPER expression in stromal fibroblasts predicted high tumor metabolic activity and potent energy transfer between the stroma and cancer cells [304].

Besides glycolysis, other energy routes are involved in the metabolic adaptation of cancer cells, namely, glutamine metabolism, and its resulting product glutamate

[305]. The glutamate transporter-1 (GLT-1) is an essential player removing the extracellular excess of glutamate. GLT-1 expression was shown to be increased by GPER activation triggered by G1 in rat primary astrocytes, which was accompanied by the increased uptake of glutamate [306]. The expression of glutamate aspartate transporter (GLAST), other transporter involved in glutamate uptake, was also increased in response to G1 [307]. Moreover, the mechanistic under this regulation was outlined with effects reported to occur via the MAPK/ERK, PI3K/Akt, TGF-α/EGFR, Src, NF-κB and CREB pathways [307]. Further studies are warranted to confirm whether GPER actions regulate glutamine and glutamate metabolism in cancer cells, and if it has impact on disease progression and aggressiveness.

GPER also seems to have a major role in the regulation of lipid metabolism, another energetic pathway widely altered in cancer cells. Besides having a glucose intolerance phenotype, GPER knockout male mice display an altered lipid profile with increased cholesterol and triglyceride levels [295], which was corroborated by *in vitro* findings. GPER activation decreased triglyceride accumulation in cultured rodent β cells and reduced lipid synthesis in pancreatic islets by reducing the expression and activity of fatty acid synthase (FAS), the master effector of *de novo* lipogenesis [308]. This study also showed that the β cell transcription factor pancreatic and duodenal homeobox 1 (Pdx1) and the downstream target genes, such as proinsulin 1 (Ins1), glucose transporter 2 (GLUT2), and glucokinase (Gckr) were down-regulated by E_2 via GPER action [308]. Furthermore, the FAS is the target of lipogenic transcriptional regulators, namely, the sterol regulatory element-binding protein 1c (SREBP1c), the carbohydrate response element binding protein (ChREBP), and the liver X receptor (LXR) [309–311], which were suppressed in response to GPER activation [312].

However, in breast (SkBr3), colorectal (LoVo), and hepatocarcinoma (HepG2) cancer cells and CAFs, GPER seems to up-regulate FAS expression and activity, through the EGFR/ERK/c-fos/AP1 transduction pathway [313]. Other studies reported that the GPER action increased the expression of low-density lipoprotein (LDL) receptor, and that G1 stimulation induced the uptake of LDL in liver cells [314, 315]. Curiously, GPER promoted adipogenesis *in vitro* and seems to be involved in the development of obesity in female mice exposed to a high-fat diet since GPER knockout mice were resistant to diet induced-obesity, glucose intolerance and insulin resistance [316].

Malate dehydrogenases (MDHs) are metabolic enzymes that function as transcriptional factors to regulate the expression of oncogenes and tumor suppressor genes. The expression of MDH2, one of the isoforms of MDHs involved in citric acid cycle in mitochondria [317], was shown to be increased in endometrial cancer cells in response to GPER activation [318]. Moreover, the augmented expression of MDH2 lead to enhanced cell proliferation, migration and invasion but inhibited apoptosis of endometrial cancer cells by suppressing PTEN expression [318].

In sum, GPER has important roles in insulin sensitivity, glucose tolerance, adiposity and energy balance in many cell types, which ignites the curiosity about the activity of GPER in the metabolic reprogramming of cancer cells. Also, it is liable to speculate that GPER selective ligands would have therapeutic value for cancer treatment targeting metabolism.

GPER in Prostate Cancer

PCa is one of the most common oncological disorders in men, being highly frequent in men over the age of 50 [319]. Indeed, aging is the main risk factor for PCa, which raises the concern about its incidence in the next years considering the population aging in consequence of the increased expectancy of life. The vast majority of PCa patients (90%) develop castration-resistant PCa (CRPC), an advanced metastatic form of disease, characterized by its aggressiveness and resistance to the classical anti-androgen therapies, which remains incurable with an average survival time of only 16–18 months [320, 321]. The search for new and effective approaches for better management and treatment of both localized and metastatic PCa remains crucial and urgent.

It is well established that the development and progression of PCa are strongly regulated by sex steroid hormones, with the male hormones androgens highly implicated in the regulation of prostate functions from the embryonic development to adulthood, but also in the onset of prostate malignancy [322]. Estrogens, classically viewed as female hormones, also seem to have a relevant role in PCa with some studies reporting their actions as causative factors in prostate carcinogenesis [323, 324]. On the other hand, estrogens have been used to treat PCa by its negative feedback actions at the hypothalamus and pituitary reducing circulating androgens levels [325]. However, has also been shown that estrogens have beneficial direct effects on PCa cells [326]. Independent studies have indicated that these steroids can be protective against PCa by their proapoptotic and antiproliferative actions over prostate cells [326, 327].

Estrogenic actions through the interaction with the classical nuclear estrogen receptors, ERα and ERβ, and involving genomic responses characterized by changes in gene transcription and protein *de novo* synthesis have been reported in both non-neoplastic and neoplastic prostate cells [328]. However, the activity of these receptors in PCa relies on the paradigm that ERα is responsible for the proliferative effects of estrogens being oncogenic whereas ERβ is considered protective, anti-carcinogenic and related with the activation of apoptosis [329].

Also, the GPER has been identified in normal human prostate, and benign and neoplastic conditions. The prostate gland mainly encompasses two compartments: the fibromuscular stroma and the glandular tissue. In the glandular epithelium, it is possible to distinguish two histological layers, the secretory luminal layer consisting of tall columnar cells responsible for the production of prostatic secretions, and an underpinning basal layer of cuboidal epithelial cells. The population of basal epithelial cells has shown to be heterogeneous and containing adult prostate stem cells, which have been proposed as cancer stem cells in the origin of PCa [330]. In the human benign prostate gland, a strong GPER expression was observed in the cytoplasm of basal epithelial cells [331]. Therefore, it is liable to assume that GPER may the mediator of estrogenic actions in this proliferative prostatic compartment. Accordingly, GPER expression was found in prostaspheres derived from human normal adult prostate stem cells [332].

GPER also was found in human prostate epithelial cells, displaying both membrane and cytoplasm localization [332–334]. Furthermore, a weak staining in the cytoplasm of stromal cells of benign prostate was also detected, suggesting the possible involvement of this receptor in estrogen signaling between stromal and epithelial prostatic compartments [331]. Accordingly, GPER was found to be expressed in human primary prostate stroma cells, derived from fresh surgical prostate specimens of benign prostate hyperplasia (BPH), and in the normal human prostate stroma cell line, WPMY-1 [335].

The reports characterizing the GPER expression in neoplastic conditions are relatively scarce, but a study found that GPER was expressed in 97.5% of all prostate tumor cases evaluated [336]. It was described that pre-neoplastic lesions (high-grade prostatic intra-epithelial neoplasia) show higher GPER expression comparatively with non-neoplastic tissues, and that GPER expression decreases from moderately to poorly differentiated PCa [331]. Moreover, GPER expression was correlated with the Gleason score of prostate adenocarcinoma being intense in Gleason patterns 2 and 3, and weak to moderate in Gleason pattern 4 [331], which indicates an inverse relationship between GPER expression and neoplastic cell differentiation. Despite the predominant, and sometimes nearly exclusive, cytoplasmic expression of GPER in PCa cases, areas with membrane and nuclear staining were also found [331, 336, 337].

Considering the androgen-sensitive stage of disease and CRPC, GPER seems to be more expressed in CRPC cases; 80% of metastatic CRPC highly express GPER, against only 54% of the primary PCa cases [338]. However, no correlation was found between GPER expression and, age, the Gleason score of primary cancer or PSA level. The type of androgen deprivation therapy, or duration of treatment also were not associated with GPER [338]. These results are not totally in agreement with the findings obtained in PCa cell lines. GPER expression levels were markedly higher in the normal progenitor prostate cells relatively to the androgen-sensitive PCa cell line LNCaP [332]. In CRPC cell line models PC3 and DU145, GPER expression was higher in PC3 cells, which mimic a more aggressive metastatic stage of disease [334]. Recent findings from our research group also demonstrated that GPER expression tends to be lost with the acquisition of the castration phenotype and aggressiveness of PCa cell line models; GPER expression in LNCaP > PC3 > DU145 cells (results to be published elsewhere).

As previously discussed for other cancer types, also in PCa, the GPER dual-mode of action has been described with opposite effects arising from separate independent studies. Some authors argue that GPER is involved in PCa development and progression based on the assumption that GPER may mediate the estrogen-initiated transformation of prostate epithelium derived from normal human prostate stem-progenitor cells, and promote the progression to invasive adenocarcinoma [332]. These data were obtained in a novel chimeric mice prostate model containing human-rat prostate tissues. Human prostate progenitor cells cytodifferentiated in chimeric prostate tissue originating epithelial hyperplasia, prostate intraepithelial neoplasia and PCa after exposure of nude mice to elevated concentrations of testosterone plus E_2 [332]. The pro-tumor action of GPER was also supported by the fact that its activation by the specific agonist G1 induced growth of normal stromal prostate cells, but not of BPH-derived cells [339]. Furthermore, GPER activation induced ERK1/2 phosphorylation, which was suggested contributing to PCa progression and hormonal independence in the PC3 cell line [333].

On the contrary, a substantial body of evidence defends the protective role of GPER in PCa. Treatment of LAPC-4 (androgen-sensitive) and PC3 (castration-resistant) PCa cells with E_2 or DES identified the growth inhibitory activity of GPER triggered by different mechanisms [340]. Also, G1 administration showed to inhibit growth of both androgen-sensitive (LNCaP) and CRPC (DU145 and PC3) cell line models [334]. The agonist G1 induced a persistent cell-cycle arrest at the G2/M phase in LNCaP and PC3 cells, which resulted in enhanced apoptotic activity, as indicated by the increase in the Annexin V–positive cell population [334]. Moreover, G1 effects mediated by GPER induced the activation of ERK1/2 and c-jun/c-fos dependent upregulation of p21, which was accompanied by the down-regulated expression of G2-checkpoint regulators, such as cyclin B1, cyclin dependent kinase 1 (cdc2) and its phosphorylated proteins, as well as by the diminished levels of cdc25C and cyclin A2 [334]. The mechanism involved in the GPER sustained activation of ERK1/2 was disclosed recently being shown that it depends on the activation of Gαi1 proteins, which are highly expressed in PCa cells [341].

Interestingly, the low GPER expression described in poorly differentiated PCa was associated with an increased expression and activity of Akt, and the transcription factor CREB [331], indicating that loss of GPER may promote PCa cell survival and proliferative activity.

Other relevant study in this field, using LNCaP (androgen-sensitive cell line) and PC3 (castration-resistant cell line) xenografts as PCa models, reported that GPER activation by G1 inhibited growth of castration-resistant, but not of androgen-sensitive tumors with no observable toxicity [338]. Growth inhibition of castration-resistant tumors by G1 was underpinned by the increased apoptosis, demonstrated by the increased number of cleaved caspase 3-positive cells [338]. Other study, similarly described the *in vivo* anti-cancer role of GPER in PC3 xenografts. G1 administration significantly suppressed tumor growth having no effect on growth and histological features in the prostates of intact mice [334]. Moreover, this study showed that G1 treatment only inhibited the growth of PCa and actively proliferating BPH cells without affecting growth of quiescent cells.

In agreement with the *in vitro* and *in vivo* anti-tumorigenic actions of GPER mediated by estrogens and G1, a study with natural compounds also found decreased proliferation and increased apoptosis of PCa cells in response to GPER activation. The SDG, a lignan extracted from flaxseed was shown to suppress the development of BPH in a rat model [151]. ENL, the metabolite of SDG inhibited prostate cell growth, which significantly restricted the enlargement of rat prostate [151].

Mechanistically ENL activated the ERK pathway causing cell cycle arrest with upregulation of p53 and p21 and downregulation of cyclin D1 [151].

The available literature unquestionably places GPER as a modulator of PCa cells growth regulating the expression of cell cycle-regulators and other cellular elements, but its influence over the control of metastatic features also was reported. G1 treatment inhibited migration and the invasion properties of both PC3 and DU145 cells by reducing the formation of filopodia and stress fibers [341]. G1 administration *in vivo* was also capable of reducing the intratumoral microvessel density in a castration-resistant xenograft tumor model [338], suggesting the anti-angiogenic role of GPER in PCa.

Other described effect of GPER in prostate is related with inflammation, with its activation leading to increased expression of COX-2 and augmented secretion of IL-8 by human prostate cells [342]. The pro-inflammatory actions of GPER were further supported by the observed marked intratumoral infiltration of neutrophils and upregulated expression of neutrophil-related chemokines and inflammation-mediated cytokines in a CRPC model after treatment with G1 [338].

In what concerns cancer cells metabolism, there are no reports implicating estrogens and GPER actions in the metabolic reprogramming. However, the emergent impact of metabolism as a therapeutic target together with others and ours recent findings highlighting the role of sex steroid hormones as metabolic regulators [343–345] strongly encourage opening this “avenue” of research.

Overall, the high expression of GPER in PCa, particularly in CRPC, and the demonstrated beneficial effects of G1 agonist counteracting the malignant behavior of PCa and BPH cells, without affecting non-proliferating cells, confer GPER the “status” of a therapeutic target. However, and despite some important advances characterizing GPER activity in prostate cells and tissues, much more investigation is needed to fully resolve the multiple GPER’ actions in PCa and its therapeutic potential.

Conclusions

GPER is a seven-transmembrane receptor with a wide mode of action activating several intracellular signaling pathways involved in the genomic and non-genomic effects of estrogens. In addition to E_2 and G1, the most potent activators of GPER, several other compounds with the ability of binding GPER acting as agonists or antagonists have been pointed out. The astonishingly increasing list of these compounds, encompass hormone ligands, synthetic compounds and environmental phytoestrogens and xenoestrogens, which places GPER in the physiological context of normal and oncological conditions, but also as a target of endocrine disruption. The diversity of biological processes under GPER control mainly includes cell proliferation, apoptosis, cell migration and invasion, angiogenesis, inflammation, and metabolism.

The multifaceted GPER is expressed in a broad range of neoplastic and non-neoplastic tissues, which further extends its broad scope of action. The development of GPER specific agonists and antagonists, as well as cell and animal knockout models have contributed greatly to ascertain its role in cancer and other diseases. GPER was associated with the development and progression of breast, ovarian, endometrial and testicular cancers. However, other strong evidence of GPER actions counteracting known hallmarks of cancer as exacerbated proliferative activity and resistance to apoptosis support its role as a tumor suppressor.

In prostatic tissues, GPER is expressed both in normal gland and PCa. Also in this case, a duality in the GPER role has been reported. Studies arguing the causative role of GPER in prostate carcinogenesis are available in the literature, co-existing with others defending the protective action of GPER in PCa development. Reports indicating the high expression of GPER in PCa cases, together with its effects suppressing cell growth, also allow speculating about its usefulness as a therapeutic target. Nevertheless, future research is needed to deeply clarify the role of GPER in PCa and establish its potential as a therapeutic point of intervention.

Overall, we can discuss that the distinct functions of GPER in different tissues, or even within the same tissue, may depend on the specific physiological/environmental conditions in each tissue at a given moment, and/or of the experimental design including the type of ligand and the concentrations tested. The question of concentration would be quite relevant. For example, in the context of testicular apoptosis and male infertility, it was shown that at low concentrations estrogens act as survival factors whereas at high concentrations have the opposite effect inducing cell death [346, 347].

Finally, GPER can be viewed as a "cell guardian", with their functions modulated by tissue specificity and environmental conditions, i.e., GPER would act protecting cells that need to be protected from harmful stimulus, or destroying those that are "dangerous" for tissue homeostasis.

References

1. Jensen EV, DeSombre ER. Estrogen-receptor interaction. Science (New York, NY). 1973;182(4108):126–34.
2. Kuiper GG, Enmark E, Pelto-Huikko M, Nilsson S, Gustafsson JA. Cloning of a novel receptor expressed in rat prostate and ovary. Proc Natl Acad Sci U S A. 1996;93(12):5925–30.
3. Wang Z, Zhang X, Shen P, Loggie BW, Chang Y, Deuel TF. Identification, cloning, and expression of human estrogen receptor-alpha36, a novel variant of human estrogen receptor-alpha66. Biochem Biophys Res Commun. 2005;336(4):1023–7. https://doi.org/10.1016/j.bbrc.2005.08.226.
4. Maruyama K, Endoh H, Sasaki-Iwaoka H, Kanou H, Shimaya E, Hashimoto S, Kato S, Kawashima H. A novel isoform of rat estrogen receptor beta with 18 amino acid insertion in the ligand binding domain as a putative dominant negative regular of estrogen action. Biochem Biophys Res Commun. 1998;246(1):142–7.
5. Mosselman S, Polman J, Dijkema R. ER beta: identification and characterization of a novel human estrogen receptor. FEBS Lett. 1996;392(1):49–53.

6. Grandien K. Determination of transcription start sites in the human estrogen receptor gene and identification of a novel, tissue-specific, estrogen receptor-mRNA isoform. Mol Cell Endocrinol. 1996;116(2):207–12.
7. Ye Q, Chung LW, Cinar B, Li S, Zhau HE. Identification and characterization of estrogen receptor variants in prostate cancer cell lines. J Steroid Biochem Mol Biol. 2000;75(1):21–31.
8. Lewandowski S, Kalita K, Kaczmarek L. Estrogen receptor beta. Potential functional significance of a variety of mRNA isoforms. FEBS Lett. 2002;524(1–3):1–5.
9. Fujimura T, Takahashi S, Urano T, Ogawa S, Ouchi Y, Kitamura T, Muramatsu M, Inoue S. Differential expression of estrogen receptor beta (ERbeta) and its C-terminal truncated splice variant ERbetacx as prognostic predictors in human prostatic cancer. Biochem Biophys Res Commun. 2001;289(3):692–9. https://doi.org/10.1006/bbrc.2001.6038.
10. Gibson DA, Saunders PT. Estrogen dependent signaling in reproductive tissues - a role for estrogen receptors and estrogen related receptors. Mol Cell Endocrinol. 2012;348(2):361–72. https://doi.org/10.1016/j.mce.2011.09.026.
11. Laurentino S, Pinto P, Correia S, Cavaco JE, Canário AVM, Socorro S. Structural variants of sex steroid hormone receptors in the testis: from molecular biology to physiological roles. OA Biotechnol. 2012;1(2):4.
12. Bjornstrom L, Sjoberg M. Mechanisms of estrogen receptor signaling: convergence of genomic and nongenomic actions on target genes. Mol Endocrinol. 2005;19(4):833–42. https://doi.org/10.1210/me.2004-0486.
13. Filardo EJ, Quinn JA, Bland KI, Frackelton AR Jr. Estrogen-induced activation of Erk-1 and Erk-2 requires the G protein-coupled receptor homolog, GPR30, and occurs via trans-activation of the epidermal growth factor receptor through release of HB-EGF. Mol Endocrinol. 2000;14(10):1649–60. https://doi.org/10.1210/mend.14.10.0532.
14. Carmeci C, Thompson DA, Ring HZ, Francke U, Weigel RJ. Identification of a gene (GPR30) with homology to the G-protein-coupled receptor superfamily associated with estrogen receptor expression in breast cancer. Genomics. 1997;45(3):607–17. https://doi.org/10.1006/geno.1997.4972.
15. Vrtacnik P, Ostanek B, Mencej-Bedrac S, Marc J. The many faces of estrogen signaling. Biochem Med. 2014;24(3):329–42. 10.11613/bm.2014.035.
16. Chavalmane AK, Comeglio P, Morelli A, Filippi S, Fibbi B, Vignozzi L, Sarchielli E, Marchetta M, Failli P, Sandner P, Saad F, Gacci M, Vannelli GB, Maggi M. Sex steroid receptors in male human bladder: expression and biological function. J Sex Med. 2010;7(8):2698–713. https://doi.org/10.1111/j.1743-6109.2010.01811.x.
17. Ge C, Yu M, Zhang C. G protein-coupled receptor 30 mediates estrogen-induced proliferation of primordial germ cells via EGFR/Akt/beta-catenin signaling pathway. Endocrinology. 2012;153(7):3504–16. https://doi.org/10.1210/en.2012-1200.
18. Szwejser E, Maciuszek M, Casanova-Nakayama A, Segner H, Verburg-van Kemenade BM, Chadzinska M. A role for multiple estrogen receptors in immune regulation of common carp. Dev Comp Immunol. 2017;66:61–72. https://doi.org/10.1016/j.dci.2016.04.003.
19. Maggioli E, McArthur S, Mauro C, Kieswich J, Kusters DH, Reutelingsperger CP, Yaqoob M, Solito E. Estrogen protects the blood-brain barrier from inflammation-induced disruption and increased lymphocyte trafficking. Brain Behav Immun. 2016;51:212–22. https://doi.org/10.1016/j.bbi.2015.08.020.
20. Cote M, Bourque M, Poirier AA, Aube B, Morissette M, Di Paolo T, Soulet D. GPER1-mediated immunomodulation and neuroprotection in the myenteric plexus of a mouse model of Parkinson's disease. Neurobiol Dis. 2015;82:99–113. https://doi.org/10.1016/j.nbd.2015.05.017.
21. Seto K, Hoang M, Santos T, Bandyopadhyay M, Kindy MS, Dasgupta S. Non-genomic oestrogen receptor signal in B lymphocytes: an approach towards therapeutic interventions for infection, autoimmunity and cancer. Int J Biochem Cell Biol. 2016;76:115–8. https://doi.org/10.1016/j.biocel.2016.04.018.

22. Meyer MR, Fredette NC, Daniel C, Sharma G, Amann K, Arterburn JB, Barton M, Prossnitz ER. Obligatory role for GPER in cardiovascular aging and disease. Sci Signal. 2016;9(452):ra105. https://doi.org/10.1126/scisignal.aag0240.
23. Zhao TZ, Ding Q, Hu J, He SM, Shi F, Ma LT. GPER expressed on microglia mediates the anti-inflammatory effect of estradiol in ischemic stroke. Brain Behav. 2016;6(4):e00449. https://doi.org/10.1002/brb3.449.
24. Barton M, Prossnitz ER. Emerging roles of GPER in diabetes and atherosclerosis. Trends Endocrinol Metab. 2015;26(4):185–92. https://doi.org/10.1016/j.tem.2015.02.003.
25. Bourque M, Morissette M, Cote M, Soulet D, Di Paolo T. Implication of GPER1 in neuroprotection in a mouse model of Parkinson's disease. Neurobiol Aging. 2013;34(3):887–901. https://doi.org/10.1016/j.neurobiolaging.2012.05.022.
26. Bourque M, Morissette M, Di Paolo T. Neuroprotection in Parkinsonian-treated mice via estrogen receptor alpha activation requires G protein-coupled estrogen receptor 1. Neuropharmacology. 2015;95:343–52. https://doi.org/10.1016/j.neuropharm.2015.04.006.
27. Kang WB, Deng YT, Wang DS, Feng D, Liu Q, Wang XS, Ru JY, Cong Y, Zhao JN, Zhao MG, Liu G. Osteoprotective effects of estrogen membrane receptor GPR30 in ovariectomized rats. J Steroid Biochem Mol Biol. 2015;154:237–44. https://doi.org/10.1016/j.jsbmb.2015.07.002.
28. Broughton BR, Brait VH, Guida E, Lee S, Arumugam TV, Gardiner-Mann CV, Miller AA, Tang SC, Drummond GR, Sobey CG. Stroke increases g protein-coupled estrogen receptor expression in the brain of male but not female mice. Neurosignals. 2013;21(3–4):229–39. https://doi.org/10.1159/000338019.
29. Broughton BR, Brait VH, Kim HA, Lee S, Chu HX, Gardiner-Mann CV, Guida E, Evans MA, Miller AA, Arumugam TV, Drummond GR, Sobey CG. Sex-dependent effects of G protein-coupled estrogen receptor activity on outcome after ischemic stroke. Stroke. 2014;45(3):835–41. https://doi.org/10.1161/strokeaha.113.001499.
30. Prabhushankar R, Krueger C, Manrique C. Membrane estrogen receptors: their role in blood pressure regulation and cardiovascular disease. Curr Hypertens Rep. 2014;16(1):408. https://doi.org/10.1007/s11906-013-0408-6.
31. Boscia F, Passaro C, Gigantino V, Perdona S, Franco R, Portella G, Chieffi S, Chieffi P. High levels of GPR30 protein in human testicular carcinoma in situ and seminomas correlate with low levels of estrogen receptor-beta and indicate a switch in estrogen responsiveness. J Cell Physiol. 2015;230(6):1290–7. https://doi.org/10.1002/jcp.24864.
32. Han G, White RE. G-protein-coupled estrogen receptor as a new therapeutic target for treating coronary artery disease. World J Cardiol. 2014;6(6):367–75. https://doi.org/10.4330/wjc.v6.i6.367.
33. Ignatov T, Modl S, Thulig M, Weissenborn C, Treeck O, Ortmann O, Zenclussen A, Costa SD, Kalinski T, Ignatov A. GPER-1 acts as a tumor suppressor in ovarian cancer. J Ovarian Res. 2013;6(1):51. https://doi.org/10.1186/1757-2215-6-51.
34. Ignatov T, Weissenborn C, Poehlmann A, Lemke A, Semczuk A, Roessner A, Costa SD, Kalinski T, Ignatov A. GPER-1 expression decreases during breast cancer tumorigenesis. Cancer Investig. 2013;31(5):309–15. https://doi.org/10.3109/07357907.2013.789901.
35. Petrie WK, Dennis MK, Hu C, Dai D, Arterburn JB, Smith HO, Hathaway HJ, Prossnitz ER. G protein-coupled estrogen receptor-selective ligands modulate endometrial tumor growth. Obstet Gynecol Int. 2013;2013:472720. https://doi.org/10.1155/2013/472720.
36. Jala VR, Radde BN, Haribabu B, Klinge CM. Enhanced expression of G-protein coupled estrogen receptor (GPER/GPR30) in lung cancer. BMC Cancer. 2012;12:624. https://doi.org/10.1186/1471-2407-12-624.
37. Chevalier N, Vega A, Bouskine A, Siddeek B, Michiels JF, Chevallier D, Fenichel P. GPR30, the non-classical membrane G protein related estrogen receptor, is overexpressed in human seminoma and promotes seminoma cell proliferation. PLoS One. 2012;7(4):e34672. https://doi.org/10.1371/journal.pone.0034672.

38. Kumar A, Klinge CM, Goldstein RE. Estradiol-induced proliferation of papillary and follicular thyroid cancer cells is mediated by estrogen receptors alpha and beta. Int J Oncol. 2010;36(5):1067–80.
39. Aiad HA, Wahed MM, Asaad NY, El-Tahmody M, Elhosary E. Immunohistochemical expression of GPR30 in breast carcinoma of Egyptian patients: an association with immunohistochemical subtypes. APMIS. 2014;122(10):976–84. https://doi.org/10.1111/apm.12241.
40. Fujiwara S, Terai Y, Kawaguchi H, Takai M, Yoo S, Tanaka Y, Tanaka T, Tsunetoh S, Sasaki H, Kanemura M, Tanabe A, Yamashita Y, Ohmichi M. GPR30 regulates the EGFR-Akt cascade and predicts lower survival in patients with ovarian cancer. J Ovarian Res. 2012;5(1):35. https://doi.org/10.1186/1757-2215-5-35.
41. Lieberherr M, Grosse B, Kachkache M, Balsan S. Cell signaling and estrogens in female rat osteoblasts: a possible involvement of unconventional nonnuclear receptors. J Bone Miner Res. 1993;8(11):1365–76. https://doi.org/10.1002/jbmr.5650081111.
42. Benten WP, Stephan C, Lieberherr M, Wunderlich F. Estradiol signaling via sequestrable surface receptors. Endocrinology. 2001;142(4):1669–77. https://doi.org/10.1210/endo.142.4.8094.
43. Evinger AJ III, Levin ER. Requirements for estrogen receptor alpha membrane localization and function. Steroids. 2005;70(5–7):361–3. https://doi.org/10.1016/j.steroids.2005.02.015.
44. Manavathi B, Kumar R. Steering estrogen signals from the plasma membrane to the nucleus: two sides of the coin. J Cell Physiol. 2006;207(3):594–604. https://doi.org/10.1002/jcp.20551.
45. Kelly MJ, Lagrange AH, Wagner EJ, Ronnekleiv OK. Rapid effects of estrogen to modulate G protein-coupled receptors via activation of protein kinase A and protein kinase C pathways. Steroids. 1999;64(1–2):64–75.
46. Prossnitz ER, Arterburn JB, Sklar LA. GPR30: a G protein-coupled receptor for estrogen. Mol Cell Endocrinol. 2007;265–266:138–42. https://doi.org/10.1016/j.mce.2006.12.010.
47. Owman C, Blay P, Nilsson C, Lolait SJ. Cloning of human cDNA encoding a novel heptahelix receptor expressed in Burkitt's lymphoma and widely distributed in brain and peripheral tissues. Biochem Biophys Res Commun. 1996;228(2):285–92. https://doi.org/10.1006/bbrc.1996.1654.
48. Takada Y, Kato C, Kondo S, Korenaga R, Ando J. Cloning of cDNAs encoding G protein-coupled receptor expressed in human endothelial cells exposed to fluid shear stress. Biochem Biophys Res Commun. 1997;240(3):737–41. https://doi.org/10.1006/bbrc.1997.7734.
49. Kvingedal AM, Smeland EB. A novel putative G-protein-coupled receptor expressed in lung, heart and lymphoid tissue. FEBS Lett. 1997;407(1):59–62.
50. O'Dowd BF, Nguyen T, Marchese A, Cheng R, Lynch KR, Heng HH, Kolakowski LF Jr, George SR. Discovery of three novel G-protein-coupled receptor genes. Genomics. 1998;47(2):310–3. https://doi.org/10.1006/geno.1998.5095.
51. Feng Y, Gregor P. Cloning of a novel member of the G protein-coupled receptor family related to peptide receptors. Biochem Biophys Res Commun. 1997;231(3):651–4. https://doi.org/10.1006/bbrc.1997.6161.
52. Revankar CM, Cimino DF, Sklar LA, Arterburn JB, Prossnitz ER. A transmembrane intracellular estrogen receptor mediates rapid cell signaling. Science (New York, NY). 2005;307(5715):1625–30. https://doi.org/10.1126/science.1106943.
53. Thomas P, Pang Y, Filardo EJ, Dong J. Identity of an estrogen membrane receptor coupled to a G protein in human breast cancer cells. Endocrinology. 2005;146(2):624–32. https://doi.org/10.1210/en.2004-1064.
54. Funakoshi T, Yanai A, Shinoda K, Kawano MM, Mizukami Y. G protein-coupled receptor 30 is an estrogen receptor in the plasma membrane. Biochem Biophys Res Commun. 2006;346(3):904–10. https://doi.org/10.1016/j.bbrc.2006.05.191.
55. Alexander SP, Benson HE, Faccenda E, Pawson AJ, Sharman JL, McGrath JC, Catterall WA, Spedding M, Peters JA, Harmar AJ, Abul-Hasn N, Anderson CM, Anderson CM, Araiksinen MS, Arita M, Arthofer E, Barker EL, Barratt C, Barnes NM, Bathgate R, Beart PM, Belelli D,

Bennett AJ, Birdsall NJ, Boison D, Bonner TI, Brailsford L, Broer S, Brown P, Calo G, Carter WG, Catterall WA, Chan SL, Chao MV, Chiang N, Christopoulos A, Chun JJ, Cidlowski J, Clapham DE, Cockcroft S, Connor MA, Cox HM, Cuthbert A, Dautzenberg FM, Davenport AP, Dawson PA, Dent G, Dijksterhuis JP, Dollery CT, Dolphin AC, Donowitz M, Dubocovich ML, Eiden L, Eidne K, Evans BA, Fabbro D, Fahlke C, Farndale R, Fitzgerald GA, Fong TM, Fowler CJ, Fry JR, Funk CD, Futerman AH, Ganapathy V, Gaisnier B, Gershengorn MA, Goldin A, Goldman ID, Gundlach AL, Hagenbuch B, Hales TG, Hammond JR, Hamon M, Hancox JC, Hauger RL, Hay DL, Hobbs AJ, Hollenberg MD, Holliday ND, Hoyer D, Hynes NA, Inui KI, Ishii S, Jacobson KA, Jarvis GE, Jarvis MF, Jensen R, Jones CE, Jones RL, Kaibuchi K, Kanai Y, Kennedy C, Kerr ID, Khan AA, Klienz MJ, Kukkonen JP, Lapoint JY, Leurs R, Lingueglia E, Lippiat J, Lolait SJ, Lummis SC, Lynch JW, MacEwan D, Maguire JJ, Marshall IL, May JM, McArdle CA, McGrath JC, Michel MC, Millar NS, Miller LJ, Mitolo V, Monk PN, Moore PK, Moorhouse AJ, Mouillac B, Murphy PM, Neubig RR, Neumaier J, Niesler B, Obaidat A, Offermanns S, Ohlstein E, Panaro MA, Parsons S, Pwrtwee RG, Petersen J, Pin JP, Poyner DR, Prigent S, Prossnitz ER, Pyne NJ, Pyne S, Quigley JG, Ramachandran R, Richelson EL, Roberts RE, Roskoski R, Ross RA, Roth M, Rudnick G, Ryan RM, Said SI, Schild L, Sanger GJ, Scholich K, Schousboe A, Schulte G, Schulz S, Serhan CN, Sexton PM, Sibley DR, Siegel JM, Singh G, Sitsapesan R, Smart TG, Smith DM, Soga T, Stahl A, Stewart G, Stoddart LA, Summers RJ, Thorens B, Thwaites DT, Toll L, Traynor JR, Usdin TB, Vandenberg RJ, Villalon C, Vore M, Waldman SA, Ward DT, Willars GB, Wonnacott SJ, Wright E, Ye RD, Yonezawa A, Zimmermann M. The concise guide to PHARMACOLOGY 2013/14: overview. Br J Pharmacol. 2013;170(8):1449–58. https://doi.org/10.1111/bph.12444.
56. Levin ER. G protein-coupled receptor 30: estrogen receptor or collaborator? Endocrinology. 2009;150(4):1563–5. https://doi.org/10.1210/en.2008-1759.
57. Otto C, Rohde-Schulz B, Schwarz G, Fuchs I, Klewer M, Brittain D, Langer G, Bader B, Prelle K, Nubbemeyer R, Fritzemeier KH. G protein-coupled receptor 30 localizes to the endoplasmic reticulum and is not activated by estradiol. Endocrinology. 2008;149(10):4846–56. https://doi.org/10.1210/en.2008-0269.
58. Lafferty AR, Torpy DJ, Stowasser M, Taymans SE, Lin JP, Huggard P, Gordon RD, Stratakis CA. A novel genetic locus for low renin hypertension: familial hyperaldosteronism type II maps to chromosome 7 (7p22). J Med Genet. 2000;37(11):831–5.
59. Gurevich EV, Tesmer JJ, Mushegian A, Gurevich VV. G protein-coupled receptor kinases: more than just kinases and not only for GPCRs. Pharmacol Ther. 2012;133(1):40–69. https://doi.org/10.1016/j.pharmthera.2011.08.001.
60. Moreira IS. Structural features of the G-protein/GPCR interactions. Biochim Biophys Acta. 2014;1840(1):16–33. https://doi.org/10.1016/j.bbagen.2013.08.027.
61. Wong SK. G protein selectivity is regulated by multiple intracellular regions of GPCRs. Neurosignals. 2003;12(1):1–12. https://doi.org/10.1159/000068914.
62. Broselid S, Berg KA, Chavera TA, Kahn R, Clarke WP, Olde B, Leeb-Lundberg LM. G protein-coupled receptor 30 (GPR30) forms a plasma membrane complex with membrane-associated guanylate kinases (MAGUKs) and protein kinase A-anchoring protein 5 (AKAP5) that constitutively inhibits cAMP production. J Biol Chem. 2014;289(32):22117–27. https://doi.org/10.1074/jbc.M114.566893.
63. Tran QK, Vermeer M. Biosensor-based approach identifies four distinct calmodulin-binding domains in the G protein-coupled estrogen receptor 1. PLoS One. 2014;9(2):e89669. https://doi.org/10.1371/journal.pone.0089669.
64. Filardo EJ. Epidermal growth factor receptor (EGFR) transactivation by estrogen via the G-protein-coupled receptor, GPR30: a novel signaling pathway with potential significance for breast cancer. J Steroid Biochem Mol Biol. 2002;80(2):231–8.
65. Filardo E, Quinn J, Pang Y, Graeber C, Shaw S, Dong J, Thomas P. Activation of the novel estrogen receptor G protein-coupled receptor 30 (GPR30) at the plasma membrane. Endocrinology. 2007;148(7):3236–45. https://doi.org/10.1210/en.2006-1605.

66. Sakamoto H, Matsuda K, Hosokawa K, Nishi M, Morris JF, Prossnitz ER, Kawata M. Expression of G protein-coupled receptor-30, a G protein-coupled membrane estrogen receptor, in oxytocin neurons of the rat paraventricular and supraoptic nuclei. Endocrinology. 2007;148(12):5842–50. https://doi.org/10.1210/en.2007-0436.
67. Matsuda K, Sakamoto H, Mori H, Hosokawa K, Kawamura A, Itose M, Nishi M, Prossnitz ER, Kawata M. Expression and intracellular distribution of the G protein-coupled receptor 30 in rat hippocampal formation. Neurosci Lett. 2008;441(1):94–9. https://doi.org/10.1016/j.neulet.2008.05.108.
68. Sanden C, Broselid S, Cornmark L, Andersson K, Daszkiewicz-Nilsson J, Martensson UE, Olde B, Leeb-Lundberg LM. G protein-coupled estrogen receptor 1/G protein-coupled receptor 30 localizes in the plasma membrane and traffics intracellularly on cytokeratin intermediate filaments. Mol Pharmacol. 2011;79(3):400–10. https://doi.org/10.1124/mol.110.069500.
69. Pupo M, Vivacqua A, Perrotta I, Pisano A, Aquila S, Abonante S, Gasperi-Campani A, Pezzi V, Maggiolini M. The nuclear localization signal is required for nuclear GPER translocation and function in breast Cancer-Associated Fibroblasts (CAFs). Mol Cell Endocrinol. 2013;376(1–2):23–32. https://doi.org/10.1016/j.mce.2013.05.023.
70. Madeo A, Maggiolini M. Nuclear alternate estrogen receptor GPR30 mediates 17beta-estradiol-induced gene expression and migration in breast cancer-associated fibroblasts. Cancer Res. 2010;70(14):6036–46. https://doi.org/10.1158/0008-5472.can-10-0408.
71. Tian R, Wang Z, Shi Z, Li D, Wang Y, Zhu Y, Lin W, Gui Y, Zheng XL. Differential expression of G-protein-coupled estrogen receptor-30 in human myometrial and uterine leiomyoma smooth muscle. Fertil Steril. 2013;99(1):256–63. https://doi.org/10.1016/j.fertnstert.2012.09.011.
72. Cheng SB, Graeber CT, Quinn JA, Filardo EJ. Retrograde transport of the transmembrane estrogen receptor, G-protein-coupled-receptor-30 (GPR30/GPER) from the plasma membrane towards the nucleus. Steroids. 2011;76(9):892–6. https://doi.org/10.1016/j.steroids.2011.02.018.
73. Gaudet HM, Cheng SB, Christensen EM, Filardo EJ. The G-protein coupled estrogen receptor. GPER. 2015;418(Pt 3):207–19. https://doi.org/10.1016/j.mce.2015.07.016.
74. Cheng SB, Quinn JA, Graeber CT, Filardo EJ. Down-modulation of the G-protein-coupled estrogen receptor, GPER, from the cell surface occurs via a trans-Golgi-proteasome pathway. J Biol Chem. 2011;286(25):22441–55. https://doi.org/10.1074/jbc.M111.224071.
75. Meng R, Qin Q, Xiong Y, Wang Y, Zheng J, Zhao Y, Tao T, Wang Q, Liu H, Wang S, Jiang WG, He J. NHERF1, a novel GPER associated protein, increases stability and activation of GPER in ER-positive breast cancer. Oncotarget. 2016;7(34):54983–97. 10.18632/oncotarget.10713.
76. Ho KJ, Liao JK. Non-nuclear actions of estrogen: new targets for prevention and treatment of cardiovascular disease. Mol Interv. 2002;2(4):219–28. https://doi.org/10.1124/mi.2.4.219.
77. Hurowitz EH, Melnyk JM, Chen YJ, Kouros-Mehr H, Simon MI, Shizuya H. Genomic characterization of the human heterotrimeric G protein alpha, beta, and gamma subunit genes. DNA Res. 2000;7(2):111–20.
78. Luttrell LM, van Biesen T, Hawes BE, Koch WJ, Krueger KM, Touhara K, Lefkowitz RJ. G-protein-coupled receptors and their regulation: activation of the MAP kinase signaling pathway by G-protein-coupled receptors. Adv Second Messenger Phosphoprotein Res. 1997;31:263–77.
79. Filardo EJ, Quinn JA, Frackelton AR Jr, Bland KI. Estrogen action via the G protein-coupled receptor, GPR30: stimulation of adenylyl cyclase and cAMP-mediated attenuation of the epidermal growth factor receptor-to-MAPK signaling axis. Mol Endocrinol. 2002;16(1):70–84. https://doi.org/10.1210/mend.16.1.0758.
80. Chimento A, Sirianni R, Delalande C, Silandre D, Bois C, Ando S, Maggiolini M, Carreau S, Pezzi V. 17 beta-estradiol activates rapid signaling pathways involved in rat pachytene spermatocytes apoptosis through GPR30 and ER alpha. Mol Cell Endocrinol. 2010;320(1–2):136–44. https://doi.org/10.1016/j.mce.2010.01.035.

81. Kleuser B, Malek D, Gust R, Pertz HH, Potteck H. 17-Beta-estradiol inhibits transforming growth factor-beta signaling and function in breast cancer cells via activation of extracellular signal-regulated kinase through the G protein-coupled receptor 30. Mol Pharmacol. 2008;74(6):1533–43. https://doi.org/10.1124/mol.108.046854.
82. Malek D, Gust R, Kleuser B. 17-Beta-estradiol inhibits transforming-growth-factor-beta-induced MCF-7 cell migration by Smad3-repression. Eur J Pharmacol. 2006;534(1–3):39–47. https://doi.org/10.1016/j.ejphar.2006.01.025.
83. De Francesco EM, Pellegrino M, Santolla MF, Lappano R, Ricchio E, Abonante S, Maggiolini M. GPER mediates activation of HIF1alpha/VEGF signaling by estrogens. Cancer Res. 2014;74(15):4053–64. https://doi.org/10.1158/0008-5472.can-13-3590.
84. De Francesco EM, Lappano R, Santolla MF, Marsico S, Caruso A, Maggiolini M. HIF-1alpha/GPER signaling mediates the expression of VEGF induced by hypoxia in breast cancer associated fibroblasts (CAFs). Breast Cancer Res. 2013;15(4):R64. https://doi.org/10.1186/bcr3458.
85. Wei Y, Zhang Z, Liao H, Wu L, Wu X, Zhou D, Xi X, Zhu Y, Feng Y. Nuclear estrogen receptor-mediated Notch signaling and GPR30-mediated PI3K/AKT signaling in the regulation of endometrial cancer cell proliferation. Oncol Rep. 2012;27(2):504–10. https://doi.org/10.3892/or.2011.1536.
86. Ge X, Guo R, Qiao Y, Zhang Y, Lei J, Wang X, Li L, Hu D. The G protein-coupled receptor GPR30 mediates the nontranscriptional effect of estrogen on the activation of PI3K/Akt pathway in endometrial cancer cells. Int J Gynecol Cancer. 2013;23(1):52–9. https://doi.org/10.1097/IGC.0b013e31827912b8.
87. Popli P, Sirohi VK, Manohar M, Shukla V, Kaushal JB, Gupta K, Dwivedi A. Regulation of cyclooxygenase-2 expression in rat oviductal epithelial cells: evidence for involvement of GPR30/Src kinase-mediated EGFR signaling. J Steroid Biochem Mol Biol. 2015;154:130–41. https://doi.org/10.1016/j.jsbmb.2015.07.019.
88. Bartella V, De Francesco EM, Perri MG, Curcio R, Dolce V, Maggiolini M, Vivacqua A. The G protein estrogen receptor (GPER) is regulated by endothelin-1 mediated signaling in cancer cells. Cell Signal. 2016;28(2):61–71. https://doi.org/10.1016/j.cellsig.2015.11.010.
89. Pupo M, Pisano A, Abonante S, Maggiolini M, Musti AM. GPER activates Notch signaling in breast cancer cells and cancer-associated fibroblasts (CAFs). Int J Biochem Cell Biol. 2014;46:56–67. https://doi.org/10.1016/j.biocel.2013.11.011.
90. Razandi M, Pedram A, Park ST, Levin ER. Proximal events in signaling by plasma membrane estrogen receptors. J Biol Chem. 2003;278(4):2701–12. https://doi.org/10.1074/jbc.M205692200.
91. Filardo EJ, Thomas P. GPR30: a seven-transmembrane-spanning estrogen receptor that triggers EGF release. Trends Endocrinol Metab. 2005;16(8):362–7. https://doi.org/10.1016/j.tem.2005.08.005.
92. Sukocheva O, Wadham C, Holmes A, Albanese N, Verrier E, Feng F, Bernal A, Derian CK, Ullrich A, Vadas MA, Xia P. Estrogen transactivates EGFR via the sphingosine 1-phosphate receptor Edg-3: the role of sphingosine kinase-1. J Cell Biol. 2006;173(2):301–10. https://doi.org/10.1083/jcb.200506033.
93. Lappano R, De Marco P, De Francesco EM, Chimento A, Pezzi V, Maggiolini M. Cross-talk between GPER and growth factor signaling. J Steroid Biochem Mol Biol. 2013;137:50–6. https://doi.org/10.1016/j.jsbmb.2013.03.005.
94. De Marco P, Bartella V, Vivacqua A, Lappano R, Santolla MF, Morcavallo A, Pezzi V, Belfiore A, Maggiolini M. Insulin-like growth factor-I regulates GPER expression and function in cancer cells. Oncogene. 2013;32(6):678–88. https://doi.org/10.1038/onc.2012.97.
95. Maggiolini M, Vivacqua A, Fasanella G, Recchia AG, Sisci D, Pezzi V, Montanaro D, Musti AM, Picard D, Ando S. The G protein-coupled receptor GPR30 mediates c-fos up-regulation by 17beta-estradiol and phytoestrogens in breast cancer cells. J Biol Chem. 2004;279(26):27008–16. https://doi.org/10.1074/jbc.M403588200.

96. Albanito L, Madeo A, Lappano R, Vivacqua A, Rago V, Carpino A, Oprea TI, Prossnitz ER, Musti AM, Ando S, Maggiolini M. G protein-coupled receptor 30 (GPR30) mediates gene expression changes and growth response to 17beta-estradiol and selective GPR30 ligand G-1 in ovarian cancer cells. Cancer Res. 2007;67(4):1859–66. https://doi.org/10.1158/0008-5472.can-06-2909.
97. Pandey DP, Lappano R, Albanito L, Madeo A, Maggiolini M, Picard D. Estrogenic GPR30 signalling induces proliferation and migration of breast cancer cells through CTGF. EMBO J. 2009;28(5):523–32. https://doi.org/10.1038/emboj.2008.304.
98. Vivacqua A, Romeo E, De Marco P, De Francesco EM, Abonante S, Maggiolini M. GPER mediates the Egr-1 expression induced by 17beta-estradiol and 4-hydroxitamoxifen in breast and endometrial cancer cells. Breast Cancer Res Treat. 2012;133(3):1025–35. https://doi.org/10.1007/s10549-011-1901-8.
99. Kanda N, Watanabe S. 17beta-estradiol inhibits oxidative stress-induced apoptosis in keratinocytes by promoting Bcl-2 expression. J Invest Dermatol. 2003;121(6):1500–9. https://doi.org/10.1111/j.1523-1747.2003.12617.x.
100. Kanda N, Watanabe S. 17beta-estradiol stimulates the growth of human keratinocytes by inducing cyclin D2 expression. J Invest Dermatol. 2004;123(2):319–28. https://doi.org/10.1111/j.0022-202X.2004.12645.x.
101. Santolla MF, Avino S, Pellegrino M, De Francesco EM, De Marco P, Lappano R, Vivacqua A, Cirillo F, Rigiracciolo DC, Scarpelli A, Abonante S, Maggiolini M. SIRT1 is involved in oncogenic signaling mediated by GPER in breast cancer. Cell Death Dis. 2015;6:e1834. https://doi.org/10.1038/cddis.2015.201.
102. Prossnitz ER, Barton M. Estrogen biology: new insights into GPER function and clinical opportunities. Mol Cell Endocrinol. 2014;389(1–2):71–83. https://doi.org/10.1016/j.mce.2014.02.002.
103. Hay DL, Pioszak AA. Receptor activity-modifying proteins (RAMPs): new insights and roles. Annu Rev Pharmacol Toxicol. 2016;56:469–87. https://doi.org/10.1146/annurev-pharmtox-010715-103120.
104. Lenhart PM, Broselid S, Barrick CJ, Leeb-Lundberg LM, Caron KM. G-protein-coupled receptor 30 interacts with receptor activity-modifying protein 3 and confers sex-dependent cardioprotection. J Mol Endocrinol. 2013;51(1):191–202. https://doi.org/10.1530/jme-13-0021.
105. Mendez-Luna D, Martinez-Archundia M, Maroun RC, Ceballos-Reyes G, Fragoso-Vazquez MJ, Gonzalez-Juarez DE, Correa-Basurto J. Deciphering the GPER/GPR30-agonist and antagonists interactions using molecular modeling studies, molecular dynamics, and docking simulations. J Biomol Struct Dyn. 2015;33(10):2161–72. https://doi.org/10.1080/07391102.2014.994102.
106. Koganti S, Snyder R, Gumaste U, Karamyan VT, Thekkumkara T. 2-methoxyestradiol binding of GPR30 down-regulates angiotensin AT(1) receptor. Eur J Pharmacol. 2014;723:131–40. https://doi.org/10.1016/j.ejphar.2013.10.064.
107. Chourasia TK, Pang Y, Thomas P. The catecholestrogen, 2-hydroxyestradiol-17beta, acts as a G protein-coupled estrogen receptor 1 (GPER/GPR30) antagonist to promote the resumption of meiosis in zebrafish oocytes. Biol Reprod. 2015;92(3):69. https://doi.org/10.1095/biolreprod.114.125674.
108. Zucchetti AE, Barosso IR, Boaglio AC, Basiglio CL, Miszczuk G, Larocca MC, Ruiz ML, Davio CA, Roma MG, Crocenzi FA, Pozzi EJ. G-protein-coupled receptor 30/adenylyl cyclase/protein kinase A pathway is involved in estradiol 17ss-D-glucuronide-induced cholestasis. Hepatology (Baltimore, MD). 2014;59(3):1016–29. https://doi.org/10.1002/hep.26752.
109. Lappano R, Rosano C, De Marco P, De Francesco EM, Pezzi V, Maggiolini M. Estriol acts as a GPR30 antagonist in estrogen receptor-negative breast cancer cells. Mol Cell Endocrinol. 2010;320(1–2):162–70. https://doi.org/10.1016/j.mce.2010.02.006.

110. Girgert R, Emons G, Grundker C. Inhibition of GPR30 by estriol prevents growth stimulation of triple-negative breast cancer cells by 17beta-estradiol. BMC Cancer. 2014;14:935. https://doi.org/10.1186/1471-2407-14-935.
111. Sandra N, Ester P, Marie-Agnes P, Robert M, Olivier H. The DHEA metabolite 7beta-hydroxy-epiandrosterone exerts anti-estrogenic effects on breast cancer cell lines. Steroids. 2012;77(5):542–51. https://doi.org/10.1016/j.steroids.2012.01.019.
112. Gros R, Ding Q, Liu B, Chorazyczewski J, Feldman RD. Aldosterone mediates its rapid effects in vascular endothelial cells through GPER activation. Am J Physiol Cell Physiol. 2013;304(6):C532–40. https://doi.org/10.1152/ajpcell.00203.2012.
113. Feldman RD, Ding Q, Hussain Y, Limbird LE, Pickering JG, Gros R. Aldosterone mediates metastatic spread of renal cancer via the G protein-coupled estrogen receptor (GPER). FASEB J. 2016. https://doi.org/10.1096/fj.15-275552.
114. Hermidorff MM, de Assis LV, Isoldi MC. Genomic and rapid effects of aldosterone: what we know and do not know thus far. Heart Fail Rev. 2017;22(1):65–89. https://doi.org/10.1007/s10741-016-9591-2.
115. Qiao H, Hu B, Zhou H, Yan J, Jia R, Lu B, Sun B, Luo X, Fan Y, Wang N. Aldosterone induces rapid sodium intake by a nongenomic mechanism in the nucleus tractus solitarius. Sci Rep. 2016;6:38631. https://doi.org/10.1038/srep38631.
116. Cheng SB, Dong J, Pang Y, LaRocca J, Hixon M, Thomas P, Filardo EJ. Anatomical location and redistribution of G protein-coupled estrogen receptor-1 during the estrus cycle in mouse kidney and specific binding to estrogens but not aldosterone. Mol Cell Endocrinol. 2014;382(2):950–9. https://doi.org/10.1016/j.mce.2013.11.005.
117. Gros R, Ding Q, Sklar LA, Prossnitz EE, Arterburn JB, Chorazyczewski J, Feldman RD. GPR30 expression is required for the mineralocorticoid receptor-independent rapid vascular effects of aldosterone. Hypertension. 2011;57(3):442–51. https://doi.org/10.1161/hypertensionaha.110.161653.
118. Heublein S, Vrekoussis T, Kuhn C, Friese K, Makrigiannakis A, Mayr D, Lenhard M, Jeschke U. Inducers of G-protein coupled estrogen receptor (GPER) in endometriosis: potential implications for macrophages and follicle maturation. J Reprod Immunol. 2013;97(1):95–103. https://doi.org/10.1016/j.jri.2012.10.013.
119. Cameron DA, Keen JC, Dixon JM, Bellamy C, Hanby A, Anderson TJ, Miller WR. Effective tamoxifen therapy of breast cancer involves both antiproliferative and pro-apoptotic changes. Eur J Cancer. 2000;36(7):845–51.
120. Ariazi EA, Ariazi JL, Cordera F, Jordan VC. Estrogen receptors as therapeutic targets in breast cancer. Curr Top Med Chem. 2006;6(3):181–202.
121. Jordan VC. SERMs: meeting the promise of multifunctional medicines. J Natl Cancer Inst. 2007;99(5):350–6. https://doi.org/10.1093/jnci/djk062.
122. Long N, Long B, Mana A, Le D, Nguyen L, Chokr S, Sinchak K. Tamoxifen and ICI 182,780 activate hypothalamic G protein-coupled estrogen receptor 1 to rapidly facilitate lordosis in female rats. Horm Behav. 2017;89:98–103. https://doi.org/10.1016/j.yhbeh.2016.12.013.
123. Zhang L, Li Y, Lan L, Liu R, Wu Y, Qu Q, Wen K. Tamoxifen has a proliferative effect in endometrial carcinoma mediated via the GPER/EGFR/ERK/cyclin D1 pathway: a retrospective study and an in vitro study. Mol Cell Endocrinol. 2016;437:51–61. https://doi.org/10.1016/j.mce.2016.08.011.
124. Qiu J, Bosch MA, Tobias SC, Grandy DK, Scanlan TS, Ronnekleiv OK, Kelly MJ. Rapid signaling of estrogen in hypothalamic neurons involves a novel G-protein-coupled estrogen receptor that activates protein kinase C. J Neurosci. 2003;23(29):9529–40.
125. Bourque M, Morissette M, Di Paolo T. Raloxifene activates G protein-coupled estrogen receptor 1/Akt signaling to protect dopamine neurons in 1-methyl-4-phenyl-1,2,3,6-tetrahydropyridine mice. Neurobiol Aging. 2014;35(10):2347–56. https://doi.org/10.1016/j.neurobiolaging.2014.03.017.
126. Vivacqua A, Bonofiglio D, Recchia AG, Musti AM, Picard D, Ando S, Maggiolini M. The G protein-coupled receptor GPR30 mediates the proliferative effects induced by 17beta-estradiol

and hydroxytamoxifen in endometrial cancer cells. Mol Endocrinol. 2006;20(3):631–46. https://doi.org/10.1210/me.2005-0280.
127. Zhang X, Li JH, Duan SX, Lin QJ, Ke S, Ma L, Huang TH, Jiang XW. G protein-coupled estrogen receptor-protein kinase A-ERK-CREB signaling pathway is involved in the regulation of mouse gubernaculum testis cells by diethylstilbestrol. Arch Environ Contam Toxicol. 2014;67(1):97–103. https://doi.org/10.1007/s00244-013-9976-3.
128. Zhang X, Ke S, Chen KH, Li JH, Ma L, Jiang XW. Diethylstilbestrol affects the expression of GPER in the gubernaculum testis. Int J Clin Exp Pathol. 2015;8(6):7217–22.
129. Chandra V, Fatima I, Saxena R, Hussain MK, Hajela K, Sankhwar P, Roy BG, Chandna S, Dwivedi A. Anti-tumorigenic action of 2-[piperidinoethoxyphenyl]-3-[4-hydroxyphenyl]-2H-benzo(b)pyran: evidence for involvement of GPR30/EGFR signaling pathway. Gynecol Oncol. 2013;129(2):433–42. https://doi.org/10.1016/j.ygyno.2013.02.005.
130. Lappano R, Santolla MF, Pupo M, Sinicropi MS, Caruso A, Rosano C, Maggiolini M. MIBE acts as antagonist ligand of both estrogen receptor alpha and GPER in breast cancer cells. Breast Cancer Res. 2012;14(1):R12. https://doi.org/10.1186/bcr3096.
131. Shaul PW. Regulation of endothelial nitric oxide synthase: location, location, location. Annu Rev Physiol. 2002;64:749–74. https://doi.org/10.1146/annurev.physiol.64.081501.155952.
132. Harrington WR, Kim SH, Funk CC, Madak-Erdogan Z, Schiff R, Katzenellenbogen JA, Katzenellenbogen BS. Estrogen dendrimer conjugates that preferentially activate extranuclear, nongenomic versus genomic pathways of estrogen action. Mol Endocrinol. 2006;20(3):491–502. https://doi.org/10.1210/me.2005-0186.
133. Revankar CM, Mitchell HD, Field AS, Burai R, Corona C, Ramesh C, Sklar LA, Arterburn JB, Prossnitz ER. Synthetic estrogen derivatives demonstrate the functionality of intracellular GPR30. ACS Chem Biol. 2007;2(8):536–44. https://doi.org/10.1021/cb700072n.
134. Sirotkin AV, Harrath AH. Phytoestrogens and their effects. Eur J Pharmacol. 2014;741:230–6. https://doi.org/10.1016/j.ejphar.2014.07.057.
135. Moreno-Ulloa A, Mendez-Luna D, Beltran-Partida E, Castillo C, Guevara G, Ramirez-Sanchez I, Correa-Basurto J, Ceballos G, Villarreal F. The effects of (−)-epicatechin on endothelial cells involve the G protein-coupled estrogen receptor (GPER). Pharmacol Res. 2015;100:309–20. https://doi.org/10.1016/j.phrs.2015.08.014.
136. Ma HR, Wang J, Chen YF, Chen H, Wang WS, Aisa HA. Icariin and icaritin stimulate the proliferation of SKBr3 cells through the GPER1-mediated modulation of the EGFR-MAPK signaling pathway. Int J Mol Med. 2014;33(6):1627–34. https://doi.org/10.3892/ijmm.2014.1722.
137. Li YC, Ding XS, Li HM, Zhang C. Icariin attenuates high glucose-induced type IV collagen and fibronectin accumulation in glomerular mesangial cells by inhibiting transforming growth factor-beta production and signalling through G protein-coupled oestrogen receptor 1. Clin Exp Pharmacol Physiol. 2013;40(9):635–43. https://doi.org/10.1111/1440-1681.12143.
138. Shang D, Li Z, Zhu Z, Chen H, Zhao L, Wang X, Chen Y. Baicalein suppresses 17-beta-estradiol-induced migration, adhesion and invasion of breast cancer cells via the G protein-coupled receptor 30 signaling pathway. Oncol Rep. 2015;33(4):2077–85. https://doi.org/10.3892/or.2015.3786.
139. Chen Y, Wang J, Hong DY, Chen L, Zhang YY, Xu YN, Pan D, Fu LY, Tao L, Luo H, Shen XC. Baicalein has protective effects on the 17beta-estradiol-induced transformation of breast epithelial cells. Oncotarget. 2017;8(6):10470–84. 10.18632/oncotarget.14433.
140. Thomas P, Dong J. Binding and activation of the seven-transmembrane estrogen receptor GPR30 by environmental estrogens: a potential novel mechanism of endocrine disruption. J Steroid Biochem Mol Biol. 2006;102(1–5):175–9. https://doi.org/10.1016/j.jsbmb.2006.09.017.
141. Pisani SL, Neese SL, Doerge DR, Helferich WG, Schantz SL, Korol DL. Acute genistein treatment mimics the effects of estradiol by enhancing place learning and impairing response learning in young adult female rats. Horm Behav. 2012;62(4):491–9. https://doi.org/10.1016/j.yhbeh.2012.08.006.

142. Lucki NC, Sewer MB. Genistein stimulates MCF-7 breast cancer cell growth by inducing acid ceramidase (ASAH1) gene expression. J Biol Chem. 2011;286(22):19399–409. https://doi.org/10.1074/jbc.M110.195826.
143. Kajta M, Rzemieniec J, Litwa E, Lason W, Lenartowicz M, Krzeptowski W, Wojtowicz AK. The key involvement of estrogen receptor beta and G-protein-coupled receptor 30 in the neuroprotective action of daidzein. Neuroscience. 2013;238:345–60. https://doi.org/10.1016/j.neuroscience.2013.02.005.
144. Rowlands DJ, Chapple S, Siow RC, Mann GE. Equol-stimulated mitochondrial reactive oxygen species activate endothelial nitric oxide synthase and redox signaling in endothelial cells: roles for F-actin and GPR30. Hypertension. 2011;57(4):833–40. https://doi.org/10.1161/hypertensionaha.110.162198.
145. Cheng YF, Zhu G, Wu QW, Xie YS, Jiang Y, Guo L, Guan YL, Liu YS, Zhang J. GPR30 activation contributes to the Puerarin-mediated neuroprotection in MPP+-induced SH-SY5Y cell death. J Mol Neurosci. 2017;61(2):227–34. https://doi.org/10.1007/s12031-016-0856-y.
146. Khan K, Pal S, Yadav M, Maurya R, Trivedi AK, Sanyal S, Chattopadhyay N. Prunetin signals via G-protein-coupled receptor, GPR30(GPER1): stimulation of adenylyl cyclase and cAMP-mediated activation of MAPK signaling induces Runx2 expression in osteoblasts to promote bone regeneration. J Nutr Biochem. 2015;26(12):1491–501. https://doi.org/10.1016/j.jnutbio.2015.07.021.
147. Kang K, Lee SB, Jung SH, Cha KH, Park WD, Sohn YC, Nho CW. Tectoridin, a poor ligand of estrogen receptor alpha, exerts its estrogenic effects via an ERK-dependent pathway. Mol Cells. 2009;27(3):351–7. https://doi.org/10.1007/s10059-009-0045-8.
148. Varoni EM, Lo Faro AF, Sharifi-Rad J, Iriti M. Anticancer molecular mechanisms of resveratrol. Front Nutr. 2016;3:8. https://doi.org/10.3389/fnut.2016.00008.
149. Dong WH, Chen JC, He YL, Xu JJ, Mei YA. Resveratrol inhibits K(v)2.2 currents through the estrogen receptor GPR30-mediated PKC pathway. Am J Physiol Cell Physiol. 2013;305(5):C547–57. https://doi.org/10.1152/ajpcell.00146.2013.
150. Chimento A, Casaburi I, Rosano C, Avena P, De Luca A, Campana C, Martire E, Santolla MF, Maggiolini M, Pezzi V, Sirianni R. Oleuropein and hydroxytyrosol activate GPER/GPR30-dependent pathways leading to apoptosis of ER-negative SKBR3 breast cancer cells. Mol Nutr Food Res. 2014;58(3):478–89. https://doi.org/10.1002/mnfr.201300323.
151. Ren GY, Chen CY, Chen WG, Huang Y, Qin LQ, Chen LH. The treatment effects of flaxseed-derived secoisolariciresinol diglycoside and its metabolite enterolactone on benign prostatic hyperplasia involve the G protein-coupled estrogen receptor 1. Appl Physiol Nutr Metab. 2016;41(12):1303–10. https://doi.org/10.1139/apnm-2016-0332.
152. Shanle EK, Xu W. Endocrine disrupting chemicals targeting estrogen receptor signaling: identification and mechanisms of action. Chem Res Toxicol. 2011;24(1):6–19. https://doi.org/10.1021/tx100231n.
153. Dong S, Terasaka S, Kiyama R. Bisphenol A induces a rapid activation of Erk1/2 through GPR30 in human breast cancer cells. Environ Pollut. 2011;159(1):212–8. https://doi.org/10.1016/j.envpol.2010.09.004.
154. Li M, Guo J, Gao W, Yu J, Han X, Zhang J, Shao B. Bisphenol AF-induced endogenous transcription is mediated by ERalpha and ERK1/2 activation in human breast cancer cells. PLoS One. 2014;9(4):e94725. https://doi.org/10.1371/journal.pone.0094725.
155. Sheng ZG, Huang W, Liu YX, Zhu BZ. Bisphenol A at a low concentration boosts mouse spermatogonial cell proliferation by inducing the G protein-coupled receptor 30 expression. Toxicol Appl Pharmacol. 2013;267(1):88–94. https://doi.org/10.1016/j.taap.2012.12.014.
156. Sheng ZG, Zhu BZ. Low concentrations of bisphenol A induce mouse spermatogonial cell proliferation by G protein-coupled receptor 30 and estrogen receptor-alpha. Environ Health Perspect. 2011;119(12):1775–80. https://doi.org/10.1289/ehp.1103781.
157. Ge LC, Chen ZJ, Liu HY, Zhang KS, Liu H, Huang HB, Zhang G, Wong CK, Giesy JP, Du J, Wang HS. Involvement of activating ERK1/2 through G protein coupled receptor 30 and

estrogen receptor alpha/beta in low doses of bisphenol A promoting growth of Sertoli TM4 cells. Toxicol Lett. 2014;226(1):81–9. https://doi.org/10.1016/j.toxlet.2014.01.035.

158. Zhang KS, Chen HQ, Chen YS, Qiu KF, Zheng XB, Li GC, Yang HD, Wen CJ. Bisphenol A stimulates human lung cancer cell migration via upregulation of matrix metalloproteinases by GPER/EGFR/ERK1/2 signal pathway. Biomed Pharmacother. 2014;68(8):1037–43. https://doi.org/10.1016/j.biopha.2014.09.003.
159. Liu J, Jin X, Zhao N, Ye X, Ying C. Bisphenol A promotes X-linked inhibitor of apoptosis protein-dependent angiogenesis via G protein-coupled estrogen receptor pathway. J Appl Toxicol. 2015;35(11):1309–17. https://doi.org/10.1002/jat.3112.
160. Ge LC, Wang HS. A commentary on "Involvement of activating ERK1/2 trough G protein coupled receptor 30 and estrogen receptor alpha/beta in low doses of bisphenol A promoting growth of Sertoli TM4 cells". Toxicol Lett. 2016;240(1):236–7. https://doi.org/10.1016/j.toxlet.2015.09.021.
161. Mahemuti L, Chen Q, Coughlan MC, Zhang M, Florian M, Mailloux RJ, Cao XL, Scoggan KA, Willmore WG, Jin X. Bisphenol A exposure alters release of immune and developmental modulators and expression of estrogen receptors in human fetal lung fibroblasts. J Environ Sci (China). 2016;48:11–23. https://doi.org/10.1016/j.jes.2016.02.013.
162. Song H, Zhang T, Yang P, Li M, Yang Y, Wang Y, Du J, Pan K, Zhang K. Low doses of bisphenol A stimulate the proliferation of breast cancer cells via ERK1/2/ERRgamma signals. Toxicol In Vitro. 2015;30(1 Pt B):521–8. https://doi.org/10.1016/j.tiv.2015.09.009.
163. Lei B, Peng W, Xu G, Wu M, Wen Y, Xu J, Yu Z, Wang Y. Activation of G protein-coupled receptor 30 by thiodiphenol promotes proliferation of estrogen receptor alpha-positive breast cancer cells. Chemosphere. 2017;169:204–11. https://doi.org/10.1016/j.chemosphere.2016.11.066.
164. Bindal RD, Carlson KE, Katzenellenbogen BS, Katzenellenbogen JA. Lipophilic impurities, not phenolsulfonphthalein, account for the estrogenic activity in commercial preparations of phenol red. J Steroid Biochem. 1988;31(3):287–93.
165. Kajta M, Litwa E, Rzemieniec J, Wnuk A, Lason W, Zelek-Molik A, Nalepa I, Grzegorzewska-Hiczwa M, Tokarski K, Golas A, Guzik E, Grochowalski A, Szychowski KA, Wojtowicz AK. Isomer-nonspecific action of dichlorodiphenyltrichloroethane on aryl hydrocarbon receptor and G-protein-coupled receptor 30 intracellular signaling in apoptotic neuronal cells. Mol Cell Endocrinol. 2014;392(1–2):90–105. https://doi.org/10.1016/j.mce.2014.05.008.
166. Tian PC, Wang HL, Chen GH, Luo Q, Chen Z, Wang Y, Liu YF. 2,2′,4,4′-Tetrabromodiphenyl ether promotes human neuroblastoma SH-SY5Y cells migration via the GPER/PI3K/Akt signal pathway. Hum Exp Toxicol. 2016;35(2):124–34. https://doi.org/10.1177/0960327115578974.
167. Florian CP, Mansfield SR, Schroeder JR. Differences in GPR30 regulation by chlorotriazine herbicides in human breast cells. Biochem Res Int. 2016;2016:2984081. https://doi.org/10.1155/2016/2984081.
168. Leblanc K, Sexton E, Parent S, Belanger G, Dery MC, Boucher V, Asselin E. Effects of 4-hydroxytamoxifen, raloxifene and ICI 182 780 on survival of uterine cancer cell lines in the presence and absence of exogenous estrogens. Int J Oncol. 2007;30(2):477–87.
169. PA. Gale, JL Sessler, Kral V (1998) Calixpyrroles. Chem Commun (1):1–8. doi:https://doi.org/10.1039/A706280J.
170. Gale PA, Anzenbacher P Jr, Sessler JL. Calixpyrroles II. Coord Chem Rev. 2001;222(1):57–102. https://doi.org/10.1016/S0010-8545(01)00346-0
171. Lappano R, Rosano C, Pisano A, Santolla MF, De Francesco EM, De Marco P, Dolce V, Ponassi M, Felli L, Cafeo G, Kohnke FH, Abonante S, Maggiolini M. A calixpyrrole derivative acts as an antagonist to GPER, a G-protein coupled receptor: mechanisms and models. Dis Model Mech. 2015;8(10):1237–46. https://doi.org/10.1242/dmm.021071.
172. Wrobel AM, Gregoraszczuk EL. Action of methyl-, propyl- and butylparaben on GPR30 gene and protein expression, cAMP levels and activation of ERK1/2 and PI3K/Akt signaling pathways in MCF-7 breast cancer cells and MCF-10A non-transformed breast epithelial cells. Toxicol Lett. 2015;238(2):110–6. https://doi.org/10.1016/j.toxlet.2015.08.001.

173. Gupte A, Mumper RJ. Elevated copper and oxidative stress in cancer cells as a target for cancer treatment. Cancer Treat Rev. 2009;35(1):32–46. https://doi.org/10.1016/j.ctrv.2008.07.004.
174. Rigiracciolo DC, Scarpelli A, Lappano R, Pisano A, Santolla MF, De Marco P, Cirillo F, Cappello AR, Dolce V, Belfiore A, Maggiolini M, De Francesco EM. Copper activates HIF-1alpha/GPER/VEGF signalling in cancer cells. Oncotarget. 2015;6(33):34158–77. 10.18632/oncotarget.5779.
175. Huff MO, Todd SL, Smith AL, Elpers JT, Smith AP, Murphy RD, Bleser-Shartzer AS, Hoerter JE, Radde BN, Klinge CM. Arsenite and cadmium activate MAPK/ERK via membrane estrogen receptors and G-protein coupled estrogen receptor signaling in human lung adenocarcinoma cells. Toxicol Sci. 2016;152(1):62–71. https://doi.org/10.1093/toxsci/kfw064.
176. Yu X, Filardo EJ, Shaikh ZA. The membrane estrogen receptor GPR30 mediates cadmium-induced proliferation of breast cancer cells. Toxicol Appl Pharmacol. 2010;245(1):83–90. https://doi.org/10.1016/j.taap.2010.02.005.
177. Liu Z, Yu X, Shaikh ZA. Rapid activation of ERK1/2 and AKT in human breast cancer cells by cadmium. Toxicol Appl Pharmacol. 2008;228(3):286–94. https://doi.org/10.1016/j.taap.2007.12.017.
178. Ariazi EA, Brailoiu E, Yerrum S, Shupp HA, Slifker MJ, Cunliffe HE, Black MA, Donato AL, Arterburn JB, Oprea TI, Prossnitz ER, Dun NJ, Jordan VC. The G protein-coupled receptor GPR30 inhibits proliferation of estrogen receptor-positive breast cancer cells. Cancer Res. 2010;70(3):1184–94. https://doi.org/10.1158/0008-5472.can-09-3068.
179. Zhu P, Liao LY, Zhao TT, Mo XM, Chen GG, Liu ZM. GPER/ERK&AKT/NF-kappaB pathway is involved in cadmium-induced proliferation, invasion and migration of GPER-positive thyroid cancer cells. Mol Cell Endocrinol. 2017;442:68–80. https://doi.org/10.1016/j.mce.2016.12.007.
180. Pisano A, Santolla MF, De Francesco EM, De Marco P, Rigiracciolo DC, Perri MG, Vivacqua A, Abonante S, Cappello AR, Dolce V, Belfiore A, Maggiolini M, Lappano R. GPER, IGF-IR, and EGFR transduction signaling are involved in stimulatory effects of zinc in breast cancer cells and cancer-associated fibroblasts. Mol Carcinog. 2017;56(2):580–93. https://doi.org/10.1002/mc.22518.
181. Santolla MF, De Francesco EM, Lappano R, Rosano C, Abonante S, Maggiolini M. Niacin activates the G protein estrogen receptor (GPER)-mediated signalling. Cell Signal. 2014;26(7):1466–75. https://doi.org/10.1016/j.cellsig.2014.03.011.
182. Bologa CG, Revankar CM, Young SM, Edwards BS, Arterburn JB, Kiselyov AS, Parker MA, Tkachenko SE, Savchuck NP, Sklar LA, Oprea TI, Prossnitz ER. Virtual and biomolecular screening converge on a selective agonist for GPR30. Nat Chem Biol. 2006;2(4):207–12. https://doi.org/10.1038/nchembio775.
183. Dennis MK, Burai R, Ramesh C, Petrie WK, Alcon SN, Nayak TK, Bologa CG, Leitao A, Brailoiu E, Deliu E, Dun NJ, Sklar LA, Hathaway HJ, Arterburn JB, Oprea TI, Prossnitz ER. In vivo effects of a GPR30 antagonist. Nat Chem Biol. 2009;5(6):421–7. https://doi.org/10.1038/nchembio.168.
184. Bai LY, Weng JR, Hu JL, Wang D, Sargeant AM, Chiu CF. G15, a GPR30 antagonist, induces apoptosis and autophagy in human oral squamous carcinoma cells. Chem Biol Interact. 2013;206(2):375–84. https://doi.org/10.1016/j.cbi.2013.10.014.
185. Dennis MK, Field AS, Burai R, Ramesh C, Petrie WK, Bologa CG, Oprea TI, Yamaguchi Y, Hayashi S, Sklar LA, Hathaway HJ, Arterburn JB, Prossnitz ER. Identification of a GPER/GPR30 antagonist with improved estrogen receptor counterselectivity. J Steroid Biochem Mol Biol. 2011;127(3–5):358–66. https://doi.org/10.1016/j.jsbmb.2011.07.002.
186. Chimento A, Sirianni R, Casaburi I, Zolea F, Rizza P, Avena P, Malivindi R, De Luca A, Campana C, Martire E, Domanico F, Fallo F, Carpinelli G, Cerquetti L, Amendola D, Stigliano A, Pezzi V. GPER agonist G-1 decreases adrenocortical carcinoma (ACC) cell growth in vitro and in vivo. Oncotarget. 2015;6(22):19190–203. 10.18632/oncotarget.4241.

187. Alencar AK, Montes GC, Montagnoli T, Silva AM, Martinez ST, Fraga AG, Wang H, Groban L, Sudo RT, Zapata-Sudo G. Activation of GPER ameliorates experimental pulmonary hypertension in male rats. Eur J Pharm Sci. 2017;97:208–17. https://doi.org/10.1016/j.ejps.2016.11.009.
188. Mendes-Oliveira J, Lopes Campos F, Videira RA, Baltazar G. GPER activation is effective in protecting against inflammation-induced nigral dopaminergic loss and motor function impairment. Brain Behav Immun. 2017;64:296–307. https://doi.org/10.1016/j.bbi.2017.04.016.
189. Lappano R, Rosano C, Santolla MF, Pupo M, De Francesco EM, De Marco P, Ponassi M, Spallarossa A, Ranise A, Maggiolini M. Two novel GPER agonists induce gene expression changes and growth effects in cancer cells. Curr Cancer Drug Targets. 2012;12(5):531–42.
190. Maggiolini M, Santolla MF, Avino S, Aiello F, Rosano C, Garofalo A, Grande F. Identification of two benzopyrroloxazines acting as selective GPER antagonists in breast cancer cells and cancer-associated fibroblasts. Future Med Chem. 2015;7(4):437–48. https://doi.org/10.4155/fmc.15.3.
191. Arias-Pulido H, Royce M, Gong Y, Joste N, Lomo L, Lee SJ, Chaher N, Verschraegen C, Lara J, Prossnitz ER, Cristofanilli M. GPR30 and estrogen receptor expression: new insights into hormone dependence of inflammatory breast cancer. Breast Cancer Res Treat. 2010;123(1):51–8. https://doi.org/10.1007/s10549-009-0631-7.
192. Filardo EJ, Graeber CT, Quinn JA, Resnick MB, Giri D, DeLellis RA, Steinhoff MM, Sabo E. Distribution of GPR30, a seven membrane-spanning estrogen receptor, in primary breast cancer and its association with clinicopathologic determinants of tumor progression. Clin Cancer Res. 2006;12(21):6359–66. https://doi.org/10.1158/1078-0432.ccr-06-0860.
193. Luo HJ, Luo P, Yang GL, Peng QL, Liu MR, Tu G. G-protein coupled estrogen receptor 1 expression in primary breast cancers and its correlation with clinicopathological variables. J Breast Cancer. 2011;14(3):185–90. https://doi.org/10.4048/jbc.2011.14.3.185.
194. Sjostrom M, Hartman L, Grabau D, Fornander T, Malmstrom P, Nordenskjold B, Sgroi DC, Skoog L, Stal O, Leeb-Lundberg LM, Ferno M. Lack of G protein-coupled estrogen receptor (GPER) in the plasma membrane is associated with excellent long-term prognosis in breast cancer. Breast Cancer Res Treat. 2014;145(1):61–71. https://doi.org/10.1007/s10549-014-2936-4.
195. Samartzis EP, Noske A, Meisel A, Varga Z, Fink D, Imesch P. The G protein-coupled estrogen receptor (GPER) is expressed in two different subcellular localizations reflecting distinct tumor properties in breast cancer. PLoS One. 2014;9(1):e83296. https://doi.org/10.1371/journal.pone.0083296.
196. Ignatov A, Ignatov T, Weissenborn C, Eggemann H, Bischoff J, Semczuk A, Roessner A, Costa SD, Kalinski T. G-protein-coupled estrogen receptor GPR30 and tamoxifen resistance in breast cancer. Breast Cancer Res Treat. 2011;128(2):457–66. https://doi.org/10.1007/s10549-011-1584-1.
197. Marjon NA, Hu C, Hathaway HJ, Prossnitz ER. G protein-coupled estrogen receptor regulates mammary tumorigenesis and metastasis. Mol Cancer Res. 2014;12(11):1644–54. https://doi.org/10.1158/1541-7786.mcr-14-0128-t.
198. Liu Q, Li JG, Zheng XY, Jin F, Dong HT. Expression of CD133, PAX2, ESA, and GPR30 in invasive ductal breast carcinomas. Chin Med J. 2009;122(22):2763–9.
199. Steiman J, Peralta EA, Louis S, Kamel O. Biology of the estrogen receptor, GPR30, in triple negative breast cancer. Am J Surg. 2013;206(5):698–703. https://doi.org/10.1016/j.amjsurg.2013.07.014.
200. Konecny G, Pauletti G, Pegram M, Untch M, Dandekar S, Aguilar Z, Wilson C, Rong HM, Bauerfeind I, Felber M, Wang HJ, Beryt M, Seshadri R, Hepp H, Slamon DJ. Quantitative association between HER-2/neu and steroid hormone receptors in hormone receptor-positive primary breast cancer. J Natl Cancer Inst. 2003;95(2):142–53.
201. Rakha EA, El-Sayed ME, Green AR, Lee AH, Robertson JF, Ellis IO. Prognostic markers in triple-negative breast cancer. Cancer. 2007;109(1):25–32. https://doi.org/10.1002/cncr.22381.

202. Girgert R, Emons G, Grundker C. Inactivation of GPR30 reduces growth of triple-negative breast cancer cells: possible application in targeted therapy. Breast Cancer Res Treat. 2012;134(1):199–205. https://doi.org/10.1007/s10549-012-1968-x.
203. Mo Z, Liu M, Yang F, Luo H, Li Z, Tu G, Yang G. GPR30 as an initiator of tamoxifen resistance in hormone-dependent breast cancer. Breast Cancer Res. 2013;15(6):R114. https://doi.org/10.1186/bcr3581.
204. Yuan J, Liu M, Yang L, Tu G, Zhu Q, Chen M, Cheng H, Luo H, Fu W, Li Z, Yang G. Acquisition of epithelial-mesenchymal transition phenotype in the tamoxifen-resistant breast cancer cell: a new role for G protein-coupled estrogen receptor in mediating tamoxifen resistance through cancer-associated fibroblast-derived fibronectin and beta1-integrin signaling pathway in tumor cells. Breast Cancer Res. 2015;17:69. https://doi.org/10.1186/s13058-015-0579-y.
205. Weissenborn C, Ignatov T, Poehlmann A, Wege AK, Costa SD, Zenclussen AC, Ignatov A. GPER functions as a tumor suppressor in MCF-7 and SK-BR-3 breast cancer cells. J Cancer Res Clin Oncol. 2014;140(4):663–71. https://doi.org/10.1007/s00432-014-1598-2.
206. Weissenborn C, Ignatov T, Nass N, Kalinski T, Dan Costa S, Zenclussen AC, Ignatov A. GPER promoter methylation controls GPER expression in breast cancer patients. Cancer Investig. 2017;35(2):100–7. https://doi.org/10.1080/07357907.2016.1271886.
207. Poola I, Abraham J, Liu A, Marshalleck JJ, Dewitty RL. The cell surface estrogen receptor, G protein-coupled receptor 30 (GPR30), is markedly down regulated during breast tumorigenesis. Breast Cancer. 2008;1:65–78.
208. Weissenborn C, Ignatov T, Ochel HJ, Costa SD, Zenclussen AC, Ignatova Z, Ignatov A. GPER functions as a tumor suppressor in triple-negative breast cancer cells. J Cancer Res Clin Oncol. 2014;140(5):713–23. https://doi.org/10.1007/s00432-014-1620-8.
209. Chen ZJ, Wei W, Jiang GM, Liu H, Wei WD, Yang X, Wu YM, Liu H, Wong CK, Du J, Wang HS. Activation of GPER suppresses epithelial mesenchymal transition of triple negative breast cancer cells via NF-kappaB signals. Mol Oncol. 2016. https://doi.org/10.1016/j.molonc.2016.01.002.
210. Plante BJ, Lessey BA, Taylor RN, Wang W, Bagchi MK, Yuan L, Scotchie J, Fritz MA, Young SL. G protein-coupled estrogen receptor (GPER) expression in normal and abnormal endometrium. Reprod Sci. 2012;19(7):684–93. https://doi.org/10.1177/1933719111431000.
211. Samartzis N, Samartzis EP, Noske A, Fedier A, Dedes KJ, Caduff R, Fink D, Imesch P. Expression of the G protein-coupled estrogen receptor (GPER) in endometriosis: a tissue microarray study. Reprod Biol Endocrinol. 2012;10:30. https://doi.org/10.1186/1477-7827-10-30.
212. He YY, Cai B, Yang YX, Liu XL, Wan XP. Estrogenic G protein-coupled receptor 30 signaling is involved in regulation of endometrial carcinoma by promoting proliferation, invasion potential, and interleukin-6 secretion via the MEK/ERK mitogen-activated protein kinase pathway. Cancer Sci. 2009;100(6):1051–61. https://doi.org/10.1111/j.1349-7006.2009.01148.x.
213. Skrzypczak M, Schuler S, Lattrich C, Ignatov A, Ortmann O, Treeck O. G protein-coupled estrogen receptor (GPER) expression in endometrial adenocarcinoma and effect of agonist G-1 on growth of endometrial adenocarcinoma cell lines. Steroids. 2013;78(11):1087–91. https://doi.org/10.1016/j.steroids.2013.07.007.
214. Tica AA, Tica OS, Georgescu CV, Pirici D, Bogdan M, Ciurea T, Mogoanta SS, Georgescu CC, Comanescu AC, Balseanu TA, Ciurea RN, Osiac E, Buga AM, Ciurea ME. GPER and ERalpha expression in abnormal endometrial proliferations. Romanian J Morphol Embryol. 2016;57(2):413–8.
215. Krakstad C, Trovik J, Wik E, Engelsen IB, Werner HM, Birkeland E, Raeder MB, Oyan AM, Stefansson IM, Kalland KH, Akslen LA, Salvesen HB. Loss of GPER identifies new targets for therapy among a subgroup of ERalpha-positive endometrial cancer patients with poor outcome. Br J Cancer. 2012;106(10):1682–8. https://doi.org/10.1038/bjc.2012.91.

216. Huang GS, Gunter MJ, Arend RC, Li M, Arias-Pulido H, Prossnitz ER, Goldberg GL, Smith HO. Co-expression of GPR30 and ERbeta and their association with disease progression in uterine carcinosarcoma. Am J Obstet Gynecol. 2010;203(3):242.e241–5. https://doi.org/10.1016/j.ajog.2010.04.046.
217. Parker LP, Taylor DD, Kesterson S, Gercel-Taylor C. Gene expression profiling in response to estradiol and genistein in ovarian cancer cells. Cancer Genomics Proteomics. 2009;6(3):189–94.
218. Heublein S, Mayr D, Friese K, Jarrin-Franco MC, Lenhard M, Mayerhofer A, Jeschke U. The G-protein-coupled estrogen receptor (GPER/GPR30) in ovarian granulosa cell tumors. Int J Mol Sci. 2014;15(9):15161–72. https://doi.org/10.3390/ijms150915161.
219. Francois CM, Wargnier R, Petit F, Goulvent T, Rimokh R, Treilleux I, Ray-Coquard I, Zazzu V, Cohen-Tannoudji J, Guigon CJ. 17beta-estradiol inhibits spreading of metastatic cells from granulosa cell tumors through a non-genomic mechanism involving GPER1. Carcinogenesis. 2015;36(5):564–73. https://doi.org/10.1093/carcin/bgv041.
220. Smith HO, Arias-Pulido H, Kuo DY, Howard T, Qualls CR, Lee SJ, Verschraegen CF, Hathaway HJ, Joste NE, Prossnitz ER. GPR30 predicts poor survival for ovarian cancer. Gynecol Oncol. 2009;114(3):465–71. https://doi.org/10.1016/j.ygyno.2009.05.015.
221. Long L, Cao Y, Tang LD. Transmembrane estrogen receptor GPR30 is more frequently expressed in malignant than benign ovarian endometriotic cysts and correlates with MMP-9 expression. Int J Gynecol Cancer. 2012;22(4):539–45. https://doi.org/10.1097/IGC.0b013e318247323d.
222. Ghazarian AA, Trabert B, Devesa SS, McGlynn KA. Recent trends in the incidence of testicular germ cell tumors in the United States. Andrology. 2015;3(1):13–8. https://doi.org/10.1111/andr.288.
223. Chevalier N, Paul-Bellon R, Camparo P, Michiels JF, Chevallier D, Fenichel P. Genetic variants of GPER/GPR30, a novel estrogen-related G protein receptor, are associated with human seminoma. Int J Mol Sci. 2014;15(1):1574–89. https://doi.org/10.3390/ijms15011574.
224. Franco R, Boscia F, Gigantino V, Marra L, Esposito F, Ferrara D, Pariante P, Botti G, Caraglia M, Minucci S, Chieffi P. GPR30 is overexpressed in post-puberal testicular germ cell tumors. Cancer Biol Ther. 2011;11(6):609–13.
225. Rago V, Romeo F, Giordano F, Maggiolini M, Carpino A. Identification of the estrogen receptor GPER in neoplastic and non-neoplastic human testes. Reprod Biol Endocrinol. 2011;9:135. https://doi.org/10.1186/1477-7827-9-135.
226. Chieffi P. An overview on predictive biomarkers of testicular germ cell tumors. J Cell Physiol. 2017;232(2):276–80. https://doi.org/10.1002/jcp.25482.
227. Liu C, Liao Y, Fan S, Tang H, Jiang Z, Zhou B, Xiong J, Zhou S, Zou M, Wang J. G protein-coupled estrogen receptor (GPER) mediates NSCLC progression induced by 17beta-estradiol (E2) and selective agonist G1. Med Oncol. 2015;32(4):104. https://doi.org/10.1007/s12032-015-0558-2.
228. Mau M, Mielenz M, Sudekum KH, Obukhov AG. Expression of GPR30 and GPR43 in oral tissues: deriving new hypotheses on the role of diet in animal physiology and the development of oral cancers. J Anim Physiol Anim Nutr. 2011;95(3):280–5. https://doi.org/10.1111/j.1439-0396.2010.01052.x.
229. Tang C, Yang L, Wang N, Li L, Xu M, Chen GG, Liu ZM. High expression of GPER1, EGFR and CXCR1 is associated with lymph node metastasis in papillary thyroid carcinoma. Int J Clin Exp Pathol. 2014;7(6):3213–23.
230. Zhao L, Zhu XY, Jiang R, Xu M, Wang N, Chen GG, Liu ZM. Role of GPER1, EGFR and CXCR1 in differentiating between malignant follicular thyroid carcinoma and benign follicular thyroid adenoma. Int J Clin Exp Pathol. 2015;8(9):11236–47.
231. Yamasaki M, Mukai A, Ohba M, Mine Y, Sakakibara Y, Suiko M, Morishita K, Nishiyama K. Genistein induced apoptotic cell death in adult T-cell leukemia cells through estrogen receptors. Biosci Biotechnol Biochem. 2010;74(10):2113–5. https://doi.org/10.1271/bbb.100359.

232. Huang W, Chen Y, Liu Y, Zhang Q, Yu Z, Mou L, Wu H, Zhao L, Long T, Qin D, Gui Y. Roles of ERbeta and GPR30 in proliferative response of human bladder cancer cell to estrogen. Biomed Res Int. 2015;2015:251780. https://doi.org/10.1155/2015/251780.
233. Glass JP, Parasher G, Arias-Pulido H, Donohue R, Prossnitz ER, Cerilli LA. Mesothelin and GPR30 staining among a spectrum of pancreatic epithelial neoplasms. Int J Surg Pathol. 2011;19(5):588–96. https://doi.org/10.1177/1066896911409575.
234. Zhang Q, Wu YZ, Zhang YM, Ji XH, Hao Q. Activation of G-protein coupled estrogen receptor inhibits the proliferation of cervical cancer cells via sustained activation of ERK1/2. Cell Biochem Funct. 2015;33(3):134–42. https://doi.org/10.1002/cbf.3097.
235. Guan BZ, Yan RL, Huang JW, Li FL, Zhong YX, Chen Y, Liu FN, Hu B, Huang SB, Yin LH. Activation of G protein coupled estrogen receptor (GPER) promotes the migration of renal cell carcinoma via the PI3K/AKT/MMP-9 signals. Cell Adhes Migr. 2015. https://doi.org/10.4161/19336918.2014.990781.
236. Watson CS, Jeng YJ, Hu G, Wozniak A, Bulayeva N, Guptarak J. Estrogen- and xenoestrogen-induced ERK signaling in pituitary tumor cells involves estrogen receptor-alpha interactions with G protein-alphai and caveolin I. Steroids. 2012;77(5):424–32. https://doi.org/10.1016/j.steroids.2011.12.025.
237. Scaling AL, Prossnitz ER, Hathaway HJ. GPER mediates estrogen-induced signaling and proliferation in human breast epithelial cells and normal and malignant breast. Horm Cancer. 2014;5(3):146–60. https://doi.org/10.1007/s12672-014 0174-1.
238. Luo H, Yang G, Yu T, Luo S, Wu C, Sun Y, Liu M, Tu G. GPER-mediated proliferation and estradiol production in breast cancer-associated fibroblasts. Endocr Relat Cancer. 2014;21(2):355–69. https://doi.org/10.1530/erc-13-0237.
239. Lappano R, Maggiolini M. GPER is involved in the functional liaison between breast tumor cells and cancer-associated fibroblasts (CAFs). J Steroid Biochem Mol Biol. 2017. https://doi.org/10.1016/j.jsbmb.2017.02.019.
240. He YY, Du GQ, Cai B, Yan Q, Zhou L, Chen XY, Lu W, Yang YX, Wan XP. Estrogenic transmembrane receptor of GPR30 mediates invasion and carcinogenesis by endometrial cancer cell line RL95-2. J Cancer Res Clin Oncol. 2012;138(5):775–83. https://doi.org/10.1007/s00432-011-1133-7.
241. Du GQ, Zhou L, Chen XY, Wan XP, He YY. The G protein-coupled receptor GPR30 mediates the proliferative and invasive effects induced by hydroxytamoxifen in endometrial cancer cells. Biochem Biophys Res Commun. 2012;420(2):343–9. https://doi.org/10.1016/j.bbrc.2012.02.161.
242. Jiang X, Ye X, Ma J, Li W, Wu R, Jun L. G protein-coupled estrogen receptor 1 (GPER 1) mediates estrogen-induced, proliferation of leiomyoma cells. Gynecol Endocrinol. 2015;31(11):894–8. https://doi.org/10.3109/09513590.2015.1092022.
243. Imesch P, Samartzis EP, Dedes KJ, Fink D, Fedier A. Histone deacetylase inhibitors down-regulate G-protein-coupled estrogen receptor and the GPER-antagonist G-15 inhibits proliferation in endometriotic cells. Fertil Steril. 2013;100(3):770–6. https://doi.org/10.1016/j.fertnstert.2013.05.008.
244. Zhang H, Wang X, Chen Z, Wang W. MicroRNA-424 suppresses estradiol-induced cell proliferation via targeting GPER in endometrial cancer cells. Cell Mol Biol (Noisy-le-Grand). 2015;61(7):96–101.
245. Liu H, Yan Y, Wen H, Jiang X, Cao X, Zhang G, Liu G. A novel estrogen receptor GPER mediates proliferation induced by 17beta-estradiol and selective GPER agonist G-1 in estrogen receptor alpha (ERalpha)-negative ovarian cancer cells. Cell Biol Int. 2014;38(5):631–8. https://doi.org/10.1002/cbin.10243.
246. Sirianni R, Chimento A, Ruggiero C, De Luca A, Lappano R, Ando S, Maggiolini M, Pezzi V. The novel estrogen receptor, G protein-coupled receptor 30, mediates the proliferative effects induced by 17beta-estradiol on mouse spermatogonial GC-1 cell line. Endocrinology. 2008;149(10):5043–51. https://doi.org/10.1210/en.2007-1593.

247. Vivacqua A, Bonofiglio D, Albanito L, Madeo A, Rago V, Carpino A, Musti AM, Picard D, Ando S, Maggiolini M. 17beta-estradiol, genistein, and 4-hydroxytamoxifen induce the proliferation of thyroid cancer cells through the g protein-coupled receptor GPR30. Mol Pharmacol. 2006;70(4):1414–23. https://doi.org/10.1124/mol.106.026344.
248. Rudelius M, Rauert-Wunderlich H, Hartmann E, Hoster E, Dreyling M, Klapper W, Ott G, Rosenwald A. The G protein-coupled estrogen receptor 1 (GPER-1) contributes to the proliferation and survival of mantle cell lymphoma cells. Haematologica. 2015;100(11):e458–61. https://doi.org/10.3324/haematol.2015.127399.
249. Pupo M, Pisano A, Lappano R, Santolla MF, De Francesco EM, Abonante S, Rosano C, Maggiolini M. Bisphenol A induces gene expression changes and proliferative effects through GPER in breast cancer cells and cancer-associated fibroblasts. Environ Health Perspect. 2012;120(8):1177–82. https://doi.org/10.1289/ehp.1104526.
250. Chevalier N, Bouskine A, Fenichel P. Bisphenol A promotes testicular seminoma cell proliferation through GPER/GPR30. Int J Cancer. 2012;130(1):241–2. https://doi.org/10.1002/ijc.25972.
251. Bouskine A, Nebout M, Brucker-Davis F, Benahmed M, Fenichel P. Low doses of bisphenol A promote human seminoma cell proliferation by activating PKA and PKG via a membrane G-protein-coupled estrogen receptor. Environ Health Perspect. 2009;117(7):1053–8. https://doi.org/10.1289/ehp.0800367.
252. Wei W, Chen ZJ, Zhang KS, Yang XL, Wu YM, Chen XH, Huang HB, Liu HL, Cai SH, Du J, Wang HS. The activation of G protein-coupled receptor 30 (GPR30) inhibits proliferation of estrogen receptor-negative breast cancer cells in vitro and in vivo. Cell Death Dis. 2014;5:e1428. https://doi.org/10.1038/cddis.2014.398.
253. Lv X, He C, Huang C, Hua G, Wang Z, Remmenga SW, Rodabaugh KJ, Karpf AR, Dong J, Davis JS, Wang C. G-1 inhibits breast cancer cell growth via targeting colchicine-binding site of tubulin to interfere with microtubule assembly. Mol Cancer Ther. 2017. https://doi.org/10.1158/1535-7163.mct-16-0626.
254. Wang C, Lv X, He C, Hua G, Tsai MY, Davis JS. The G-protein-coupled estrogen receptor agonist G-1 suppresses proliferation of ovarian cancer cells by blocking tubulin polymerization. Cell Death Dis. 2013;4:e869. https://doi.org/10.1038/cddis.2013.397.
255. Holm A, Baldetorp B, Olde B, Leeb-Lundberg LM, Nilsson BO. The GPER1 agonist G-1 attenuates endothelial cell proliferation by inhibiting DNA synthesis and accumulating cells in the S and G2 phases of the cell cycle. J Vasc Res. 2011;48(4):327–35. https://doi.org/10.1159/000322578.
256. Duarte-Guterman P, Lieblich SE, Chow C, Galea LA. Estradiol and GPER activation differentially affect cell proliferation but not GPER expression in the hippocampus of adult female rats. PLoS One. 2015;10(6):e0129880. https://doi.org/10.1371/journal.pone.0129880.
257. Correia S, Cardoso HJ, Cavaco JE, Socorro S. Oestrogens as apoptosis regulators in mammalian testis: angels or devils? Expert Rev Mol Med. 2015;17. https://doi.org/10.1017/erm.2014.25.
258. Lawen A. Apoptosis-an introduction. BioEssays. 2003;25(9):888–96. https://doi.org/10.1002/bies.10329.
259. Royer C, Lucas TF, Lazari MF, Porto CS. 17Beta-estradiol signaling and regulation of proliferation and apoptosis of rat Sertoli cells. Biol Reprod. 2012;86(4):108. https://doi.org/10.1095/biolreprod.111.096891.
260. Li WL, Xiang W, Ping Y. Activation of novel estrogen receptor GPER results in inhibition of cardiocyte apoptosis and cardioprotection. Mol Med Rep. 2015;12(2):2425–30. https://doi.org/10.3892/mmr.2015.3674.
261. Bopassa JC, Eghbali M, Toro L, Stefani E. A novel estrogen receptor GPER inhibits mitochondria permeability transition pore opening and protects the heart against ischemia-reperfusion injury. Am J Phys Heart Circ Phys. 2010;298(1):H16–23. https://doi.org/10.1152/ajpheart.00588.2009.

262. Sbert-Roig M, Bauza-Thorbrugge M, Galmes-Pascual BM, Capllonch-Amer G, Garcia-Palmer FJ, Llado I, Proenza AM, Gianotti M. GPER mediates the effects of 17beta-estradiol in cardiac mitochondrial biogenesis and function. Mol Cell Endocrinol. 2016;420:116–24. https://doi.org/10.1016/j.mce.2015.11.027.
263. Chen J, Hu R, Ge H, Duanmu W, Li Y, Xue X, Hu S, Feng H. G-protein-coupled receptor 30-mediated antiapoptotic effect of estrogen on spinal motor neurons following injury and its underlying mechanisms. Mol Med Rep. 2015;12(2):1733–40. https://doi.org/10.3892/mmr.2015.3601.
264. Zekas E, Prossnitz ER. Estrogen-mediated inactivation of FOXO3a by the G protein-coupled estrogen receptor GPER. BMC Cancer. 2015;15:702. https://doi.org/10.1186/s12885-015-1699-6.
265. Calnan DR, Brunet A. The FoxO code. Oncogene. 2008;27(16):2276–88. https://doi.org/10.1038/onc.2008.21.
266. Broselid S, Cheng B, Sjostrom M, Lovgren K, Klug-De Santiago HL, Belting M, Jirstrom K, Malmstrom P, Olde B, Bendahl PO, Hartman L, Ferno M, Leeb-Lundberg LM. G protein-coupled estrogen receptor is apoptotic and correlates with increased distant disease-free survival of estrogen receptor-positive breast cancer patients. Clin Cancer Res. 2013;19(7):1681–92. https://doi.org/10.1158/1078-0432.ccr-12-2376.
267. Chimento A, Casaburi I, Bartucci M, Patrizii M, Dattilo R, Avena P, Ando S, Pezzi V, Sirianni R. Selective GPER activation decreases proliferation and activates apoptosis in tumor Leydig cells. Cell Death Dis. 2013;4:e747. https://doi.org/10.1038/cddis.2013.275.
268. Mori T, Ito F, Matsushima H, Takaoka O, Tanaka Y, Koshiba A, Kusuki I, Kitawaki J. G protein-coupled estrogen receptor 1 agonist G-1 induces cell cycle arrest in the mitotic phase, leading to apoptosis in endometriosis. Fertil Steril. 2015;103(5):1228–1235.e1221. https://doi.org/10.1016/j.fertnstert.2015.01.026.
269. Kurt AH, Celik A, Kelleci BM. Oxidative/antioxidative enzyme-mediated antiproliferative and proapoptotic effects of the GPER1 agonist G-1 on lung cancer cells. Oncol Lett. 2015;10(5):3177–82. https://doi.org/10.3892/ol.2015.3711.
270. Chimento A, Sirianni R, Casaburi I, Ruggiero C, Maggiolini M, Ando S, Pezzi V. 17beta-Estradiol activates GPER- and ESR1-dependent pathways inducing apoptosis in GC-2 cells, a mouse spermatocyte-derived cell line. Mol Cell Endocrinol. 2012;355(1):49–59. https://doi.org/10.1016/j.mce.2012.01.017.
271. Altmann JB, Yan G, Meeks JF, Abood ME, Brailoiu E, Brailoiu GC. G protein-coupled estrogen receptor-mediated effects on cytosolic calcium and nanomechanics in brain microvascular endothelial cells. J Neurochem. 2015;133(5):629–39. https://doi.org/10.1111/jnc.13066.
272. Yue J, Zhang Y, Li X, Gong S, Tao J, Jiang X. Activation of G-protein-coupled receptor 30 increases T-type calcium currents in trigeminal ganglion neurons via the cholera toxin-sensitive protein kinase A pathway. Pharmazie. 2014;69(11):804–8.
273. Hao J, Bao X, Jin B, Wang X, Mao Z, Li X, Wei L, Shen D, Wang JL. Ca2+ channel subunit alpha 1D promotes proliferation and migration of endometrial cancer cells mediated by 17beta-estradiol via the G protein-coupled estrogen receptor. FASEB J. 2015;29(7):2883–93. https://doi.org/10.1096/fj.14-265603.
274. Ji Y, Han Z, Shao L, Zhao Y. Ultrasound-targeted microbubble destruction of calcium channel subunit alpha 1D siRNA inhibits breast cancer via G protein-coupled receptor 30. Oncol Rep. 2016;36(4):1886–92. https://doi.org/10.3892/or.2016.5031.
275. Okegawa T, Pong RC, Li Y, Hsieh JT. The role of cell adhesion molecule in cancer progression and its application in cancer therapy. Acta Biochim Pol. 2004;51(2):445–57. doi: 035001445
276. Mehlen P, Puisieux A. Metastasis: a question of life or death. Nat Rev Cancer. 2006;6(6):449–58.
277. Jiang QF, Wu TT, Yang JY, Dong CR, Wang N, Liu XH, Liu ZM. 17beta-estradiol promotes the invasion and migration of nuclear estrogen receptor-negative breast cancer cells through

cross-talk between GPER1 and CXCR1. J Steroid Biochem Mol Biol. 2013;138:314–24. https://doi.org/10.1016/j.jsbmb.2013.07.011.
278. Xie K. Interleukin-8 and human cancer biology. Cytokine Growth Factor Rev. 2001;12(4):375–91.
279. Ohshiro K, Schwartz AM, Levine PH, Kumar R. Alternate estrogen receptors promote invasion of inflammatory breast cancer cells via non-genomic signaling. PLoS One. 2012;7(1):e30725. https://doi.org/10.1371/journal.pone.0030725.
280. Yu T, Liu M, Luo H, Wu C, Tang X, Tang S, Hu P, Yan Y, Wang Z, Tu G. GPER mediates enhanced cell viability and motility via non-genomic signaling induced by 17beta-estradiol in triple-negative breast cancer cells. J Steroid Biochem Mol Biol. 2014;143:392–403. https://doi.org/10.1016/j.jsbmb.2014.05.003.
281. Li Y, Chen Y, Zhu ZX, Liu XH, Yang L, Wan L, Lei TW, Wang XD. 4-Hydroxytamoxifen-stimulated processing of cyclin E is mediated via G protein-coupled receptor 30 (GPR30) and accompanied by enhanced migration in MCF-7 breast cancer cells. Toxicology. 2013;309:61–5. https://doi.org/10.1016/j.tox.2013.04.012.
282. De Marco P, Lappano R, De Francesco EM, Cirillo F, Pupo M, Avino S, Vivacqua A, Abonante S, Picard D, Maggiolini M. GPER signalling in both cancer-associated fibroblasts and breast cancer cells mediates a feedforward IL1beta/IL1R1 response. Sci Rep. 2016;6:24354. https://doi.org/10.1038/srep24354.
283. Liu Y, Du FY, Chen W, Fu PF, Yao MY, Zheng SS. G15 sensitizes epithelial breast cancer cells to doxorubicin by preventing epithelial-mesenchymal transition through inhibition of GPR30. Am J Transl Res. 2015;7(5):967–75.
284. Ruan SQ, Wang SW, Wang ZH, Zhang SZ. Regulation of HRG-beta1-induced proliferation, migration and invasion of MCF-7 cells by upregulation of GPR30 expression. Mol Med Rep. 2012;6(1):131–8. https://doi.org/10.3892/mmr.2012.874.
285. Chen Y, Li Z, He Y, Shang D, Pan J, Wang H, Chen H, Zhu Z, Wan L, Wang X. Estrogen and pure antiestrogen fulvestrant (ICI 182 780) augment cell-matrigel adhesion of MCF-7 breast cancer cells through a novel G protein coupled estrogen receptor (GPR30)-to-calpain signaling axis. Toxicol Appl Pharmacol. 2014;275(2):176–81. https://doi.org/10.1016/j.taap.2014.01.005.
286. Castillo Sanchez R, Gomez R, Perez Salazar E. Bisphenol A induces migration through a GPER-, FAK-, Src-, and ERK2-dependent pathway in MDA-MB-231 breast cancer cells. Chem Res Toxicol. 2016;29(3):285–95. https://doi.org/10.1021/acs.chemrestox.5b00457.
287. Yan Y, Liu H, Wen H, Jiang X, Cao X, Zhang G, Liu G. The novel estrogen receptor GPER regulates the migration and invasion of ovarian cancer cells. Mol Cell Biochem. 2013;378(1–2):1–7. https://doi.org/10.1007/s11010-013-1579-9.
288. Yan Y, Jiang X, Zhao Y, Wen H, Liu G. Role of GPER on proliferation, migration and invasion in ligand-independent manner in human ovarian cancer cell line SKOV3. Cell Biochem Funct. 2015;33(8):552–9. https://doi.org/10.1002/cbf.3154.
289. Siesser PM, Hanks SK. The signaling and biological implications of FAK overexpression in cancer. Clin Cancer Res. 2006;12(11 Pt 1):3233–7. https://doi.org/10.1158/1078-0432.ccr-06-0456.
290. Tsai CL, Wu HM, Lin CY, Lin YJ, Chao A, Wang TH, Hsueh S, Lai CH, Wang HS. Estradiol and tamoxifen induce cell migration through GPR30 and activation of focal adhesion kinase (FAK) in endometrial cancers with low or without nuclear estrogen receptor alpha (ERalpha). PLoS One. 2013;8(9):e72999. https://doi.org/10.1371/journal.pone.0072999.
291. Avino S, De Marco P, Cirillo F, Santolla MF, De Francesco EM, Perri MG, Rigiracciolo D, Dolce V, Belfiore A, Maggiolini M, Lappano R, Vivacqua A. Stimulatory actions of IGF-I are mediated by IGF-IR cross-talk with GPER and DDR1 in mesothelioma and lung cancer cells. Oncotarget. 2016;7(33):52710–28. 10.18632/oncotarget.10348.
292. Fraser SP, Ozerlat-Gunduz I, Onkal R, Diss JK, Latchman DS, Djamgoz MB. Estrogen and non-genomic upregulation of voltage-gated Na(+) channel activity in MDA-MB-231

human breast cancer cells: role in adhesion. J Cell Physiol. 2010;224(2):527–39. https://doi.org/10.1002/jcp.22154.

293. Henic E, Noskova V, Hoyer-Hansen G, Hansson S, Casslen B. Estradiol attenuates EGF-induced rapid uPAR mobilization and cell migration via the G-protein-coupled receptor 30 in ovarian cancer cells. Int J Gynecol Cancer. 2009;19(2):214–22. https://doi.org/10.1111/IGC.0b013e31819bcb75.
294. Zhu G, Huang Y, Wu C, Wei D, Shi Y. Activation of G-protein-coupled estrogen receptor inhibits the migration of human nonsmall cell lung cancer cells via IKK-beta/NF-kappaB signals. DNA Cell Biol. 2016;35(8):434–42. https://doi.org/10.1089/dna.2016.3235.
295. Sharma G, Hu C, Brigman JL, Zhu G, Hathaway HJ, Prossnitz ER. GPER deficiency in male mice results in insulin resistance, dyslipidemia, and a proinflammatory state. Endocrinology. 2013;154(11):4136–45. https://doi.org/10.1210/en.2013-1357.
296. Chakrabarti S, Davidge ST. G-protein coupled receptor 30 (GPR30): a novel regulator of endothelial inflammation. PLoS One. 2012;7(12):e52357. https://doi.org/10.1371/journal.pone.0052357.
297. Itoga M, Konno Y, Moritoki Y, Saito Y, Ito W, Tamaki M, Kobayashi Y, Kayaba H, Kikuchi Y, Chihara J, Takeda M, Ueki S, Hirokawa M. G-protein-coupled estrogen receptor agonist suppresses airway inflammation in a mouse model of asthma through IL-10. PLoS One. 2015;10(3):e0123210. https://doi.org/10.1371/journal.pone.0123210.
298. Okamoto M, Mizukami Y. GPER negatively regulates TNFalpha-induced IL-6 production in human breast cancer cells via NF-kappaB pathway. Endocr J. 2016. https://doi.org/10.1507/endocrj.EJ15-0571.
299. Hanahan D, Weinberg RA. Hallmarks of cancer: the next generation. Cell. 2011;144(5):646–74. https://doi.org/10.1016/j.cell.2011.02.013.
300. Warburg O, Wind F, Negelein E. The metabolism of tumors in the body. J Gen Physiol. 1927;8(6):519–30.
301. Martensson UE, Salehi SA, Windahl S, Gomez MF, Sward K, Daszkiewicz-Nilsson J, Wendt A, Andersson N, Hellstrand P, Grande PO, Owman C, Rosen CJ, Adamo ML, Lundquist I, Rorsman P, Nilsson BO, Ohlsson C, Olde B, Leeb-Lundberg LM. Deletion of the G protein-coupled receptor 30 impairs glucose tolerance, reduces bone growth, increases blood pressure, and eliminates estradiol-stimulated insulin release in female mice. Endocrinology. 2009;150(2):687–98. https://doi.org/10.1210/en.2008-0623.
302. Sharma G, Prossnitz ER. Mechanisms of estradiol-induced insulin secretion by the G protein-coupled estrogen receptor GPR30/GPER in pancreatic beta-cells. Endocrinology. 2011;152(8):3030–9. https://doi.org/10.1210/en.2011-0091.
303. Xie BY, Lv QY, Ning CC, Yang BY, Shan WW, Cheng YL, Gu C, Luo XZ, Zhang ZB, Chen XJ, Xi XW, Feng YJ. TET1-GPER-PI3K/AKT pathway is involved in insulin-driven endometrial cancer cell proliferation. Biochem Biophys Res Commun. 2017;482(4):857–62. https://doi.org/10.1016/j.bbrc.2016.11.124.
304. Yu T, Yang G, Hou Y, Tang X, Wu C, Wu XA, Guo L, Zhu Q, Luo H, Du YE, Wen S, Xu L, Yin J, Tu G, Liu M. Cytoplasmic GPER translocation in cancer-associated fibroblasts mediates cAMP/PKA/CREB/glycolytic axis to confer tumor cells with multidrug resistance. Oncogene. 2016. https://doi.org/10.1038/onc.2016.370.
305. Altman BJ, Stine ZE, Dang CV. From Krebs to clinic: glutamine metabolism to cancer therapy. Nat Rev Cancer. 2016;16(10):619–34. https://doi.org/10.1038/nrc.2016.71.
306. Rothstein JD, Dykes-Hoberg M, Pardo CA, Bristol LA, Jin L, Kuncl RW, Kanai Y, Hediger MA, Wang Y, Schielke JP, Welty DF. Knockout of glutamate transporters reveals a major role for astroglial transport in excitotoxicity and clearance of glutamate. Neuron. 1996;16(3):675–86.
307. Lee E, Sidoryk-Wegrzynowicz M, Wang N, Webb A, Son DS, Lee K, Aschner M. GPR30 regulates glutamate transporter GLT-1 expression in rat primary astrocytes. J Biol Chem. 2012;287(32):26817–28. https://doi.org/10.1074/jbc.M112.341867.

308. Tiano JP, Delghingaro-Augusto V, Le May C, Liu S, Kaw MK, Khuder SS, Latour MG, Bhatt SA, Korach KS, Najjar SM, Prentki M, Mauvais-Jarvis F. Estrogen receptor activation reduces lipid synthesis in pancreatic islets and prevents beta cell failure in rodent models of type 2 diabetes. J Clin Invest. 2011;121(8):3331–42. https://doi.org/10.1172/jci44564.
309. Cha JY, Repa JJ. The liver X receptor (LXR) and hepatic lipogenesis. The carbohydrate-response element-binding protein is a target gene of LXR. J Biol Chem. 2007;282(1):743–51. https://doi.org/10.1074/jbc.M605023200.
310. Repa JJ, Liang G, Ou J, Bashmakov Y, Lobaccaro JM, Shimomura I, Shan B, Brown MS, Goldstein JL, Mangelsdorf DJ. Regulation of mouse sterol regulatory element-binding protein-1c gene (SREBP-1c) by oxysterol receptors, LXRalpha and LXRbeta. Genes Dev. 2000;14(22):2819–30.
311. Uyeda K, Repa JJ. Carbohydrate response element binding protein, ChREBP, a transcription factor coupling hepatic glucose utilization and lipid synthesis. Cell Metab. 2006;4(2):107–10. https://doi.org/10.1016/j.cmet.2006.06.008.
312. Tiano JP, Mauvais-Jarvis F. Molecular mechanisms of estrogen receptors' suppression of lipogenesis in pancreatic beta-cells. Endocrinology. 2012;153(7):2997–3005. https://doi.org/10.1210/en.2011-1980.
313. Santolla MF, Lappano R, De Marco P, Pupo M, Vivacqua A, Sisci D, Abonante S, Iacopetta D, Cappello AR, Dolce V, Maggiolini M. G protein-coupled estrogen receptor mediates the up-regulation of fatty acid synthase induced by 17beta-estradiol in cancer cells and cancer-associated fibroblasts. J Biol Chem. 2012;287(52):43234–45. https://doi.org/10.1074/jbc.M112.417303.
314. Horton JD, Cohen JC, Hobbs HH. Molecular biology of PCSK9: its role in LDL metabolism. Trends Biochem Sci. 2007;32(2):71–7. https://doi.org/10.1016/j.tibs.2006.12.008.
315. Hussain Y, Ding Q, Connelly PW, Brunt JH, Ban MR, McIntyre AD, Huff MW, Gros R, Hegele RA, Feldman RD. G-protein estrogen receptor as a regulator of low-density lipoprotein cholesterol metabolism: cellular and population genetic studies. Arterioscler Thromb Vasc Biol. 2015;35(1):213–21. https://doi.org/10.1161/atvbaha.114.304326.
316. Wang A, Luo J, Moore W, Alkhalidy H, Wu L, Zhang J, Zhen W, Wang Y, Clegg DJ, Bin X, Cheng Z, McMillan RP, Hulver MW, Liu D. GPR30 regulates diet-induced adiposity in female mice and adipogenesis in vitro. Sci Rep. 2016;6:34302. https://doi.org/10.1038/srep34302.
317. Hall DA, Zhu H, Zhu X, Royce T, Gerstein M, Snyder M. Regulation of gene expression by a metabolic enzyme. Science (New York, NY). 2004;306(5695):482–4. https://doi.org/10.1126/science.1096773.
318. Zhuang Y, Xiang J, Bao W, Sun Y, Wang L, Tan M, He Y, Xi X. MDH2 stimulated by estrogen-GPR30 pathway down-regulated PTEN expression promoting the proliferation and invasion of cells in endometrial cancer. Transl Oncol. 2017;10(2):203–10. https://doi.org/10.1016/j.tranon.2017.01.009.
319. Siegel R, Ma J, Zou Z, Jemal A. Cancer statistics, 2014. CA. 2014;64(1):9–29. https://doi.org/10.3322/caac.21208.
320. Ribal MJ, Martínez-Salamanca JI, García Freire C (2015) Current patterns of management of advanced prostate cancer in routine clinical practice in Spain. Prostate Cancer. 2015:6. doi:https://doi.org/10.1155/2015/186740.
321. Karantanos T, Corn PG, Thompson TC. Prostate cancer progression after androgen deprivation therapy: mechanisms of castrate resistance and novel therapeutic approaches. Oncogene. 2013;32(49):5501–11. https://doi.org/10.1038/onc.2013.206.
322. Huggins C. Endocrine control of prostatic cancer. Science (New York, NY). 1943;97(2529):541–4. https://doi.org/10.1126/science.97.2529.541.
323. Ricke WA, McPherson SJ, Bianco JJ, Cunha GR, Wang Y, Risbridger GP. Prostatic hormonal carcinogenesis is mediated by in situ estrogen production and estrogen receptor alpha signaling. FASEB J. 2008;22(5):1512–20. https://doi.org/10.1096/fj.07-9526com.

324. Yang M, Wang J, Wang L, Shen C, Su B, Qi M, Hu J, Gao W, Tan W, Han B. Estrogen induces androgen-repressed SOX4 expression to promote progression of prostate cancer cells. Prostate. 2015;75(13):1363–75. https://doi.org/10.1002/pros.23017.
325. Denmeade S, Isaacs J. Androgen deprivation strategies in the treatment of advanced prostate cancer. In: Kufe DW, Pollock RE, Weichselbaum RR, et al., editors. Holland-Frei cancer medicine. Hamilton, ON: BC Decker; 2000.
326. Montgomery B, Nelson PS, Vessella R, Kalhorn T, Hess D, Corey E. Estradiol suppresses tissue androgens and prostate cancer growth in castration resistant prostate cancer. BMC Cancer. 2010;10:244. https://doi.org/10.1186/1471-2407-10-244.
327. Figueira MI, Correia S, Vaz CV, Cardoso HJ, Gomes IM, Marques R, Maia CJ, Socorro S. Estrogens down-regulate the stem cell factor (SCF)/c-KIT system in prostate cells: evidence of antiproliferative and proapoptotic effects. Biochem Pharmacol. 2016;99:73–87. https://doi.org/10.1016/j.bcp.2015.11.016.
328. Levin ER. Integration of the extranuclear and nuclear actions of estrogen. Mol Endocrinol. 2005;19(8):1951–9. https://doi.org/10.1210/me.2004-0390.
329. Nelson AW, Tilley WD, Neal DE, Carroll JS. Estrogen receptor beta in prostate cancer: friend or foe? Endocr Relat Cancer. 2014;21(4):T219–34. https://doi.org/10.1530/erc-13-0508.
330. Signoretti S, Loda M. Prostate stem cells: from development to cancer. Semin Cancer Biol. 2007;17(3):219–24. https://doi.org/10.1016/j.semcancer.2006.04.004.
331. Rago V, Romeo F, Giordano F, Ferraro A, Carpino A. Identification of the G protein coupled estrogen receptor (GPER) in human prostate: expression site of the estrogen receptor in the benign and neoplastic gland. Andrology. 2016;4(1):121–7. https://doi.org/10.1111/andr.12131.
332. Hu WY, Shi GB, Lam HM, Hu DP, Ho SM, Madueke IC, Kajdacsy-Balla A, Prins GS. Estrogen-initiated transformation of prostate epithelium derived from normal human prostate stem-progenitor cells. Endocrinology. 2011;152(6):2150–63. https://doi.org/10.1210/en.2010-1377.
333. Pisolato R, Lombardi AP, Vicente CM, Lucas TF, Lazari MF, Porto CS. Expression and regulation of the estrogen receptors in PC-3 human prostate cancer cells. Steroids. 2015. https://doi.org/10.1016/j.steroids.2015.12.021.
334. Chan QK, Lam HM, Ng CF, Lee AY, Chan ES, Ng HK, Ho SM, Lau KM. Activation of GPR30 inhibits the growth of prostate cancer cells through sustained activation of Erk1/2, c-jun/c-fos-dependent upregulation of p21, and induction of G(2) cell-cycle arrest. Cell Death Differ. 2010;17(9):1511–23. https://doi.org/10.1038/cdd.2010.20.
335. Zhang Z, Duan L, Du X, Ma H, Park I, Lee C, Zhang J, Shi J. The proliferative effect of estradiol on human prostate stromal cells is mediated through activation of ERK. Prostate. 2008;68(5):508–16. https://doi.org/10.1002/pros.20722.
336. Schade GR, Holt S, Zhang X, Song D, Wright JL, Zhao S, Kolb S, Lam HM, Levin L, Leung YK, Ho SM, Stanford JL. Prostate cancer expression profiles of cytoplasmic ERbeta1 and nuclear ERbeta2 are associated with poor outcomes following radical prostatectomy. J Urol. 2016. https://doi.org/10.1016/j.juro.2015.12.101.
337. Gan WD, Zhou M, Hu Y, Li DM, Jia RP. Low-dose nonylphenol promotes the proliferation of DU-145 cells and expression of membrane estrogen receptor GPR30 in DU-145 cells. Zhonghua Nan Ke Xue. 2014;20(5):405–9.
338. Lam HM, Ouyang B, Chen J, Ying J, Wang J, Wu CL, Jia L, Medvedovic M, Vessella RL, Ho SM. Targeting GPR30 with G-1: a new therapeutic target for castration-resistant prostate cancer. Endocr Relat Cancer. 2014;21(6):903–14. https://doi.org/10.1530/erc-14-0402.
339. Park II, Zhang Q, Liu V, Kozlowski JM, Zhang J, Lee C. 17Beta-estradiol at low concentrations acts through distinct pathways in normal versus benign prostatic hyperplasia-derived prostate stromal cells. Endocrinology. 2009;150(10):4594–605. https://doi.org/10.1210/en.2008-1591.
340. Koong LY, Watson CS. Direct estradiol and diethylstilbestrol actions on early- versus late-stage prostate cancer cells. Prostate. 2014;74(16):1589–603. https://doi.org/10.1002/pros.22875.

341. Lau KM, Ma FM, Xia JT, Chan QK, Ng CF, To KF. Activation of GPR30 stimulates GTP-binding of Galphai1 protein to sustain activation of Erk1/2 in inhibition of prostate cancer cell growth and modulates metastatic properties. Exp Cell Res. 2017;350(1):199–209. https://doi.org/10.1016/j.yexcr.2016.11.022.
342. Comeglio P, Morelli A, Cellai I, Vignozzi L, Sarchielli E, Filippi S, Maneschi E, Corcetto F, Corno C, Gacci M, Vannelli GB, Maggi M. Opposite effects of tamoxifen on metabolic syndrome-induced bladder and prostate alterations: a role for GPR30/GPER? Prostate. 2014;74(1):10–28. https://doi.org/10.1002/pros.22723.
343. Vaz CV, Alves MG, Marques R, Moreira PI, Oliveira PF, Maia CJ, Socorro S. Androgen-responsive and nonresponsive prostate cancer cells present a distinct glycolytic metabolism profile. Int J Biochem Cell Biol. 2012;44(11):2077–84. https://doi.org/10.1016/j.biocel.2012.08.013.
344. Vaz CV, Marques R, Alves MG, Oliveira PF, Cavaco JE, Maia CJ, Socorro S. Androgens enhance the glycolytic metabolism and lactate export in prostate cancer cells by modulating the expression of GLUT1, GLUT3, PFK, LDH and MCT4 genes. J Cancer Res Clin Oncol. 2016;142(1):5–16. https://doi.org/10.1007/s00432-015-1992-4.
345. Massie CE, Lynch A, Ramos-Montoya A, Boren J, Stark R, Fazli L, Warren A, Scott H, Madhu B, Sharma N, Bon H, Zecchini V, Smith DM, Denicola GM, Mathews N, Osborne M, Hadfield J, Macarthur S, Adryan B, Lyons SK, Brindle KM, Griffiths J, Gleave ME, Rennie PS, Neal DE, Mills IG. The androgen receptor fuels prostate cancer by regulating central metabolism and biosynthesis. EMBO J. 2011;30(13):2719–33. https://doi.org/10.1038/emboj.2011.158.
346. Correia S, Cardoso HJ, Cavaco JE, Socorro S. Oestrogens as apoptosis regulators in mammalian testis: angels or devils? Expert Rev Mol Med. 2015;17:e2. https://doi.org/10.1017/erm.2014.25.
347. Correia S, Alves MR, Cavaco JE, Oliveira PF, Socorro S. Estrogenic regulation of testicular expression of stem cell factor and c-kit: implications in germ cell survival and male fertility. Fertil Steril. 2014;102(1):299–306. https://doi.org/10.1016/j.fertnstert.2014.04.009.

Chapter 6
Bone Cancer: Dysregulation of Signaling Cascades by microRNAs

Janaina de Andréa Dernowsek

Abstract Osteogenesis is a biological process involving the specification of mesenchymal stem cells and their maturation and differentiation into osteoblasts and osteocyte cells. Complex regulatory networks are required during these events in osteogenesis, and their imbalance can cause disastrous disordering and consequently diseases, such as cancer. Bone-related diseases involve a deregulation of genes, microRNAs, transcription factors (TF) or proteins implicates in the biological process. However, the mechanism of action of miRNAs in bone cancers is not understood clearly. Key elements of the intrinsic molecular processes that modulate essential cell phenomena are studied continuously and with increasingly advanced tools. Nevertheless, such processes, such as cell cycle control, cell proliferation, metabolism, and apoptosis, remain a challenge in the development of new targeted therapies to treat heterogeneous and complex diseases. These new meta-analysis methods—omics technologies—will be useful for more efficient analyses of a vast number of data generated by the multidisciplinary areas involved in the translational studies for medicine. The information about miRNAs associations with bone cancer were results of the networks visualizations of the independent studies found in the literature. Therefore, this chapter mainly summarizes how deregulation of different miRNAs contributes to bone cancer, such as Osteosarcoma, Chondrosarcoma and Ewing sarcoma. Overall, finding new methods and biomarkers as miRNAs could be useful to improve the diagnosis and create advanced treatments.

Keywords Bone cancer · MicroRNAs · Signaling cascades · Osteosarcoma · Chondrosarcoma · Ewing sarcoma

---

J. de Andréa Dernowsek (✉)
Center for Information Technology Renato Archer (CTI), 3D Technologies Research Group (NT3D), Campinas, SP, Brazil
e-mail: janaina.dernowsek@cti.gov.br

S. Fayyaz, A. A. Farooqi (eds.), *Recent Trends in Cancer Biology: Spotlight on Signaling Cascades and microRNAs*, https://doi.org/10.1007/978-3-319-71553-7_6

Introduction

Bone cancer is therapeutically challenging and technological advancements have transformed our understanding of diseases profiling, and the data may help the identification of novel diagnostics and therapeutic approaches, as well as new biomarkers.

Recent progress in Multi-Omics data integration methods can help us in developing a better understanding of the biology of disease with results that are more accurate by considering multiple approaches. Researches involving data integrations of the genome, transcriptome, miRnome, proteome, and metabolome can answer previously unanswered questions. In fact, these results are due to the advance in computational power and the availability of next-generation technologies. This scenery influencing the future of the molecular cancer biology and needs to be frequently updated. Therefore, the focus of this book chapter is a review of recent data of the dysregulation of signaling cascade by microRNAs in cancer, specifically in bone cancer.

Bone cancer biology is characterized by the complex phenomenon that implicates many factors including the dysregulation of microRNAs (miRNAs) expression, which may contribute to loss of the normal regulatory controls of the signaling cascades [1–3]. Deregulation of genes and miRNAs is predominantly involved in cell proliferation, differentiation, and apoptosis [2, 4].

Unquestionably, field of the miRNAs research has attracted considerable appreciation, with the identification in 1993 of the first of these small single-stranded non-coding ribonucleic acids (RNAs) in the nematode *Caenorhabditis elegans* [5]. Since then more than 28,000 entries representing precursor miRNAs, expressing 35,828 mature miRNAs in 223 species have been identified with 2588 in humans at the time of writing, but this number may rise (miRBase Release 21: June 2017) [6].

MicroRNAs are involved in a myriad of normal and abnormal biological processes. Based on the insights gleaned from decades of research, it seems more understandable that miRNAs tactfully modulate myriad of different genes in different cancers [7–9]. Rapidly emerging scientific evidence is demystifying the mechanisms of dysregulation of miRNAs which has been related to several cellular developments, proliferation, and differentiation, as well as osteogenesis and bone disease [10–13]. In recent years, studies on profiling of the miRnome (global miRNA expression levels) and post-transcriptional interactions between mRNA and miRNA, have significantly improved our concepts related to miRNA regulation of transducers of multiple cell signaling pathways in different cancers [14].

Cancer becomes the leading cause of deaths worldwide and reducing the patient quality of life, including bone cancer that is an abnormal growth that is found in the native tissues. Depending on the case, the growth may be malignant or may be benign. Besides, the bone tumor can be divided into the primary, which originates from the bone tissue, and a secondary tumor that originates from other areas of the body [15].

A myriad of signaling pathways is associated with the developing bone cancer, such as Wnt/ß, MAPK, Notch, JNK, IGF, mTOR among other [16–22]. Despite several efforts, specific molecular interactions involved in bone cancer remain unclear [3, 23]. Therefore, this chapter focuses on recent findings of the dysregulated microRNAs in signaling cascades of the bone cancer mainly the most common forms, including osteosarcoma (OS), chondrosarcoma (COS), and Ewing sarcoma (ES).

Bone Cancer

Osteogenic differentiation, a crucial biological process in bone development, involves the activation of multiple signaling pathways, including TGFb, BMP, Wnt as well transcription factors, which are rigidly regulated by miRNAs [24, 25]. Until now, many studies have showed a huge number of miRNAs that play their role as posttranscriptional regulators during osteogenic differentiation, such as miR-20a, miR-22, miR-27, miR-28-5p, miR-29a, miR-29b, miR-29c, miR-133, miR-139-5p, miR-141, miR-196a, miR-200a, miR-204, miR-210, miR-211, miR-378, miR-450a-5p among others [1, 14, 24, 26–29]. Additionally, the dysregulation of signaling cascades by microRNAs is the cause of many cancers, including the different bone cancer types (OS, COS, and ES). The miRNA-related changes have been shown to affect the control of many cellular mechanisms, including proliferative, control cell growth, motility, survival, invasion, metastasis, angiogenesis signaling, among others [30–33]. The significant increase of miRNA data provides a comprehensive understanding of the cancer biology and its association with signaling cascades. The use of databases that contain miRNA information is appropriate for a thorough review. Therefore, resort to the miRCancer database [34] to create a summarized diagram of miRNAs associated in normal osteogenesis and bone cancers. As previously mentioned, miRNA expression implicated in normal osteogenesis can be altered in malignancies and play a fundamental role in tumor progression. Thus, Fig. 6.1 sums up all recent researches in OS, COS, and ES, which have been conducted in recent years regarding the miRNAs alterations and their targets.

In the systematic analysis present in Fig. 6.1, the miR-125b was associated with all bone cancers and with the normal osteogenesis. MiR-125b/miR-125b-5p is broadly conserved and presents 65 validated targets found in miRecords database [35]. Furthermore, the dysregulated miR-125b expression was associated with numerous types of tumor, including OS [36], gastric cancer [37], renal carcinoma [38], hepatocellular carcinoma [39], and breast cancer [40]. Further studies identified the role of miR-125b act as potential biomarker in the progression and metastasis of OS [41].

Interestingly, the compare data set does not contain any shared miRNAs between COS, ES, and normal osteogenesis. Moreover, the topics below contain recent data with more details on bone cancers.

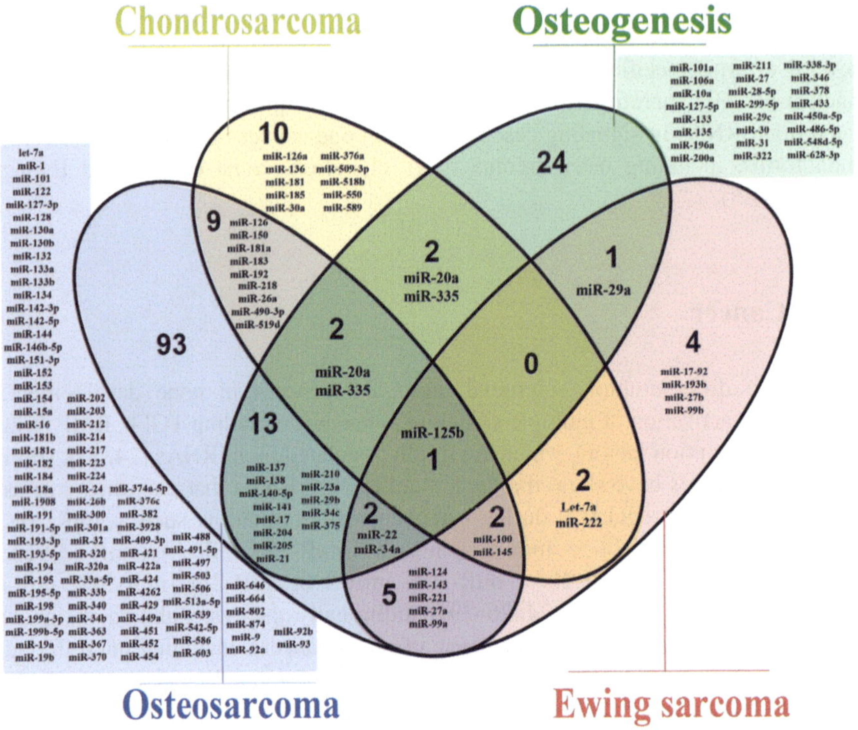

Fig. 6.1 Recent discoveries about the biology of miRNAs attributed to bone cancers—Osteosarcoma, Chondrosarcoma and Ewing sarcoma

Osteosarcoma

The first, and by far the most common bone cancer is the osteosarcoma (OS), that affects many more children and adolescent than adults.

Despite advances in OS studies, the complex mechanism of miRNA actions is not well understood. Until now, several miRNAs are known to function as oncogenes in OS, including the miR-27a, miR-21, miR-9, miR-92, miR-382, and miR135b, miR-335, miR-489-3p [42, 43]. On the other hand, the miR-183, miR133b, miR-138, miR-195 miR199a-3p, miR-124, miR-646, miR-100, miR-101, miR-1, miR-409-3p, miR-4262 and miR-33b are known to function as a tumor suppressor [43–48]. About these recent studies, the miRNAs-targets interactions and signaling pathways were summarized in Fig. 6.2.

Forty-eight validated miRNAs-target interactions were found in literature associated with OS, COS, and EW. In Fig. 6.2a, a visualization of the interactions networks was created using Cytoscape 3.5.1 software. Cytoscape is an open-source software that is used to view, analyze and integrate a significant amount of data, as well as to create functional network clusters with biological molecular information

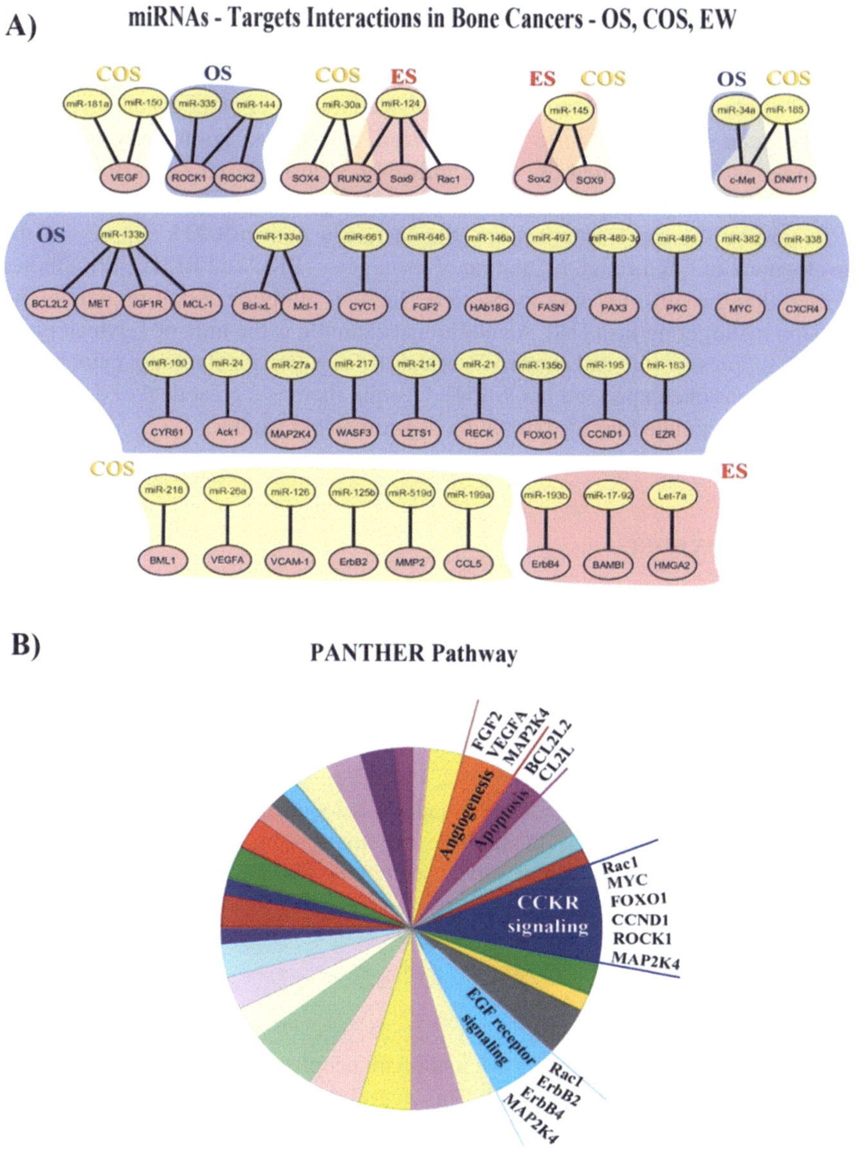

Fig. 6.2 (**a**) Interaction networks of the miRNA-targets in bone cancers based on Cytoscape 3.5.1. *OS* osteosarcoma, *COS* chondrosarcoma, *EW* Ewing sarcoma. (**b**) Pie chart of PANTHER Pathways of 48 miRNA-targets found in the literature [26, 43–45, 47–49]

[50]. Protein Analysis Through Evolutionary Relationships (PANTHER) was used to analyze the signaling pathways of the miRNA targets found by biology experiments described in the literature (Fig. 6.2b) [51].

In identifying the latest studies on miRNAs in OS, we found that with the last year, new miRNAs involved in OS have been identified, such as the miRNA-335 and its rock1 target that influence tumor progression and consequently the prognosis in osteosarcoma [52]. Wang et al. investigated the effect of the connective tissue growth factor (CTGF) on miR-543 regulation, and it was significantly correlated with tumor stage [53]. The Loss of the miR-489-3p also played a contributory role in the OS metastasis by activating the PAX3-MET pathway [48].

Regarding the tumor suppressors, the MiR-486 and miR-218 suppressed the development of OS through regulation of its targets—PKC and BMI1 [49]. Studies showed that the miR-497 was found to be such an inhibitor of cell migration and invasion by targeting fasn [18]. An additional example is the miR-661, which plays a tumor suppressor role and exerted its effects by negative regulation of cytochrome c1, which was overexpressed in OS [54]. Overall, there is a great deal of data about miRNAs dysregulated in OS, but the focus is on recent studies.

Chondrosarcoma

Chondrosarcoma (COS), is cancer of the cartilage, and it is characterized by the abnormal production of a specific cartilage matrix [7]. Although COS can manifest in any bone, this cancer preferentially develops in the pelvis, humerus, femur, scapula, and ribs and can present at any age [55].

The most recent literature has reported 28 miRNAs (Let-7a, MiRs 100, 125b, 126, 136, 145, 150, 181a, 183, 185, 192, 199a, 20a, 218, 222, 26a, 30a, 335, 376a, 490-3p, 509-3p, 518b, 519d, 550, 589, and 96) associated with COS [7, 17, 43, 56, 57]. However, in this review made, little information on validated miRNA-target interactions was found, and some can be seen in the Fig. 6.2a.

Ewing Sarcoma

Ewing Sarcoma (ES) is an aggressive malignant neoplasm, and it is the second most common primary bone cancer [58]. This aggressive cancer bone affects children and young adults predominantly [7].

Unlike OS and CH, the ES has little information about miRNA biogenesis deregulations in signaling cascades. Recent findings indicate that alterations in miR expression are widespread and can be involved in EWS gene, aside from independent mechanisms [22, 59].

Throughout the past few years, some studies about miRNAs expression profile have demonstrated miRNA dysregulation in ES. It was found 51 miRNAs, but some data exhibited discordant expression [58]. Thus, these discrepancies need to be re-assessed.

Other studies have revelated others miRNAs (Fig. 6.1), including the miRNAs (miR-124, miR-143, miR-221, miR-27a, miR-99a, miR-22, miR-34a, miR-125b), which are were associated with OS. The miRNAs (Let-7a, miR-222, miR-miR-100, miR-145, miR125b) were disclosed with COS. These last information—miRNAs associations with bone cancer—were results of the visualizations networks of the independent studies found in the literature.

Conclusions

Every day, significant numbers of studies associating miRNAs and diseases are accumulated to the data bases, and molecular biologists have gained success in putting together missing pieces of an incomplete jig-saw puzzle to comprehend the regulatory capacities of oncogenic and tumor suppressor miRNAs. It is hoped these studies, and scientific approaches will improve current technologies for the treatments and detection of new biomarkers.

Despite the increase in knowledge regarding bone cancers, commonly in children and adolescents, the survival time did not increase. However, results from the cellular and molecular biology of bone cancers, targeting the various key pathways that control the aggressive behavior of the diseases, led to new and efficient therapeutic opportunities.

Future studies of miRNA biology and the signaling cascades stand to expand the understanding of the bone cancers, which may identify new biomarkers and treatment options.

References

1. Sera SR, zur Nieden NI. MicroRNA regulation of skeletal development. Curr Osteoporos Rep. 2017;14:1–14. https://doi.org/10.1007/s11914-017-0379-7.
2. Zoni E, van der Pluijm G. The role of microRNAs in bone metastasis. J Bone Oncol. 2016;5:104–8. https://doi.org/10.1016/j.jbo.2016.04.002.
3. Ibrahim T, Mercatali L, Amadori D. Bone and cancer: the osteoncology. Clin Cases Miner Bone Metab. 2013;10(2):121–3.
4. Sun M, Zhou X, Chen L, Huang S, Leung V, Wu N, et al. The regulatory roles of microRNAs in bone remodeling and perspectives as biomarkers in osteoporosis. Biomed Res Int. 2016;5:104–8.
5. Lee RC, Feinbaum RL, Ambros V. The C. elegans heterochronic gene lin-4 encodes small RNAs with antisense comlementarity to lin-14. Cell. 1993;75:843–54.
6. Griffiths-Jones S. MiRBase: microRNA sequences and annotation. Curr Protoc Bioinformatics. 2010;34(SUPPL. 29):1291–12910.
7. Nugent M. MicroRNA cancer regulation, vol. 774. Cham: Springer; 2013. p. 201–30. https://doi.org/10.1007/978-94-007-5590-1.
8. Mahdian-shakib A, Dorostkar R, Tat M, Hashemzadeh MS, Saidi N. Differential role of microRNAs in prognosis, diagnosis, and therapy of ovarian cancer. Biomed Pharmacother. 2016;84:592–600.

9. Lee YS, Dutta A. MicroRNAs in cancer. Annu Rev Pathol Mech Dis. 2009;4:199–227.
10. Farooqi AA, Li Y, Sarkar FH. The biological complexity of RKIP signaling in human cancers. Exp Mol Med. 2015;47(9):e185. https://doi.org/10.1038/emm.2015.70.
11. Misso G, Di Martino MT, De Rosa G, Farooqi AA, Lombardi A, Campani V, et al. Mir-34: a new weapon against cancer? Mol Ther Nucleic Acids. 2014;3(August):e195. http://linkinghub.elsevier.com/retrieve/pii/S2162253116303341
12. Huang G, Nishimoto K, Yang Y, Kleinerman ES. Current advances in osteosarcoma. Adv Exp Med Biol. 2014;804:203–17. https://doi.org/10.1007/978-3-319-04843-7.
13. Manca P, Pantano F, Iuliani M, Ribelli G, De Lisi D, Danesi R, et al. Determinants of bone specific metastasis in prostate cancer. Crit Rev Oncol Hematol. 2017;112:59.
14. Dernowsek JA, Pereira MC, Fornari TA, Macedo C, Assis AF, Donate PB, et al. Posttranscriptional interaction between miR-450a-5p and miR-28-5p and STAT1 mRNA triggers osteoblastic differentiation of human mesenchymal stem cells. J Cell Biochem. 2017. https://doi.org/10.1002/jcb.26060.
15. Habeeb A, HafsaSania H, Shaik M, Baig M, Ali S, Amer S. Bone cancer - current pharmacotherapeutic approach. Int J Adv Res. 2017;5(1):378–91. http://www.journalijar.com/article/14397/bone-cancer--current-pharmacotherapeutic-approach
16. Fernandes G, Barone AW, Dziak R. The effect of ascorbic acid on bone cancer cells in vitro. Cogent Biol. 2017;3(1):1288335. https://doi.org/10.1080/23312025.2017.1288335.
17. Tian J, He H, Lei G. Wnt/β-catenin pathway in bone cancers. Tumor Biol. 2014;35(10):9439–45.
18. Jiang Y, Yin X, Wu L, Qin Q, Xu J. MAPK/P53-mediated FASN expression in bone tumors. Oncol Lett. 2017;13:4035–8. https://doi.org/10.3892/ol.2017.6015.
19. Sethi N, Kang Y. Notch signalling in cancer progression and bone metastasis. Br J Cancer. 2011;105(12):1805–10. https://doi.org/10.1038/bjc.2011.497.
20. Sasaki K, Hitora T, Nakamura O, Kono R, Yamamoto T. The role of MAPK pathway in bone and soft tissue tumors. Anticancer Res. 2011;31(2):549–53.
21. Bertoldo F, Silvestris F, Ibrahim T, Cognetti F, Generali D, Ripamonti CI, et al. Targeting bone metastatic cancer: role of the mTOR pathway. Biochim Biophys Acta. 2014;1845(2):248–54. https://doi.org/10.1016/j.bbcan.2014.01.009.
22. Dylla L, Moore C, Jedlicka P. MicroRNAs in ewing sarcoma. Front Oncol. 2013;3(March):65. http://www.pubmedcentral.nih.gov/articlerender.fcgi?artid=3610014&tool=pmcentrez&rendertype=abstract
23. Moonan DJ, Kumar R, Biermann JS. Disclosure of relevant financial relationships CE planners: individuals who provided content development and/or authorship assistance. J Natl Compr Canc Netw. 2017;15(2):155–67. http://www.jnccn.org/content/15/2/155.full?sid=f3568af2-cd0a-4b07-a6b2-f44009a17c36
24. Zhang Y, Xie RL, Croce CM, Stein JL, Lian JB, van Wijnen AJ, et al. A program of microRNAs controls osteogenic lineage progression by targeting transcription factor Runx2. Proc Natl Acad Sci. 2011;108(24):9863–8. https://doi.org/10.1073/pnas.1018493108.
25. Vimalraj S, Selvamurugan N. MicroRNAs: synthesis, gene regulation and osteoblast differentiation. Curr Issues Mol Biol. 2013;15(1):7–18.
26. Zhu W, Ma X, Gou M, Mei D, Zhang K, Chen S. 3D printing of functional biomaterials for tissue engineering. Curr Opin Biotechnol. 2016;40:103–12. https://doi.org/10.1016/j.copbio.2016.03.014.
27. Zhang J, Fu W, He M, Xie W, Lv Q, Wan G, et al. MiRNA-20a promotes osteogenic differentiation of human mesenchymal stem cells by co-regulating BMP signaling. RNA Biol. 2011;8(5):829–38. https://doi.org/10.4161/rna.8.5.16043.
28. Peng S, Gao D, Gao C, Wei P, Niu M, Shuai C. MicroRNAs regulate signaling pathways in osteogenic differentiation of mesenchymal stem cells (Review). Mol Med Rep. 2016;14(1):623–9.
29. Long H, Sun B, Cheng L, Zhao S, Zhu Y, Zhao R, et al. miR-139-5p represses BMSC osteogenesis via targeting Wnt/β-catenin signaling pathway. DNA Cell Biol. 2017;36(8):715–24. https://doi.org/10.1089/dna.2017.3657.
30. Nugent M. MicroRNA function and dysregulation in bone tumors: the evidence to date. Cancer Manag Res. 2014;6(1):15–25.

31. Wang C, Zhang X, Zhang C, Zhai F, Li Y, Huang Z. MicroRNA-155 targets MAP3K10 and regulates osteosarcoma cell growth. Pathol Res Pract. 2017;213(4):389–93. https://doi.org/10.1016/j.prp.2016.12.028.
32. Peng Y, Croce CM. The role of MicroRNAs in human cancer. Signal Transduct Target Ther. 2016;1:15004.
33. Qu Q, Chu X, Wang P. MicroRNA-195-5p suppresses osteosarcoma cell proliferation and invasion by suppressing naked cuticle homolog 1. Cell Biol Int. 2017;41(3):287–95.
34. Xie B, Ding Q, Han H, Wu D. MiRCancer: a microRNA-cancer association database constructed by text mining on literature. Bioinformatics. 2013;29(5):638–44.
35. Xiao F, Zuo Z, Cai G, Kang S, Gao X, Li T. miRecords: an integrated resource for microRNA-target interactions. Nucleic Acids Res. 2009;37(SUPPL. 1):105–10.
36. Wang F, Yu D, Liu Z, Wang R, Xu Y, Cui H, et al. MiR-125b functions as a tumor suppressor and enhances chemosensitivity to cisplatin in osteosarcoma. Technol Cancer Res Treat. 2016;15(6):NP105.
37. JG W, Wang JJ, Jiang X, Lan JP, He XJ, Wang HJ, et al. MiR-125b promotes cell migration and invasion by targeting PPP1CA-Rb signal pathways in gastric cancer, resulting in a poor prognosis. Gastric Cancer. 2015;18(4):729–39. https://doi.org/10.1007/s10120-014-0421-8.
38. Jin L, Zhang Z, Li Y, He T, Hu J, Liu J, et al. miR-125b is associated with renal cell carcinoma cell migration, invasion and apoptosis. Oncol Lett. 2017;13:4512–20. https://doi.org/10.3892/ol.2017.5985.
39. Tsang FH, Au V, Lu W-J, Shek FH, Liu AM, Luk JM, et al. Prognostic marker microRNA-125b inhibits tumorigenic properties of hepatocellular carcinoma cells via suppressing tumorigenic molecule eIF5A2. Dig Dis Sci. 2014;59(10):2477–87. https://doi.org/10.1007/s10620-014-3184-5.
40. Akhavantabasi S, Sapmaz A, Tuna S, Erson-bensan AE. miR-125b targets ARID3B in breast cancer cells. Cell Struct Funct. 2012;38:27–38.
41. Karbasy SH, Taheriazam A, Mirghasemi A, Sedaghati F, Shakeri M, Yahaghi E, et al. RETRACTED ARTICLE: upregulation of miR-300 and downregulation of miR-125b act as potential predictor biomarkers in progression, metastasis, and poor prognosis of osteosarcoma. Tumor Biol. 2016;37(12):16385. https://doi.org/10.1007/s13277-015-4000-3.
42. Jiang X, Li X, Wu F, Gao H, Wang G, Zheng H, et al. Overexpression of miR-92a promotes the tumor growth of osteosarcoma by suppressing F-box and WD repeat-containing protein 7. Gene. 2017;606:10–6. https://doi.org/10.1016/j.gene.2017.01.002.
43. Palmini G, Marini F, Brandi ML. What is new in the miRNA world regarding osteosarcoma and chondrosarcoma? Molecules. 2017;22(3):pii: E417.
44. Song K, Liu N, Yang Y, Qiu X. Regulation of osteosarcoma cell invasion through osteopontin modification by miR-4262. Tumor Biol. 2016;37(5):6493–9. https://doi.org/10.1007/s13277-015-4530-8.
45. Jiang R, Zhang C, Liu G, Gu R, Wu H. MicroRNA-101 inhibits proliferation , migration and invasion in osteosarcoma cells by targeting ROCK1. Am J Cancer Res. 2017;7(1):88–97.
46. Zhu Z, Tang J, Wang J, Duan G, Zhou L, Zhou X. MIR-138 acts as a tumor suppressor by targeting EZH2 and enhances cisplatin-induced apoptosis in osteosarcoma cells. PLoS One. 2016;11(3):1–12.
47. Wu S, Du X, Wu M, Du H, Shi X, Zhang T. MicroRNA-409-3p inhibits osteosarcoma cell migration and invasion by targeting catenin-??1. Gene. 2016;584(1):83–9. https://doi.org/10.1016/j.gene.2016.03.021.
48. Liu Q, Yang G, Qian Y. Loss of MicroRNA-489-3p promotes osteosarcoma metastasis by activating PAX3-MET pathway. Mol Carcinog. 2017;56(4):1312–21. https://doi.org/10.1002/mc.22593.
49. Ming H, Wang G, Jiang L, Qiu C, Li B, Wang J, et al. MiR-486 suppresses the development of osteosarcoma by regulating PKC-δ pathway. Int J Oncol. 2017;50(5):1590–600.
50. Shannon P, Markiel A, Ozier O, Baliga NS, Wang JT, Ramage D, et al. Cytoscape: a software environment for integrated models of biomolecular interaction networks. Genome Res. 2003;13:2498–504.

51. Mi H, Huang X, Muruganujan A, Tang H, Mills C, Kang D, et al. PANTHER version 11: expanded annotation data from Gene Ontology and Reactome pathways, and data analysis tool enhancements. Nucleic Acids Res. 2017;45(D1):D183–9.
52. Wang Y, Wang N, Zeng X, Sun J, Wang G, Xu H, et al. MicroRNA-335 and its target Rock1 synergistically influence tumor progression and prognosis in osteosarcoma. Oncol Lett. 2017;13:3057–65. https://doi.org/10.3892/ol.2017.5818.
53. Wang LH, Tsai HC, Cheng YC, Lin CY, Huang YL, Tsai CH, et al. CTGF promotes osteosarcoma angiogenesis by regulating miR-543/angiopoietin 2 signaling. Cancer Lett. 2017;391:28–37. https://doi.org/10.1016/j.canlet.2017.01.013.
54. Fan L, Zhu C, Qiu R, Zan P, Zheng Z, Xu T, et al. MicroRNA-661 enhances TRAIL or STS induced osteosarcoma cell apoptosis by modulating the expression of cytochrome c1. Cell Physiol Biochem. 2017;41(5):1935–46. https://doi.org/10.1159/000472380.
55. Leddy LR, Holmes RE. Chondrosarcoma of bone. In: Peabody TD, Attar S, editors. Orthopaedic oncology: primary and metastatic tumors of the skeletal system. Cham: Springer International Publishing; 2014. p. 117–30. https://doi.org/10.1007/978-3-319-07323-1_6.
56. Möller T, Amoroso M, Hägg D, Brantsing C, Rotter N, Apelgren P, et al. Vivo chondrogenesis in 3D bioprinted human cell-laden hydrogel constructs. Plast Reconstr Surg Glob Open. 2017;5(2):e1227. http://insights.ovid.com/crossref?an=01720096-201702000-00013
57. Chandhanayingyong C, Kim Y, Staples JR, Hahn C, Lee FY. MAPK/ERK signaling in osteosarcomas, Ewing sarcomas and chondrosarcomas: therapeutic implications and future directions. Sarcoma. 2012;2012:404810.
58. Li Z, Yu X, Shen J, WKK W, Chan MTV. MicroRNA expression and its clinical implications in Ewing's sarcoma. Cell Prolif. 2015;48(1):1–6.
59. Moore C, Parrish JK, Jedlicka P. MiR-193b, downregulated in Ewing Sarcoma, targets the ErbB4 oncogene to inhibit anchorage-independent growth. PLoS One. 2017;12(5):e0178028. https://doi.org/10.1371/journal.pone.0178028.

Chapter 7
The Roles of miR-25 and Its Targeted Genes in Human Cancer

Carmen Caiazza, Palmiro Poltronieri, and Massimo Mallardo

Abstract Since of their discovery in mammalians, microRNAs (miRNAs) have been associated to almost every physiological function within cells, tissues and organs. miRNAs are small non-coding RNAs that regulate gene expression by targeting messenger RNAs for translational repression or, at lesser extent, mRNAs degradation. Within the several functions controlled by miRNAs there are the control of cell proliferation, apoptosis, differentiation, cell migration, autophagy and metabolism. Thus, the uncontrolled expression of miRNAs has been associated with cancer onset, progression and cancer spreading into metastasis. miRNAs up- or down-regulation has been linked to oncogenic and tumor-suppressive roles in all type of cancers. Altered expression of many miRNAs has been reported many human malignant tumors, participating in various cellular processes accordingly with its broad range of potential mRNAs target. Here, we want to briefly discuss the mechanisms underlying miRNA-mediated tumorigenesis in different human cancers, presenting, as an example, the oncogenic and tumor-suppressive function of miR-25. Moreover, we summarize the possible future as a potential diagnostic and prognostic parameter as well as therapeutic target in clinical applications and the main techniques to study miRNAs.

Keywords Non-coding RNAs · miR-25 · Cancer · Oncogenic miRNAs · Oncosuppressor miRNAs

C. Caiazza · M. Mallardo (✉)
Department of Molecular Medicine and Medical Biotechnology, University of Naples Federico II, Napoli, Italy
e-mail: massimo.mallardo@unina.it

P. Poltronieri
CNR-ISPA, Institute of Sciences of Food Productions, National Research Council, Lecce, Italy

S. Fayyaz, A. A. Farooqi (eds.), *Recent Trends in Cancer Biology: Spotlight on Signaling Cascades and microRNAs*, https://doi.org/10.1007/978-3-319-71553-7_7

Introduction

miRNAs constitute a class of small, endogenous, noncoding RNAs (ncRNAs) of ~22 nucleotides in length mainly transcribed by RNA polymerases (RNA-pol) II and at minor extent by RNA-pol III [1]. miRNAs regulate gene expression at the posttranscriptional level, acting as negative regulators of targeted specific mRNA translation. They act by binding to complementary sequences in the 3′ untranslated region (3′ UTR) of their target mRNAs. miRNAs can recognize more than one mRNA as specific target and, on the other hand, mRNAs may contain multiple binding sites for different miRNAs. This results in a complex regulatory network [2] that can control almost every cellular function. MiRNAs play important roles in cell proliferation, differentiation and cell death [2–4]. Thus, miRNAs play crucial roles in oncogenesis by regulating cell proliferation, apoptosis, cell migration, autophagy and metabolism and they can act as either oncogenes (oncomirs) or tumor suppressors [5, 6].

mir-25 constitutes, with mir-106b and mir-93, the mir-106b/25 cluster, located on chromosome 7 (7q22.1) in the 13th intron of the MCM7 gene [7]. Both MCM7 and mir-25 are induced by the transcription factor E2F1 and, at the same time, mir-25 directly regulates E2F1 expression through interaction with its 3′UTR. This negative loop may be used from cancer cells as a way to escape E2F1-induced apoptosis [8]. In last years, much interest has been aroused by mir-25, supported by several evidences of its central role in both physiological and pathological mechanism. It seems capable of regulating, through different targets, some of most important pathway for cell survival, which regulation may lead to insurgence of diseases, most of all cancers. mir-25 seems to be capable of inducing reprogramming by directly repressing Wwp2 and Fbx7 [9], two components of E3 and SCF ubiquitin ligases respectively, which cause degradation of Oct4, Notch and Klf5 [10–12]. Moreover, mir-25 seems to silence BIM expression so inhibiting apoptosis and regulating indirectly TGFβ signaling [7]. At least, it has a central role in regulating p53 activity. In 293T cells, overexpression of mir-25 seems to downregulate p53 causing a reduced expression of genes downstream [13]. In glioblastoma multiforme cell lines, instead, there are evidences of a negative feedback: mir-25 stabilizes p53 protein, so inducing cell cycle arrest; at the same time, p53, inhibiting E2F1 causes a reduction of mir-25 levels [14]. This dual role of mir25 in regulating p53 could be explained with a different mechanism of action depending on cell lines and diseases: the downregulation of p53 mir-25-dependent could have a fundamental role in malignancy in which there are not somatic mutation in p53 gene [13].

Examples of miRNAs Involved in Different Human Cancers with Oncogenic and Tumor-Suppressive Function

miR-221 and 222 were described as tumor suppressing factors promoting senescence of lung HDFs [15] and EC cells [16]. Moreover they inhibit angiogenesis in response to SCF by targeting the SCF receptor c-kit. On the other hand, both were found overexpressed in many human cancers and are considered as oncogenes acting via down-regulation of the tumor suppressors p27, p57 and PTEN [17–19]. One of the most studies miRNA related to cancer is miR-155. It act as tumor suppressor gene inhibiting the transformation of human breast epithelial cells induced by ErbB2. Treatment with drugs such as Trastuzumab may results in the up-regulation of miR-155 and in a marked reduction of ErbB2 expression in ErbB2-positive breast cancer cells [20]. However, miR-155 represent one of the first example of oncogenic miRNA involved in myeloid and lymphoid malignancies and other cancer types [21]. Another example of miRNA considered as double edge sword, is represented by miRNA-146. While it is up-regulated during replicative and DNA damage-induced senescence of skin in HCA2, BJ and HDF cell lines, miR-146 may act as an oncomir by suppressing SASP via a negative feedback loop. Again, miR-130b may targets Zeb-1 and inhibits Zeb1-dependent EMT cancer cell invasion [22] while it acts as oncomir in glioblastoma by inhibiting Hippo signaling [23] and in esophageal squamous cell carcinoma by inhibiting PTEN [24].

All together, these few example clearly show how they can act in different way accordingly to the tissue and the mRNAs that can be targeted.

miR-25 in Human Cancer

miR-25 is dysregulated in many different types of human cancers. Up to date, it has been shown that miR-25 can have several mRNA targets that are involved in multiple biological pathways, including proliferation, invasion, differentiation, apoptosis, and autophagy. Characteristic of this miRNA is to possess a dual oncogenic role, functioning principally as oncogene and in some cases as tumor suppressor.

Oncogenic Role of mir-25

Breast (BC) and ovarian cancer (OC) are diffuses malignancy in women [25, 26]. The role of mir-25 in breast cancer seems to be correlated with onset of chemoresistance. In fact it appear capable of negative regulate ULK1, a protein of ULK complex that triggers the autophagy [27]. Treatments with isoliquiritigenin (ISL), a natural flavonoid isolated from the root of licorice [28] which acts inhibiting mir-25 in drug-resistant BC cells [29], cause a decreased expression of mir-25 and a consequently upregulation of ULK1. It leads to induction of autophagic cell death and accelerated degradation, via the lysosome pathway, of ABCG2 [29], an ATP-binding cassette transporter highly correlated with drug resistance [30]. mir-25 expression levels result higher in both OC tissues and cells (OVCAR3, SKOV3, ES-2) compared to normal ones [31]. Moreover, mir-25 direct targets and down-regulate LATS2 [31], a member tumor suppressors family LATS that plays an essential role in mediating Hippo (Hpo) growth inhibitory signaling [32]. The reduced expression of LATS2 promotes proliferation, migration and invasion of OC cells.

Treatments with mir-25 inhibitor restore normal expression of LATS2 suggesting a potential therapeutic strategy for OC treatment [31]. Moreover, there are evidences of mir-25 role on regulating apoptosis in OC: particularly it seems to act directly on Bim with an inverse correlation. Depletion of mir-25 explicates with increase of the intrinsic pathway of apoptosis with upregulated expression of both Bax and caspase3 and downregulation of Bcl2 [33]. Gastric cancer (GC) has the second highest mortality rate among cancers [34]. Principals risk factors are both genetic predisposition and environmental (diet rich in salt and nitrate, tobacco and alcohol consumption) and also helicobacter pylori infection [35, 36]. mir-25 expression results increased in both GC tissues and cell lines compared to controls [37–41]. Its increases directly correlate with tumor phenotype. Through different targets, in fact, mir-25 cause increased cell proliferation, motility and invasion. Among several mRNAs target we focused on: RECK [37], LATS2 [38], TOB1 [39], FBXW7 [40]. Tumor suppressor RECK, a membrane anchored glycoprotein, act suppressing matrix metalloproteinases (MMPs) such as MMP-2, MMP-9 and MMP-14, which are directly involved in tumor invasion and metastasis [41]. Mir-25 overexpression, in MKN28 GC cell line, significantly suppressed both the mRNA and protein levels of RECK: transfecting SGC7901 GC cell line with its inhibitor, both mRNA and protein levels of RECK result increased [37]. A member of LATS tumor suppressor family, LATS2, is known for its negative regulation on cell cycle; particularly, it acts inhibiting G1/S [42] and G2/M transition [43]. Mir-25 action on LATS2 has been found in gastric adenocarcinoma (GAC). It seems to be capable of negative modulating LATS2 expression causing, with increased proliferation and reduction of apoptosis, cancer progression. Particularly, the effect of mir-25 on LATS2 seems to be stronger when acting simultaneously to mir-107 [38]. TOB1 (transducer of ERBB2) is a member of Tob/BTG family which negative regulate cell growth [44]. Loss of TOB1, induced by mir-25, promotes cell proliferation, invasion and metastasis in GC cells [39]. Tumor suppressor FBXW7 regulates a network of protein directly involved in regulation of cell division, growth and differentiation [45]. Also in GC, according with what seen in lung cancer, FBXW7 result downregulated in mir-25 dependent way, with consequently tumor progression [40].

Hepatocellular carcinoma (HCC) is the second leading cause of cancer-related mortality, following lung cancer [46]. In most of cases, it develops consequently to principal risk factors: HBV or HCV infection, alcoholism, cirrhosis, non-alcoholic fatty liver disease and genetic predisposition [47]. According to several studies, mir-25 results involved in progression, development and prognosis of HCC. Expression levels of mir-25 results increased not only in HCC tissues and in cell lines (HepG2 and HuH7 infected with a lentiviral vector overexpressing mir-25) compared to controls; however, its expression levels range in tissues connected from patients according to presence/absence of metastasis: mir-25 results more expressed with metastasis expression compared to the absence [48]. Experiments on mice evince the same behavior also in vivo [48]. Its overexpression in HCC cells enhances proliferation, migration, invasion and activation of EMT. These effects results from mir-25 ability of targeting, on its 3′UTR, the tumor suppressor RhoGDI1 blocking its function [48]. When activated, RhoGDI1 acts as a negative regulator of Rho GTPases such as Rac1 and Cdc42, thus modulating cell cycle [49]. By targeting and inhibiting RhoGDI1, mir-25 acts as a positive regulator of cell cycle inducing tumor proliferation. Moreover, still through RhoGDI1, mir-25 regulate, among several transcription factors, Snail, responsible of E-cadherin repression, then inducing EMT and promoting metastasis [48]. Furthermore, experimental evidences show a direct correlation between mir-25 and β-catenin levels in HCC: from these evidences, mir-25 seems to be positively regulated by Wnt/β-catenin pathway [48]. A recent work report a different grade of methylation of genic region containing microRNA in HCC cell lines. Among several microRNA analyzed, mir-25 result hypomethylated and thus up-regulated, coinciding with above [50].

Lung cancer is often represented by Non-small cell lung cancer (NSCLC) , which accounts for 85% of all pulmonary carcinomas. A distinct form of lung cancer (~15% of the cases), associated with intensive history of smoking, is represented by Small-cell lung cancer (SCLC) [26]. In last years, many microRNA have been associated with lung cancer with both roles, tumor suppressors or oncogenes [51]. Several studies have showed a direct link between lung cancer and mir-25 particularly displaying its oncogenic role. In both SCLC cell lines (H146, H209, H446, H510A, H889) and tissues, mir-25 expression results upregulated compared to normal fetal lung fibroblast cell line (MRC5) and non-tumor tissues [52]. Moreover, its downregulation, through mir-25 anti-sense oligonucleotides inhibitor, reduces H510A cell proliferation, invasion and chemioresistence to cisplatin [52]. These effects are linked to mir-25 ability of targeting cyclin E2 on its 3′ UTR causing an increase of its activity. Indeed, mir-25 downregulation induces G0/G1 cell cycle arrest in H510A cells [52]. The role of mir-25 in NSCLC has been widely investigated. Even in this case mir-25 expression levels results higher in NSCLC cell lines (A549, SPC-A-1, H460, H661) and tissues compared with normal bronchial epithelial cell line (16 HBE) and normal tissues [53]. Also in NSCLC, mir-25 is correlated with tumor growth: its downregulation (with mir-25 inhibitor) lead to suppression of cell proliferation, migration and invasion [53], and increases chemiosensitivity to cisplatin [54]. Several target of mir-25 are described to explain its oncogenic role, particularly it seems to act regulating: modulators of cell cycle (FBXW7 [53] and Cdc42 [54]), EMT transition (Snail1 [55]) and modulators of apoptosis (RGS3 [56], MOAP1 [57]). Tumor suppressor FBXW7 (F-box and WD repeat domain-containing 7), a protein in SFC (SKP1-CUL1-F-box protein) E3 ligase complex, determines target specificity identifying and targeting proteins for ubiquitination and degradation [45]. Cdc42, a small GTPase of the Rho family, regulate, between many cell function, the cycle progression [58]. Both FBXW7 and Cdc42 are targeted by mir-25 on their 3′UTR with opposite effects: reduced protein level of FBXW7 [53] and increase of Cdc42 protein [54]. In both events, mir-25 shows an oncogenic role promoting cell proliferation, migration and invasion although opposed mechanisms. Supporting these evidences, treatment with mir-25 inhibitor in NSCLC cell lines, explicates with increase of FBXW7 protein levels [53] and reduction of Cdc42 [54], and thus, with a consequently decrease in proliferation, migration, and invasion. Between others targets, Snail1, a zinc finger protein that promotes EMT, has been showed increased in mir-25 dependent way [55]. RGS3, a negative regulator of G-protein signaling [59], result downregulated by mir-25 in NSCLCs cells [56]. Besides, inhibition of mir-25 evinces not only a reduction in proliferation but also an increase of apoptosis through regulation of its modulators: stimulated expression of caspase-3, -8, -9, and Bax, and suppressed of Bcl-2 [56]. Another regulator of apoptosis, MOAP1, is a further target of mir-25. It binds to Bax to initiate the caspase cascade leading to apoptosis [60]. In NSCLSs cells, MOAP1 results downregulated by mir-25 with a consequent decrease of apoptosis [57].

Tumor Suppressor Role of miR-25

Prostate cancer is the most common malignancy and the second leading cause of cancer related death in men worldwide. Despite many progress in detection and treatment, the prognosis remain poor because of the high risk of developing metastasis [26]. A key role in tumor initiation, metastasis and chemoresistance seems to be performed by prostate cancer stem or progenitor-like cells [61]. In this case, mir-25 expression seems to be directly correlated with differentiation: it result downregulated in stem cells compared to committed ones [62]. Among several targets of mir-25 in prostate cancer, much interest has been aroused by ITGAV and ITGA6, two integrins required for both stemness maintenance and acquisition of migratory and invasive phenotype [63]. Acting on cytoskeleton dynamics through downregulation of ITGAV and ITGA6 both at protein and mRNA levels, mir-25 causes a reduction of cell migration, invasion and metastatic ability, so operating, differently with what seen until this point, as a tumor suppressor. Thus, mir-25 seems to have a key role in inducing a less tumorigenic subpopulation of human prostate cancer cells [62].

A similar role of mir-25 has been found in colon cancer. Targeting and downregulating Smad7, a protein directly involved in metastasis and proliferation promotion [63], mir-25 acts as a tumor suppressor, so explaining its low expression in colon cancer tissues and cells [37].

At the same time, in prostatic small cell neuroendocrine carcinoma (SCNC) an aggressive form of prostatic cancer, mir-25 display an oncogenic role [39]. In this case, mir-25 seems downregulated by p53. This negative regulation leads to a high expression of Fbxw7 and a consequent ubiquitination of its substrates included Aurora kinase A, a serine/threonine kinase regulating cell cycle [40], which results overexpressed in SCNC and seems to play a key role in cell proliferation and invasion [41]. P53 mutated, often found in SCNC, enabled to targeting mir-25 causes indirectly a decreased degradation of Aurora kinase A, so resulting in an aggressive phenotype [39].

miR-25 as Possible Diagnostic and Prognostic Marker in Cancer

Emerging evidences suggest that microRNAs may improve diagnosis and prognosis accuracy for several human pathologies including cancer. Analyzing mir-25 expression levels in OC and normal patients there are evidences of its possible role as biomarker. However, its trend seems to be opposite in tissues and cells. Serum levels of mir-25 result downregulated in OC patients compared to normal ones [64] while in tissues not only there is higher expression in OC patient, but also this is correlated to poor prognosis [65]. The large involvement of mir-25 in GC pathogenesis allowed a deep analysis of its potential detection in blood. Several studies found its levels quite high in plasma so inducing to suppose its future potential role as biomarker for early detection of GC [66, 67]. Moreover, through these studies, it has been found a Single Nucleotide Polymorphism in mir-25, rs41274221, which may protect GC patients from tumor growth and metastasis [68]. The central role of mir-25 in HCC has promoted a succession of investigation for its role as a biomarker. There are evidences that both in tissues [69] and in plasma [70, 71] mir-25 detection could allow an early diagnosis. Besides, studies on patients' tissue reveal a negative prognostic role of mir-25: its overexpression is of predictive value on poor prognosis [72]. Direct links between mir-25 and lung cancer cells and tissues, encourage the interest on its application as a biomarker for early detection [73] of the disease. Finally, it has been shown plasma levels of miR-25 might be a clinically useful biomarker for cancer detection and the monitoring of tumour dynamics in oesophageal squamous cell carcinoma patients [74].

Main Techniques Used to Study miRNAs

In the last decade, several methods have been set up in order to profile and to quantify microRNA (miRNA) expression in cells, tissues and even in body fluids. Although bioinformatics plays a big part in identifying putative miRNAs, they also need to be experimentally verified in the lab. A broad range of techniques, have been developed to overcome the challenges of miRNA profiling. These techniques, range from measuring single miRNAs by conventional or multiplexed quantitative PCR in purified T cell subsets, to the genome-wide analysis of miRNA expression by microarrays and RNA sequencing. One of the most popular techniques for validating and accurately quantifying miRNAs is the quantitative real time PCR (qPCR). As well as being sensitive and quantitative, qPCR is also relatively inexpensive and flexible making it the preferred choice for validating novel miRNAs and for use in relatively small experiments. Arrays are typically chosen for larger studies covering multiple miRNA targets. While they are the least quantitative of the miRNA assay methods, conventional DNA oligonucleotide arrays are a relatively inexpensive way to measure hundreds of targets at once. RNA-seq uses the high-throughput capability of next-generation sequencing (NGS) platforms. While it cannot quantify miRNA levels with the molar resolution of qPCR, deep sequencing of miRNA have the advantage of being able to sample all miRNAs present in a sample, whether the researcher knows the sequence or not, making it an ideal discovery tool. Finally, Multiplex miRNA profiling assays using Firefly particle technology are a more recent addition to the range of tools available to assay miRNAs. A key benefit of this technique, is its ability to allow the validation of multiple miRNAs across a range of samples, without the labor intensive workflow or large sample requirement of other techniques.

References

1. Lin S, Gregory RI. MicroRNA biogenesis pathways in cancer. Nat Rev Cancer. 2015;15:321–33.
2. Esteller M. Non-coding RNAs in human disease. Nat Rev Genet. 2011;12:861–74.
3. Mendell JT. MicroRNAs: critical regulators of development, cellular physiology and malignancy. Cell Cycle. 2005;4:1179–84.
4. Tiscornia G, Izpisua Belmonte JC. MicroRNAs in embryonic stem cell function and fate. Genes Dev. 2010;24:2732–41.
5. Ruan K, Fang X, Ouyang G. MicroRNAs: novel regulators in the hallmarks of human cancer. Cancer Lett. 2009;285:116–26.
6. Voorhoeve PM. MicroRNAs: oncogenes, tumor suppressors or master regulators of cancer heterogeneity? Biochim Biophys Acta. 2010;1805:72–86.
7. Petrocca F, Vecchione A, Croce CM. Emerging role of mir-106b-25/mir-17-92 clusters in the control of transforming growth factor β signaling. Cancer Res. 2008;68(20):8191–4.
8. O'Donnel KA, Wentzel EA, Zeller KI, Dang CV, Mendell JT. c-MYC-regulated microRNAs modulate E2F expression. Nature. 2005;435:839–43.
9. Lu D, Davis MPA, Abreu-Goodger C, Wang W, Campos LS, Siede J, Vigorito E, Skarnes WC, Dunham I, Enright AJ, Liu P. mir-25 regulates Wwp2 and Fbxw7 and promotes reprogramming of mouse fibroblasts cells to IPSCs. PLoS One. 2012;7(8):e40938.
10. HM X, Liao B, Zhang QJ, Wang BB, Li H, Zhong XM, Sheng HZ, Zhao YX, Zhao YM, Jin Y. Wwp2, an E3 ubiquitin ligase that targets transcription factor Oct-4 for ubiquitination. J Biol Chem. 2004;279:23495–503.
11. Welcker M, Clurman BE. FBW7 ubiquitin ligase: a tumor suppressor at the crossroads of cell division, growth and differentiation. Nat Rev Cancer. 2008;8:83–93.
12. Liu N, Li H, Li S, Shen M, Xiao N, Chen Y, Wang Y, Wang W, Wang R, Wang Q, Sun J, Wang P. The Fbw7/hCDC4 tumor suppressor targets pro-proliferative factor KLF5 for ubiquitination and degradation through multiple phosphodegron motifs. J Biol Chem. 2010;285(24):18858–67.
13. Kumar M, Lu Z, Takwi AAL, Chen W, Callander NS, Ramos KS, Young KH, Li Y. Negative regulation of the tumor suppressor p53 gene by microRNAs. Oncogene. 2011;30(7):843–53.
14. Suh SS, Yoo JY, Nuovo GJ, Jeon YJ, Kim S, Lee TJ, Kim T, Bakàcs A, Alder H, Kaur B, Aqeilan RI, Pichiorri F, Croce CM. MicroRNAs/TP53 feedback circuitry in glioblastoma multiforme. Proc Natl Acad Sci U S A. 2012;109(14):5316–21.
15. Markopoulos GS, Roupakia E, Tokamani M, Vartholomatos G, Tzavaras T, Hatziapostolou M, Fackelmayer FO, Sandaltzopoulos R, Polytarchou C, Kolettas E. Senescence-associated microRNAs target cell cycle regulatory genes in normal human lung fibroblasts. Exp Gerontol. 2017;96:110–22.
16. Vasa-Nicotera M, Chen H, Tucci P, Yang AL, Saintigny G, Menghini R, Mahe C, Agostini M, Knight RA, Melino G, Federici M. miR-146a is modulated in human endothelial cell with aging. Atherosclerosis. 2011;217:326–30.
17. le Sage C, Nagel R, Egan DA, Schrier M, Mesman E, Mangiola A, Anile C, Maira G, Mercatelli N, Ciafre SA, Farace MG, Agami R. Regulation of the p27(Kip1) tumor suppressor by miR-221 and miR-222 promotes cancer cell proliferation. EMBO J. 2007;26:3699–708.
18. Garofalo M, Di Leva G, Romano G, Nuovo G, Suh SS, Ngankeu A, Taccioli C, Pichiorri F, Alder H, Secchiero P, Gasparini P, Gonelli A, Costinean S, Acunzo M, Condorelli G, Croce CM. miR-221&222 regulate TRAIL resistance and enhance tumorigenicity through PTEN and TIMP3 downregulation. Cancer Cell. 2009;16:498–509.
19. Ciafre SA, Galardi S, Mangiola A, Ferracin M, Liu CG, Sabatino G, Negrini M, Maira G, Croce CM, Farace MG. Extensive modulation of a set of microRNAs in primary glioblastoma. Biochem Biophys Res Commun. 2005;334:1351–8.
20. He XH, Zhu W, Yuan P, Jiang S, Li D, Zhang HW, Liu MF. miR-155 downregulates ErbB2 and suppresses ErbB2-induced malignant transformation of breast epithelial cells. Oncogene. 2016;35:6015–25.
21. Sayed D, Abdellatif M. MicroRNAs in development and disease. Physiol Rev. 2011;91:827–87.

22. Dong P, Karaayvaz M, Jia N, Kaneuchi M, Hamada J, Watari H, Sudo S, Ju J, Sakuragi N. Mutant p53 gain-of-function induces epithelial-mesenchymal transition through modulation of the miR-130b-ZEB1 axis. Oncogene. 2013;32:3286–95.
23. Zhu G, Wang Y, Mijiti M, Wang Z, Wu PF, Jiafu D. Upregulation of miR-130b enhances stem cell-like phenotype in glioblastoma by inactivating the hippo signaling pathway. Biochem Biophys Res Commun. 2015;465:194–9.
24. Yu T, Cao R, Li S, Fu M, Ren L, Chen W, Zhu H, Zhan Q, Shi R. MiR-130b plays an oncogenic role by repressing PTEN expression in esophageal squamous cell carcinoma cells. BMC Cancer. 2015;15:29.
25. DeSantis C, Ma J, Bryan L, Jemal A. Breast cancer statistics, 2013. CA Cancer J Clin. 2014;61:52–62.
26. Siegel R, Naishadham D, Jemal A. Cancer statistics, 2013. CA Cancer J Clin. 2013;63:11–30.
27. Wong PM, Puente C, Ganley IG, Jiang X. The ULK1 complex: sensing nutrient signals for autophagy activation. Autophagy. 2013;9:124–37.
28. Ye L, Gho WM, Chan FL, Chen S, Leung LK. Dietary administration of the licorice flavonoid isoliquiritigenin deters the growth of MCF-7 cells overexpressing aromatase. Int J Cancer. 2009;124:1028–36.
29. Wang Z, Wang N, Liu P, Chen Q, Situ H, Xie T, Zhang J, Peng C, Lin Y, Chen J. MicroRNA-25 regulates chemoresistance-associated autophagy in breast cancer cells, a process modulated by the natural autophagy inducer isoliquiritigenin. Oncotarget. 2014;5(16):7013–26.
30. Szakàcs G, Paterson JK, Ludwig JA, Booth-Genthe C, Gottesman MM. Targeting multidrug resistance in cancer. Nat Rev Drug Discov. 2006;5:219–34.
31. Feng S, Pan W, Jin Y, Zheng J. MIR-25 promotes ovarian cancer proliferation and motility by targeting LATS2. Tumor Biol. 2014;35:12339–44.
32. Visser S, Yang X. LATS turor suppressor: a new governor of cellular homeostasis. Cell Cycle. 2010;9:3892–903.
33. Zhang H, Zuo Z, Lu X, Wang L, Wang H, Zhu Z. MIR-25 regulates apoptosis by targeting Bim in human ovarian cancer. Oncol Rep. 2012;27(2):594–8.
34. Siegel R, Ma J, Zou Z, Jemal A. Cancer statistics. CA Cancer J Clin. 2014;64(1):9–29.
35. Graham DY. Helicobacter pylori update: gastric cancer, reliable therapy, and possible benefits. Gastroenterology. 2015;148:719–31.e713.
36. Sung H, Yang HH, Hu N, Su H, Taylor PR, Hyland PL. Functional annotation of high-quality SNP biomarkers of gastric cancer susceptibility: the Yin Yang of PSCA rs2294008. Gut. 2015;65(2):361–4.
37. Zhao H, Wang Y, Yang L, Jiang R, Li W. mir-25 promotes gastric cancer cell growth and motility by targeting RECK. Mol Cell Biochem. 2014;385:207–13.
38. Zhang M, Wang X, Li W, Cui Y. miR-107 and miR-25 simultaneously target LATS2 and regulate proliferation and invasion of gastric adenocarcinoma (GAC) cells. Biochem Biophys Res Commun. 2015;460(3):806–12.
39. Li BS, Zuo QF, Zhao YL, Xiao B, Zhuang Y, Mao XH, Wu C, Yang SM, Zeng H, Zou QM, Guo G. MicroRNA-25 promotes gastric cancer migration, invasion and proliferation by directly targeting transducer of ERBB2, 1 and correlates with poor survival. Oncogene. 2015;34(20):2556–65.
40. Gong J, Cui Z, Li L, Ma Q, Wang Q, Gao Y, Sun H. MicroRNA-25 promotes gastric cancer proliferation, invasion, and migration by directly targeting F-box and WD-40 Domain Protein 7, FBXW7. Tumor Biol. 2015;36(10):7831–40.
41. Clark JC, Thomas DM, Choong PF, Dass CR. RECK a newly discovered inhibitor of metastasis with prognostic significance in multiple forms of cancer. Cancer Metastasis Rev. 2007;26(3-4):675–83.
42. Li Y, Pei J, Xia H, Ke H, Wang H, Tao W. Lats2, a putative tumor suppressor, inhibits G1/S transition. Oncogene. 2003;22(28):4398–405.
43. Ke H, Pei J, Ni Z, Xia H, Qi H, Woods T, Kelekar A, Tao W. Putative tumor suppressor Lats2 induces apoptosis through downregulation of Bcl-2 and Bcl-x(L). Exp Cell Res. 2004;298(2):329–38.

44. Kundu J, Wahab SM, Kundu JK, Choi YL, Erkin OC, Lee HS, Park SG, Shin YK. Tob1 induces apoptosis and inhibits proliferation, migration and invasion of gastric cancer cells by activating Smad4 and inhibiting β-catenin signaling. Int J Oncol. 2012;41(3):839–48.
45. Wang Z, Inuzuka H, Zhong J, Wan L, Fukushima H, Sarkar FH, Wei W. Tumor suppressor functions of FBW7 in cancer development and progression. FEBS Lett. 2012;586:1409–18.
46. Jemal A, Bray F, Center MM, Ferlay J, Ward E, Forman D. Global Cancer statistics. CA Cancer J Clin. 2011;61(2):69–90.
47. Llovet JM, Burroughs A, Bruix J. Hepatocellular carcinoma. Lancet. 2003;362(9399):1907–17.
48. Wang C, Wang X, Su Z, Fei H, Liu X, Pan Q. mir-25 promotes hepatocellular carcinoma cell growth, migration and invasion by inhibiting RhoGDI1. Oncotarget. 2015;6(34):36231–44.
49. Hancock JF, Hall AA. novel role for RhoGDI as an inhibitor of GAP proteins. EMBO J. 1993;12(5):1915–21.
50. He XX, Kuang SZ, Liao JZ, CR X, Chang Y, YL W, Gong J, Tian DA, Guo AY, Lin JS. The regulation of microRNA expression by DNA methylation in hepatocellular carcinoma. Mol Biosyst. 2015;11(2):532–9.
51. Frixa T, Donzelli S, Blandino G. Oncogenic MicroRNAs: kay players in malignant transformation. Cancers (Basel). 2015;7(4):2466–85.
52. Zhao Z, Liu J, Wang C, Wang Y, Jiang Y, Guo M. MicroRNA-25 regulates small cell lung cancer cell development and cell cycle through cyclin E2. Int J Clin Exp Pathol. 2014;7(11):7726–34.
53. Xiang J, Hang JB, Che JM, Li HC. Mir-25 is upregolated in non-small cell lung cancer and promotes cell proliferation and motility by targeting FBXW7. Int J Clin Exp Pathol. 2015;8(8):9147–53.
54. Yang T, Chen T, Li Y, Gao L, Zhang S, Wang T, Chen M. Downregulation of mir-25 modulates non-small cell lung cancer cells by targeting CDC42. Tumor Biol. 2015;36:1903–11.
55. Savita U, Karunagaran D. MicroRNA-106b-25 cluster targets β-TRCP2, increases the expression of Snail and enhances cell migration and invasion in H1299 (non small cell lung cancer) cells. Biochem Biophys Res Commun. 2013;434:841–7.
56. Chen Z, Wu Y, Meng Q, Xia Z. Elevated microRNA-25 inhibits cell apoptosis in lung cancer by targeting RGS3. In Vitro Cell Dev Biol Anim. 2016;52:62–7.
57. Wu T, Chen W, Kong D, Li X, Lu H, Liu S, Wang J, Du L, Kong Q, Huang X, Lu Z. mir-25 targets the modulator of apoptosis 1 gene in lung cancer. Carcinogenesis. 2015;36(8):925–35.
58. Olson MF, Ashworth A, Hall A. An essential role for Rho, Rac, and Cdc42 GTPases in cell cycle progression through G1. Science. 1995;269:1270–2.
59. Ishii M, Kurachi Y. Physiological actions of regulators of G-protein signaling (RGS) proteins. Life Sci. 2003;74(2-3):163–71.
60. Tan KO, Tan KM, Chan SL, Yee KS, Bevort M, Ang KC, Yu VC. MAP-1, a novel proapoptotic protein containing a BH3-like motif that associates with Bax through its Bcl-2 homology domains. J Biol Chem. 2003;276:2802–7.
61. Maitland NJ, Collins AT. Prostate cancer stem cells: a new target for therapy. J Clin Oncol. 2008;26(17):2862–70.
62. Zoni E, van der Horst G, van de Merbel AF, Chen L, Rane JK, Pelger RC, Collins AT, Visakorpi T, Snaar-Jagalska BE, Maitland NJ, van der Pluijm G. mir-25 modulates invasiveness and dissemination of human prostate cancer cells via regulation of αv- and α6-integrin expression. Cancer Res. 2015;75(11):2326–36.
63. van der Hoogen C, van der Horst G, Cheung H, Buijs JT, Pelger RC, van der Pluijm G. Integrin alphav expression is required for the acquisition of a metastatic stem/progenitor cell phenotype in human prostate cancer. Am J Pathol. 2011;179(5):2559–68.
64. Meng X, Joosse SA, Muller V, Trillsch F, Milde-Langosch K, Mahner S, Geffken M, Pantel K, Schwarzenbach H. Diagnostic and prognostic potential of serum miR-7, miR-16, miR-25, miR93, miR-182, miR-376a and miR429 in ovarian cancer patients. Br J Cancer. 2015;113(9):1358–66.
65. Wang X, Meng X, Li H, Liu W, Shen S, Gao Z. MicroRNA-25 expression level is an independent prognostic factor in epithelial ovarian cancer. Clin Transl Oncol. 2014;16:954–8.

66. Zhang HH, GL G, Zhang XY, Li FZ, Ding L, Fan Q, Wu R, Shi W, Wang XY, Chen L, Wei XM, Yuan XY. Primary analysis and screening of microRNAs in gastric cancer side population cells. World J Gastroenterol. 2015;21(12):3519–26.
67. Zhu C, Ren C, Han J, Ding Y, Du J, Dai N, Dai J, Ma H, Hu Z, Shen H, Xu Y, Jin G. A five-microRNA panel in plasma was identified as potential biomarker for early detection of gastric cancer. Br J Cancer. 2014;110(9):2291–9.
68. Zhou J, Zhou J, Wang W, Li W, Wu L, Li G, Shi J, Zhou S. The polymorphism in mir-25 attenuated the oncogenic function in gastric cancer. Tumor Biol. 2016;37:5515.
69. Sadeghian Y, Kamyabi-Moghaddam Z, Nodushan SM, Khoshbakht S, Pedram B, Yahaghi E, Mokarizadeh A, Mohebbi M. Profiles of tissue microRNA; miR-148b and miR-25 serve as potential prognostic biomarkers for hepatocellular carcinoma. Tumor Biol. 2016;37:16379.
70. Wen Y, Han J, Chen J, Dong J, Xia Y, Liu J, Jiang Y, Dai J, Lu J, Jin G, Han J, Wei Q, Shen H, Sun B, Hu Z. Plasma miRNAs as early biomarkers for detecting hepatocellular carcinoma. Int J Cancer. 2015;137(7):1679–90.
71. Li LM, Hu ZB, Zhou ZX, Chen X, Liu FY, Zhang JF, Shen HB, Zhang CY, Zen K. Serum microRNA profiles serve as novel biomarkers for HBV infection and diagnosis of HBV-positive hepatocarcinoma. Cancer Res. 2010;70(23):9798–807.
72. Su ZX, Zhao J, Rong ZH, Geng WM, Wu YG, Qin CK. Upregulation of microRNA-25 associates with prognosis in hepatocellular carcinoma. Diagn Pathol. 2014;9:47.
73. Wang P, Yang D, Zhang H, Wei X, Ma T, Cheng Z, Hong Q, Hu J, Zhuo H, Song Y, Jia C, Jing F, Jin Q, Bai C, Mao H, Zhao J. Early detection of lung cancer in serum by a panel of microRNA biomarkers. Clin Lung Cancer. 2015;16(4):313–9.
74. Komatsu S, Ichikawa D, Hirajima S, Kawaguchi T, Miyamae M, Okajima W, Ohashi T, Arita T, Konishi H, Shiozaki A, Fujiwara H, Okamoto K, Yagi N, Otsuji E. Plasma microRNA profiles: identification of miR-25 as a novel diagnostic and monitoring biomarker in oesophageal squamous cell carcinoma. Br J Cancer. 2014;111(8):1614–24.

Chapter 8
Notch Signaling in Lung Cancer Initiation and Development

Kayla C. Lewis and Yi Liu

Abstract Cancer stem cells (CSCs), or tumorigenic cancer cells, are termed as a small proportion of tumor cells that are responsible for cancer differentiation, initiation, development and drug resistance. Accumulating evidence has shown a wide range of cancers, such as lung cancer, is initiated from CSCs. Understanding the tumorigenic mechanisms of CSCs is one of the most pressing problems in modern biology. The Notch signaling pathway is characterized as one of the three core stem cell signaling pathways and plays an instrumental role in activating mutations and amplifications in lung cancer. In this chapter, we will briefly review the experimental model and clinical progress in lung cancer stem cell research and discuss the therapeutically potential target of Notch signaling in lung cancer stem cell research.

Keywords Notch signaling · Lung cancer · Cancer stem cells

Introduction

The concept of the cancer stem cell hypothesis was proposed back to nineteenth century by Rudolph Virchow and Julius Cohnheim [1]. Cancer stem cells (CSCs) are defined as a small population of self-renewed cancer cells that give rise to different types of cancers and are responsible for tumor maintenance and metastasis. Existing chemical synthesis of drugs, radiation and chemotherapy can destroy large portions of tumors but not CSCs, which can potentially result in tumor recurrence and drug resistance. Current conventional therapies are not efficient and effective at rooting out cancer cells completely; initially reduce tumor in smaller size but

---

Kayla C. Lewis and Yi Liu contributed equally to this work.

K. C. Lewis · Y. Liu (✉)
Department of Cancer Biology, Mayo Clinic, Jacksonville, FL, USA
e-mail: Liu.Yi1@mayo.edu

S. Fayyaz, A. A. Farooqi (eds.), *Recent Trends in Cancer Biology: Spotlight on Signaling Cascades and microRNAs*, https://doi.org/10.1007/978-3-319-71553-7_8

possess the potential to grow back due to its self-renewal characteristic. Therefore, it is imperative to target the unique characteristics of CSCs and may provide a new therapeutic insight to future cancer treatment.

Lung cancer is most common cause of cancer-related deaths worldwide [2]. According to the data from Surveillance, Epidemiology, and End Results (SEER) Program in 2016, it was estimated that new cases and death of lung and bronchus cancer will be 224,390 and 158,080 people in USA, which ranking second among all the cancer types. In lung cancer development, Notch signaling pathway plays pleiotropic roles in regulation of cell-cell communication and cellular fate determination. Deregulation of Notch signaling has been reported from series types of cancers including breast cancer [3], esophageal cancer [4] and T-cell acute lymphoblastic leukemia (T-ALL) [5]. Development of Notch inhibitors as a pharmacological and chemotherapeutic strategy to treat lung cancer has been documented. In this chapter, we will first summarize recent lung CSC research in context of experimental and clinical evidence. Next, we will discuss the therapeutic roles of targeting Notch signaling in lung cancer tumorigenic cell research.

Experimental Model and Clinical Progress in Lung Cancer Stem Cell Research

To identify CSCs, fluorescence-activated cell sorting (FACS) technology is used to detect cells with specific proteins on their surface. These specific proteins are less found on the surface of regular cancer cells than found in CSCs. For example, CD133 and CD44 are known as epithelial specific antigens (ESA) whose expressions are thought to indicate stem cell like properties in lung cancer stem cell research [6, 7]. Human lung cancer stem cells isolated either from solid tumors or parental cell lines have been first identified with CD133, which has been suggested as a potential stem cell surface marker in several human lung cancer cell lines as well [7–9]. Sarvi et al. characterized CD133 expression, as a CSC surface marker, in small cell lung cancer (SCLC) cell lines and mouse models and showed the CD133 expression was positively correlated with chemotherapy in mouse and human SCLC cell lines [10].

CD44 is responsible for tumor growth and anti-apoptosis [11]. In a research of analyzing 195 lung tumor samples, higher expression of CD44 was found in higher grade of tumors and CD44 expression was higher in non-small cell lung cancer (NSCLC) compared to SCLC [12]. Up-regulated expressions of CD133 and CD44 increase tumor formation and self-renewal properties and $CD133^{+}/CD44^{high}$ cells are indicated to form spheres *in vitro* [13]. Another CSC marker in lung cancer is aldehyde dehydrogenase (ALDH) [14]. ALDH activity is an indicator of tumor progression and higher degree of metabolism. $ALDH1^{high}/CD133^{+}$ cells showed resistance to common chemotherapy agents [14]. Roudi et al. also suggested CD133 and ALDH1 can be considered as stem cell marker in lung cancer patients and found that ALDH1 and CD133 had higher expression in NSCLC compared to SCLC [15]. Increased epithelial cell adhesion molecule (EpCAM) is an indicator of tumor progression and higher degree of metabolism. The roles of EpCAM and CD44 are more like Co-CSC markers than CD133 because tumor-initiating population with $EpCAM^{high}/CD44^{high}$ fractions is usually $CD133^{+}$ [16].

Purified/enriched cancer stem cells that isolated from cell surface markers through FACS technology are able to grow as non-adherent spheres/oncospheres in defined medium cultures, which would be efficiently selected differentiated progeny and continually re-passaged at certain clonal density [2]. Besides cell surface markers, another prospective strategy that will be used to identify CSCs is Hoechst exclusion assay, which is based on a fluorescent dye that is used to stain nuclear DNA but not CSCs due to the expression of the drug resistance transporter.

The second test to prove CSCs is to inject cells with specific cell surface markers into host mice that lack a cancer-fighting immune system or immune-deficient. For tumorigenicity experiments, 6–8 weeks old non-obese diabetic, severe combined immuno-deficient (NOD/SCID) mice were used. NOD/SCID mouse, which was known as lacking functional B and T cells, is characterized by complete defects to natural killer (NK) cells, macrophage and granulocyte numbers and functions. NOD/SCID mouse model allows tumor growth after tumor implantation of human xenografts in different sites or cancer cell lines [17] and has been shown to be a preclinical model for childhood acute lymphoblastic leukemia [18], cervix [19], colon [16], and prostate cancer [17]. In order to examine tumorigenicity, CSCs are more tumorigenic than parental cells. In tumor weight/size comparison CSC group form larger and heavier tumors than parental cell group injected with the same amount of cells. The tumor volume is estimated using $V = (1/2)LH^2$, while L equals to the longest diameter and H equals to the shortest diameter.

In clinical studies, cell surface molecular signatures were used as diagnostic implications and therapeutic approaches. As mentioned before, CD133 was one of the most representative cancer stem cell surface markers. High expression of CD133 was found significantly lower disease-free survival and higher incidences in 177 patients examined with stage I [20]. Jiang et al. showed that ALDH1 expression is positively correlated with the clinical stage and malignant grade of lung tumors in 303 clinical specimens from lung cancer patients [14]. Additionally, ALDH-positive patients showed chemo-resistant characteristic and have worse overall survival. ALDH1 was demonstrated as a prognostic indicator for lung cancer cells in another study of analyzing more than 200 clinically annotated NSCLC samples [21]. Immuno-histochemical staining showed significantly opposite correlation between ALDH1A1 and prognosis in those patients with stage I and N0. A systematic conclusion and comparison of prognostic value of CSC biomarkers in lung cancer tissues was reviewed in Koren et al. [22].

Despite experimental evidence in support of the expression of CD133 has an early prognostic value in NSCLC tumors, controversies surrounding the CSC model still remain. Salnikov et al. has confirmed the chemo-resistant role of CD133-positive in 88 cases of previously untreated NSCLC, however, there was no difference found between CD133-positive and CD133-negative patients [23]. On other hand, the prognostic role of CD133 was not clearly reported in lung tumors as well [20, 23]. Another controversy issue of tumorigenic characteristic about CSC is the sensitivity of the cell surface marker. For example, CD133-negative cells showed similar tumorigenic features, including colony formation, self-renewal, proliferation, differentiation, and invasion, as well as chemotherapy resistance, as CD133-positve in human lung cancer adenocarcinoma cell lines A549 and H446 [24]. In addition, Ucar et al. injected ALDH-positive and ALDH-negative cells into NOD/SCID mice and found both cells were tumorigenic [25]. Those controversies and uncertainties suggested more robust markers are needed for CSC purification and isolation in the future direction.

Besides lung CSC, several CSC models have been isolated and identified from solid tumors based on cell-surface marker expression currently, such as acute myeloid leukemia [26], breast [27], brain [28, 29], head and neck [30], colon [31, 32] and pancreatic cancer [33]. The worldwide acceptance of CSC model keeps it an important role of cancer biology and clinical implications for lung cancer. There are considerable experimental data support and confirm the concept of CSC that a subpopulation of cancer cells that potentially result in resistance to therapeutic strategies and tumor regrowth. A lung CSC working model in lung cancer-derived cell lines (parental cell line) or lung tumors has been widely studied. The CSC experimental model was used to not only identifying and isolate lung CSCs with specific stem cell surface markers but also targeting subpopulation of tumorigenic cells through major stem cell regulated pathways by blocking their CSC-like activity. Since the capacities of self-renewal, tumorigenicity, and differentiation of lung CSC are governed by their genetic makeup, the gene expression changes and stem cells regulated pathways during the biological processes are important in understanding the characteristics of lung CSC and developing in therapeutic strategies for patients with lung cancer.

Notch Signaling in Lung Cancer Stem Cells

In human tumors, there are several developmental signaling pathways responsible for maintaining stemness, self-renewal and cell-fate determination including Wnt/β--catenin signaling, Hedgehog (Hh) signaling and Notch signaling pathways. In brief, Wnt/β-catenin signaling pathway regulates pluripotency and cell fate decisions, although this may occur through a non-canonical mechanism [34]. Hedgehog signaling pathway is essential for pluripotency development and plays critical roles in tissue maintenance, growth and differentiation [35]. Notch signaling is a highly conserved embryonic pathway in in multicellular organisms that frequently regulates cellular fate determination [36]. Signaling through these pathways results in the expression of three key transcription factors: octamer-binding transcription factor 4 (OCT-4), SRY (sex determining region Y)-box 2 (SOX2), and Nanog homeobox (NANOG) [37]. Several research and reviews have addressed the roles of those pathways and genes in tumorigenesis and highlighted the potential therapeutic strategies for eliminating CSCs by targeting the signaling pathways [38–40].

The oncogenic role of Notch signaling has been shown to be critical for cell proliferation, differentiation, cell-cell fate determination and homeostasis [41] in variety of human tumors including lymphoma, breast cancer, colorectal cancer, head and neck cancer, lung cancer, pancreatic cancer, and prostate cancer [42–45] as we mentioned above. Notch activity has been shown to be a distinguishing marker of NSCLC CSCs [46]. To determine the presence of Notch activity in the NSCLC adenocarcinoma, a Notch GFP reporter was used in combination with a γ-secretase inhibitor, a known inhibitor of the Notch pathway activity [47]. The results of this showed a subset of cells with high Notch activity. These high Notch activity cells also exhibited chemotherapeutic resistance, tumorigenity in xenograft models, and the ability to produce either cells with high Notch activity or cells with low Notch activity, suggesting a stem-like phenotype. Additionally for NSCLC, it was determined that high Notch activity causes an increase in sphere formation as well as self-renewal. Another determining factor of CSCs is chemotherapeutic agent resistance [48]. When cells are treated with chemotherapy drugs, such as cisplatin, doxorubicin or etoposide, a subset of these cells will survive (drug surviving cells, DSCs) [49]. These remaining chemotherapeutic resistant cells show high tumorgenetic potential, ability to self-renewal and express known CSC cell surface markers in the lung such as CD44, CD133, ALDH, and CD166 [50]. *In vivo* studies verify that lung CSCs have similar properties to lung CSCs generated from in vitro studies. These properties include undifferentiation of cells and self-renewal. In a study conducted by Kajstura, et al., the lungs of mice were exposed to damage and then injected with human CSCs [51]. After 2 weeks *in vivo*, lung CSCs were identified via the surface markers mentioned previously. It was verified that the cells present were in fact CSCs. Furthermore, verification of stemness was seen in these in vivo cells as they showed undifferentiation and self-renewal, known phenotypic characteristics of lung CSCs.

The study of Notch signaling in lung cancer stem cells is limited. The Notch pathway is involved in the lung tumorigenesis has been demonstrated in elevated levels of Notch transcripts in non-small cell lung cancer (NSCLC) cell lines [52, 53]. Inhibition of Notch signaling by MRK-003, a well-studied γ-secretase inhibitor, resulted in lowering of tumor cell proliferation, induction of apoptosis and inhibition of serum independence of lung cancer cell lines *in vitro* and mouse models *in vivo*, suggestion a potential clinical therapy [54]. Further study showed decreased clonogenic capacity due to inhibition of Notch3 by γ-secretase inhibitor in NSCLC cell lines [53]. As we mentioned the importance of Notch3 in the maintenance of cancer stemness, observations of knocking down Notch3 showed decreased oncosphere growth, clonal expansion, cell viability and soft agar growth in lung adenocarcinoma (LADC) cell lines [55].

To conclude, Notch signaling pathway plays different roles, either tumorigenic or tumor suppressive, in corresponding tissues. For example, a few studies have been showed that activation of Notch1 induces differentiation in glioblastoma cell lines [56, 57]. In contrast to the preventive tumorigenesis role of Notch, it's been thoroughly investigated and highlightly by the presence of physiological and oncogenic transformation in T-ALL [58]. Taken together, to promote/maintain or repress tumorigenesis that is relevant to Notch pathway is highly context-dependent in cell-type specific and tissue-type specific manner. Given the specificity of Notch ligands/receptors and the high recurrent rate of lung cancer, identification of lung cancer stem cell markers/pathways and selective Notch inhibitors in therapeutic strategies will be beneficial for patients with lung cancer.

References

1. Huntly BJ, Gilliland DG. Leukaemia stem cells and the evolution of cancer-stem-cell research. Nat Rev Cancer. 2005;5(4):311–21.
2. Alamgeer M, Peacock CD, Matsui W, Ganju V, Watkins DN. Cancer stem cells in lung cancer: evidence and controversies. Respirology. 2013;18(5):757–64.
3. D'Angelo RC, Ouzounova M, Davis A, Choi D, Tchuenkam SM, Kim G, Luther T, Quraishi AA, Senbabaoglu Y, Conley SJ, Clouthier SG, Hassan KA, Wicha MS, Korkaya H. Notch reporter activity in breast cancer cell lines identifies a subset of cells with stem cell activity. Mol Cancer Ther. 2015;14(3):779–87.
4. Wang Z, Da Silva TG, Jin K, Han X, Ranganathan P, Zhu X, Sanchez-Mejias A, Bai F, Li B, Fei DL, Weaver K, Carpio RV, Moscowitz AE, Koshenkov VP, Sanchez L, Sparling L, Pei XH, Franceschi D, Ribeiro A, Robbins DJ, Livingstone AS, Capobianco AJ. Notch signaling drives stemness and tumorigenicity of esophageal adenocarcinoma. Cancer Res. 2014;74(21):6364–74.
5. Zou J, Li P, Lu F, Liu N, Dai J, Ye J, Qu X, Sun X, Ma D, Park J, Ji C. Notch1 is required for hypoxia-induced proliferation, invasion and chemoresistance of T-cell acute lymphoblastic leukemia cells. J Hematol Oncol. 2013;6:3.
6. Leung EL, Fiscus RR, Tung JW, Tin VP, Cheng LC, Sihoe AD, Fink LM, Ma Y, Wong MP. Non-small cell lung cancer cells expressing CD44 are enriched for stem cell-like properties. PLoS One. 2010;5(11):e14062.
7. Qiu X, Wang Z, Li Y, Miao Y, Ren Y, Luan Y. Characterization of sphere-forming cells with stem-like properties from the small cell lung cancer cell line H446. Cancer Lett. 2012;323(2):161–70.
8. Eramo A, Lotti F, Sette G, Pilozzi E, Biffoni M, Di Virgilio A, Conticello C, Ruco L, Peschle C, De Maria R. Identification and expansion of the tumorigenic lung cancer stem cell population. Cell Death Differ. 2008;15(3):504–14.
9. Wang P, Gao Q, Suo Z, Munthe E, Solberg S, Ma L, Wang M, Westerdaal NA, Kvalheim G, Gaudernack G. Identification and characterization of cells with cancer stem cell properties in human primary lung cancer cell lines. PLoS One. 2013;8(3):e57020.
10. Sarvi S, Mackinnon AC, Avlonitis N, Bradley M, Rintoul RC, Rassl DM, Wang W, Forbes SJ, Gregory CD, Sethi T. CD133+ cancer stem-like cells in small cell lung cancer are highly tumorigenic and chemoresistant but sensitive to a novel neuropeptide antagonist. Cancer Res. 2014;74(5):1554–65.
11. Schneider M, Huber J, Hadaschik B, Siegers GM, Fiebig HH, Schuler J. Characterization of colon cancer cells: a functional approach characterizing CD133 as a potential stem cell marker. BMC Cancer. 2012;12:96.
12. Roudi R, Madjd Z, Korourian A, Mehrazma M, Molanae S, Sabet MN, Shariftabrizi A. Clinical significance of putative cancer stem cell marker CD44 in different histological subtypes of lung cancer. Cancer Biomark. 2014;14(6):457–67.

13. Karamboulas C, Ailles L. Developmental signaling pathways in cancer stem cells of solid tumors. Biochim Biophys Acta. 2013;1830(2):2481–95.
14. Jiang F, Qiu Q, Khanna A, Todd NW, Deepak J, Xing L, Wang H, Liu Z, Su Y, Stass SA, Katz RL. Aldehyde dehydrogenase 1 is a tumor stem cell-associated marker in lung cancer. Mol Cancer Res. 2009;7(3):330–8.
15. Roudi R, Korourian A, Shariftabrizi A, Madjd Z. Differential expression of cancer stem cell markers ALDH1 and CD133 in various lung cancer subtypes. Cancer Invest. 2015;33(7):294–302.
16. Dalerba P, Dylla SJ, Park IK, Liu R, Wang X, Cho RW, Hoey T, Gurney A, Huang EH, Simeone DM, Shelton AA, Parmiani G, Castelli C, Clarke MF. Phenotypic characterization of human colorectal cancer stem cells. Proc Natl Acad Sci U S A. 2007;104(24):10158–63.
17. Bastide C, Bagnis C, Mannoni P, Hassoun J, Bladou F. A Nod Scid mouse model to study human prostate cancer. Prostate Cancer Prostatic Dis. 2002;5(4):311–5.
18. Lock RB, Liem N, Farnsworth ML, Milross CG, Xue C, Tajbakhsh M, Haber M, Norris MD, Marshall GM, Rice AM. The nonobese diabetic/severe combined immunodeficient (NOD/SCID) mouse model of childhood acute lymphoblastic leukemia reveals intrinsic differences in biologic characteristics at diagnosis and relapse. Blood. 2002;99(11):4100–8.
19. Gu W, Yeo E, McMillan N, Yu C. Silencing oncogene expression in cervical cancer stem-like cells inhibits their cell growth and self-renewal ability. Cancer Gene Ther. 2011;18(12):897–905.
20. Woo T, Okudela K, Mitsui H, Yazawa T, Ogawa N, Tajiri M, Yamamoto T, Rino Y, Kitamura H, Masuda M. Prognostic value of CD133 expression in stage I lung adenocarcinomas. Int J Clin Exp Pathol. 2010;4(1):32–42.
21. Sullivan JP, Spinola M, Dodge M, Raso MG, Behrens C, Gao B, Schuster K, Shao C, Larsen JE, Sullivan LA, Honorio S, Xie Y, Scaglioni PP, DiMaio JM, Gazdar AF, Shay JW, Wistuba II, Minna JD. Aldehyde dehydrogenase activity selects for lung adenocarcinoma stem cells dependent on notch signaling. Cancer Res. 2010;70(23):9937–48.
22. Koren A, Motaln H, Cufer T. Lung cancer stem cells: a biological and clinical perspective. Cell Oncol (Dordr). 2013;36(4):265–75.
23. Salnikov AV, Gladkich J, Moldenhauer G, Volm M, Mattern J, Herr I. CD133 is indicative for a resistance phenotype but does not represent a prognostic marker for survival of non-small cell lung cancer patients. Int J Cancer. 2010;126(4):950–8.
24. Meng X, Li M, Wang X, Wang Y, Ma D. Both CD133+ and CD133- subpopulations of A549 and H446 cells contain cancer-initiating cells. Cancer Sci. 2009;100(6):1040–6.
25. Ucar D, Cogle CR, Zucali JR, Ostmark B, Scott EW, Zori R, Gray BA, Moreb JS. Aldehyde dehydrogenase activity as a functional marker for lung cancer. Chem Biol Interact. 2009;178(1-3):48–55.
26. Bonnet D, Dick JE. Human acute myeloid leukemia is organized as a hierarchy that originates from a primitive hematopoietic cell. Nat Med. 1997;3(7):730–7.
27. Al-Hajj M, Wicha MS, Benito-Hernandez A, Morrison SJ, Clarke MF. Prospective identification of tumorigenic breast cancer cells. Proc Natl Acad Sci U S A. 2003;100(7):3983–8.
28. Galli R, Binda E, Orfanelli U, Cipelletti B, Gritti A, De Vitis S, Fiocco R, Foroni C, Dimeco F, Vescovi A. Isolation and characterization of tumorigenic, stem-like neural precursors from human glioblastoma. Cancer Res. 2004;64(19):7011–21.
29. Singh SK, Hawkins C, Clarke ID, Squire JA, Bayani J, Hide T, Henkelman RM, Cusimano MD, Dirks PB. Identification of human brain tumour initiating cells. Nature. 2004;432(7015):396–401.
30. Allegra E, Trapasso S. Cancer stem cells in head and neck cancer. Onco Targets Ther. 2012;5:375–83.
31. O'Brien CA, Pollett A, Gallinger S, Dick JE. A human colon cancer cell capable of initiating tumour growth in immunodeficient mice. Nature. 2007;445(7123):106–10.
32. Ricci-Vitiani L, Lombardi DG, Pilozzi E, Biffoni M, Todaro M, Peschle C, De Maria R. Identification and expansion of human colon-cancer-initiating cells. Nature. 2007;445(7123):111–5.

33. Li C, Heidt DG, Dalerba P, Burant CF, Zhang L, Adsay V, Wicha M, Clarke MF, Simeone DM. Identification of pancreatic cancer stem cells. Cancer Res. 2007;67(3):1030–7.
34. Clevers H, Nusse R. Wnt/beta-catenin signaling and disease. Cell. 2012;149(6):1192–205.
35. Briscoe J, Therond PP. The mechanisms of Hedgehog signalling and its roles in development and disease. Nat Rev Mol Cell Biol. 2013;14(7):416–29.
36. Ables JL, Breunig JJ, Eisch AJ, Rakic P. Not(ch) just development: notch signalling in the adult brain. Nat Rev Neurosci. 2011;12(5):269–83.
37. Liu A, Yu X, Liu S. Pluripotency transcription factors and cancer stem cells: small genes make a big difference. Chin J Cancer. 2013;32(9):483–7.
38. Park KS, Martelotto LG, Peifer M, Sos ML, Karnezis AN, Mahjoub MR, Bernard K, Conklin JF, Szczepny A, Yuan J, Guo R, Ospina B, Falzon J, Bennett S, Brown TJ, Markovic A, Devereux WL, Ocasio CA, Chen JK, Stearns T, Thomas RK, Dorsch M, Buonamici S, Watkins DN, Peacock CD, Sage J. A crucial requirement for Hedgehog signaling in small cell lung cancer. Nat Med. 2011;17(11):1504–8.
39. Takebe N, Ivy SP. Controversies in cancer stem cells: targeting embryonic signaling pathways. Clin Cancer Res. 2010;16(12):3106–12.
40. You L, He B, Xu Z, Uematsu K, Mazieres J, Mikami I, Reguart N, Moody TW, Kitajewski J, McCormick F, Jablons DM. Inhibition of Wnt-2-mediated signaling induces programmed cell death in non-small-cell lung cancer cells. Oncogene. 2004;23(36):6170–4.
41. Ranganathan P, Weaver KL, Capobianco AJ. Notch signalling in solid tumours: a little bit of everything but not all the time. Nat Rev Cancer. 2011;11(5):338–51.
42. Carvalho FL, Simons BW, Eberhart CG, Berman DM. Notch signaling in prostate cancer: a moving target. Prostate. 2014;74(9):933–45.
43. Ma J, Xia J, Miele L, Sarkar FH, Wang Z. Notch Signaling Pathway in Pancreatic Cancer Progression. Pancreat Disord Ther. 2013;3(114):pii: 1000114.
44. Reedijk M. Notch signaling and breast cancer. Adv Exp Med Biol. 2012;727:241–57.
45. Sun W, Gaykalova DA, Ochs MF, Mambo E, Arnaoutakis D, Liu Y, Loyo M, Agrawal N, Howard J, Li R, Ahn S, Fertig E, Sidransky D, Houghton J, Buddavarapu K, Sanford T, Choudhary A, Darden W, Adai A, Latham G, Bishop J, Sharma R, Westra WH, Hennessey P, Chung CH, Califano JA. Activation of the NOTCH pathway in head and neck cancer. Cancer Res. 2014;74(4):1091–104.
46. Hassan KA, Wang L, Korkaya H, Chen G, Maillard I, Beer DG, Kalemkerian GP, Wicha MS. Notch pathway activity identifies cells with cancer stem cell-like properties and correlates with worse survival in lung adenocarcinoma. Clin Cancer Res. 2013;19(8):1972–80.
47. Olsauskas-Kuprys R, Zlobin A, Osipo C. Gamma secretase inhibitors of Notch signaling. Onco Targets Ther. 2013;6:943–55.
48. Tehranchi R, Woll PS, Anderson K, Buza-Vidas N, Mizukami T, Mead AJ, Astrand-Grundstrom I, Strombeck B, Horvat A, Ferry H, Dhanda RS, Hast R, Ryden T, Vyas P, Gohring G, Schlegelberger B, Johansson B, Hellstrom-Lindberg E, List A, Nilsson L, Jacobsen SE. Persistent malignant stem cells in del(5q) myelodysplasia in remission. N Engl J Med. 2010;363(11):1025–37.
49. Levina V, Marrangoni AM, DeMarco R, Gorelik E, Lokshin AE. Drug-selected human lung cancer stem cells: cytokine network, tumorigenic and metastatic properties. PLoS One. 2008;3(8):e3077.
50. Hardavella G, George R, Sethi T. Lung cancer stem cells-characteristics, phenotype. Transl Lung Cancer Res. 2016;5(3):272–9.
51. Kajstura J, Rota M, Hall SR, Hosoda T, D'Amario D, Sanada F, Zheng H, Ogorek B, Rondon-Clavo C, Ferreira-Martins J, Matsuda A, Arranto C, Goichberg P, Giordano G, Haley KJ, Bardelli S, Rayatzadeh H, Liu X, Quaini F, Liao R, Leri A, Perrella MA, Loscalzo J, Anversa P. Evidence for human lung stem cells. N Engl J Med. 2011;364(19):1795–806.
52. Chen H, Thiagalingam A, Chopra H, Borges MW, Feder JN, Nelkin BD, Baylin SB, Ball DW. Conservation of the Drosophila lateral inhibition pathway in human lung cancer: a hairy-

related protein (HES-1) directly represses achaete-scute homolog-1 expression. Proc Natl Acad Sci U S A. 1997;94(10):5355–60.
53. Osanyingbemi-Obidi J, Dobromilskaya I, Illei PB, Hann CL, Rudin CM. Notch signaling contributes to lung cancer clonogenic capacity in vitro but may be circumvented in tumorigenesis in vivo. Mol Cancer Res. 2011;9(12):1746–54.
54. Konishi J, Kawaguchi KS, Vo H, Haruki N, Gonzalez A, Carbone DP, Dang TP. Gamma-secretase inhibitor prevents Notch3 activation and reduces proliferation in human lung cancers. Cancer Res. 2007;67(17):8051–7.
55. Ali SA, Justilien V, Jamieson L, Murray NR, Fields AP. Protein kinase ciota drives a NOTCH3-dependent stem-like phenotype in mutant KRAS lung adenocarcinoma. Cancer Cell. 2016;29(3):367–78.
56. Schroeder T, Just U. Notch signalling via RBP-J promotes myeloid differentiation. EMBO J. 2000;19(11):2558–68.
57. Schroeder T, Kohlhof H, Rieber N, Just U. Notch signaling induces multilineage myeloid differentiation and up-regulates PU.1 expression. J Immunol. 2003;170(11):5538–48.
58. Weng AP, Ferrando AA, Lee W, Morris JP t, Silverman LB, Sanchez-Irizarry C, Blacklow SC, Look AT, Aster JC. Activating mutations of NOTCH1 in human T cell acute lymphoblastic leukemia. Science. 2004;306(5694):269–71.

Chapter 9
TGF-β/Smad Signalling Pathway in Cancer

Mohadeseh Hasanpourghadi and Mohd. Rais Mustafa

Abstract Transforming growth factor (TGF)-β signalling pathway recently received attention as putative therapeutic target in cancer therapy. TGF-β is dual functional in development of cancer. In early stages of tumour development, TGF-β plays tumour suppressive, and as tumour progressed, TGF-β is an oncogenic factor. Functioning as tumour suppressor, TGF-β is a potential growth inhibitor. Cancerous cells show high sensitivity to TGF-β inhibition either by depletion of TGF-β signalling components in genetic level or by perturbation of downstream signalling proteins in protein level. Intense investigations had revealed that Smad proteins constitute as core components of TGF-β intracellular cascades. Typical development of cancer often contains production of excess TGF-β which accelerates the invasion and metastasis as well as inhibiting the anti-tumoural immune responses. In order to design optimal approaches in cancer therapeutic regimes, comprehending the oncogenic function of TGF-β and function of its downstream proteins (Smads) are required. The approaches must be in the direction of inhibition of those TGF-β functions which induce metastasis phenotypes, but at the same time preserve its growth inhibitory effects. To date several anti-cancer drugs such as Genistein and several microRNAs such as microRNA-452, demonstrated the new insights in induction of tumour suppressive potential of TGF-β signalling pathways among various cancer types.

Keywords Cancer therapy · TGF-β/Smad signalling · Tumor suppressive · Anti-cancer drug · MicroRNAs

M. Hasanpourghadi · M.R. Mustafa (✉)
Department of Pharmacology, Faculty of Medicine, University of Malaya,
Kuala Lumpur, Malaysia
e-mail: rais@um.edu.my

S. Fayyaz, A. A. Farooqi (eds.), *Recent Trends in Cancer Biology: Spotlight on Signaling Cascades and microRNAs*, https://doi.org/10.1007/978-3-319-71553-7_9

Introduction

Transforming growth factor (TGF)-β is a cytokine with multiple functions, basically was an isolated components from rat kidney fibroblast that could cause transforming phenotypes [1]. Eventually came to light that TGF-β has growth inhibitory effects against many cell-lines [2]. Since then researchers began to investigate the molecular mechanism of autocrine tumour suppressive properties of TGF-β. There are several downstream signalling pathways that accelerate the growth inhibition property of TGF-β. Numerous proteins involved in these pathways are, namely: Smad2, Smad3, Smad4, TβRI, TβRII, which receive, produce or carrying the signals from cell membrane into nucleus. Endogenous or exogenous factors that cause loss of TGF-β, results in abnormal proliferation of tumour cells. In parallel direction, scientists found that the *In vitro* application of TGF-β enhances cancer cells invasion, metastasis and migration [3]. Over the time, several additional studies demonstrated that not only TGF-β is involved with transformation, but also is involved in metastasis and migration of cancerous cells. Today pile of evidences clarified that TGF-β plays as both pro- and anti-oncogenic factor [4]. At the embark of tumourigenesis, TGF-β suppresses the maturing of benign tumour, but as tumour develops, by biochemical or genetic changes, TGF-β rather to enhance tumour growth. TGF-β as an enhancer in more developed level of tumourigenesis, effect in both cancer cells and non-cancerous stromal cells exist in the tumour tissue. The greatest challenge regarding TGF-β in researches is not TGF-β tumour suppressive properties, but its oncogenic properties that causes outgrowth and metastases of tumour. The reason is that the most human cancer types retain an oncogenic function of TGF-β and its downstream pathways. Therefore, most successful strategies are those that beside considering the bifunctional properties of TGF-β, yet manage to focus on inhibition of pro-oncogenic properties of TGF-β [4]. Hence, comprehending the TGF-β's exact mechanisms of activity in molecular level, and function of its downstream proteins that enhance oncogenic properties is crucial. In this chapter we elaborate TGF-β's dual function, TGF-β receptors and their functions, the role of core components of TGF-β signalling (Smad proteins) and their regulation, Smads' subcellular localization, the effect of this localization in pro- or anti-oncogenic potency of TGF-β/Smad signalling, and current therapeutic regimes in control of TGF-β/Smad signalling in cancer therapy.

TGF-β and Smads as Members of Signalling Pathway

TGF-β signalling pathway most likely has emerged in the first animal species to positively or negatively regulate multiple cellular gene expressions in regards to cell survival [5]. For instance, TGF-β suppresses the Inhibitor of Differentiation 1 (ID1) in epithelial cells of mammary glands. In another circumstance, TGF-β induces the same gene in invasive breast cancer cells [6, 7]. TGF-β controls embryonic development and maintains tissue homeostasis, by regulating cell proliferation, migration, morphogenesis, cytokine secretion, extra cellular matrix productions and apoptosis. TGF-β modes of response depends on the signals that cell receives [8]. TGF-β in most cell types is produced in the formation of latent complex consists of an active TGF-β, and 24-KDA cytokine with 80-KDA dimer formations of pre-pro region, also known as Latency-Associated Peptide (LAP) segment. Active TGF-β and LAP together are termed as Small Latent Complex (SLC). Two monomers of these complexes bind together by disulfide bridges and form a dimer complex. Next, SLC dimers bind to Latent TGF-β Binding Protein (LTBP) covalently and shape a Large Latent Complex (LLC). LLC has high affinity for Extracellular Matrix (ECM) [9, 10]. Extracellular spaces are rich in latent complexes which are secreted by cells. To release TGF-β from LAP, latent complex must be matured and activated. The matured TGF-β is required to be released from LAP. The process of latent TGF-β conversion into an active TGF-β differs in different cell types. In some cases proteases cleave LAP and release the mature TGF-β. In some other cases, a multifunctional secreted thrombospondin-1 (THBS1) attaches to LAP and interrupts the non-covalent interaction of mature TGF-β and LAP, and causes the liberation and exposure of active TGF-β [11]. Activated signals of TGF-β via two classes of receptors, the TGF-β type I (TβRI) and the TGF-β type II (TβR II). These two receptors are serine/threonine kinase receptors that upon binding to TGF-β form heterodimeric formation. Although TGF-β 1 and TGF-β 2 bind to TβRII receptor, TGF-β 2 only binds to TβRII receptor in the presence of TβRIII receptor. Any TGF-β bound to TβRII, further transactivates TβRI receptor and this is the rise of downstream signalling. If instead an atypical signalling occurs, such as phosphorylation of tyrosine residues, TGF-β receptors still can be attributed and activated [12]. TβRII phosphorylates TβRI, and phosphorylated TβRI binds to and phosphorylates Smad proteins. Smad proteins function by carrying received signals from cell membrane to nucleus and also are transcriptional factors. Therefore, TGF-β signalling occurs through Smad proteins. In genetic level, Smads are involved in transcriptional activities and repression. To be more specific, the actual result of Smad activation is absence or presence of other transcriptional factors [13]. There is another receptor known as accessory receptor (TβRIII) which binds to TGF-β to accelerate its interaction with TβRI and TβRII [14]. Smad2 and Smad3 which are transcriptional factors, are phosphorylated through TβRI. These Smads are generally known as receptor phosphorylated Smads or receptor regulated Smads (R-Smads). Smad4 is a Co-Smad and R-Smad's partner, plus it is not competent to be phosphorylated by TβRI. Two additional Smads known as Smad6 and Smad7, act as inhibitors of TGF-β signals. Inhibitory Smads act in Smad-Smad interactions and also facilitate proteasomes that target degradation of TGF-β receptors [15]. In normal condition, Smads are continuously moving in and out of nucleus. Upon phosphorylation by TβRI, R-Smads are accumulated in cell's nucleus wherein they accompany a cofactor, land on particular gene promoters and regulate transcription [16].

TGF-β Signalling Activation and Intracellular Factors

TGF-β family has above 30 members in human cells. Four receptors and four Smad proteins equip the TGF-β signalling pathway [5]. TGF-β activin-nodal and Bone Morphogenetic Protein (BMP) are two ligand subfamilies which are defined by sequences similarities [14]. These ligands are disulphide-linked dimers which their dimerization is crucial for activation of receptor. The family members of TGF-β mostly play a role like paracrine factors. Upon ligand assembly to receptor, a complex consists of two type I receptor and two type II receptors forms. Type I receptor propagates the signals, whereas type II receptor are activator. This complex has a short cytoplasmic segment which continued by a kinase region. This segment in type II BMP receptor (BMPR2) has a C-terminal extension. Within this complex, the type I receptor is phosphorylated by type II receptor in advance to signal propagation [17]. By phosphorylation, a region inside type I receptor is switched from a inhibitor-binding domain that attaches to a specific inhibitor, FK506-binding protein (FKBP12), into a domain that binds to Smad substrate in order to phosphorylate and activate it [18].

In human cells, seven distinct type I and five contrasting type II receptors exist. Either the ligand is contiguous or non-contiguous determines the specific ligand and receptor pairing [19]. The TGF-β-nodal receptors are controlled by Bα and Bδ, which are two isoforms of regulatory subunit B, parts of protein phosphatase (PP2A). The PP2A which consists of Bα subunit, regulates the receptor positively, while PP2A which has Bδ subunit negatively regulates the receptor signalling [20]. Moreover in mammalian cells, TβRI can be sumoylated by a SUMO ligase which facilitates recruitment of Smad3 to the receptor to be phosphorylated [21]. Sumoylation of TβRI provides a docking domain for an adaptor which facilitates binding of Smad3 and the receptor. Upon TβRI activation, this receptor is internalized through clathrin-coated pits into early endosome wherein TβRI attaches to Smad anchors such as SARA (Smad Anchor for Receptor Activation) and ZFYVE9 which are necessary for the activation of receptor and mediates Smad2 and Smad3 recruitment for their phosphorylation by activated receptor [22]. There is a homolog of SARA, an endofin known as ZFYVE16 with the same responsibility, but in BMP signalling pathway [23]. The only role of ZFYVE16 in TGF-β signalling pathway is scaffolding Smad4 and facilitating a complex formation made of Smad2, Smad3 and Smad4 bound to receptor [24]. In mammalian cells SARA and Smad in complex formation are stabilized by promyelocytic leukemia (cPML) tumour suppressor protein [25]. Recently many additional regulators have been introduced which one way or another mediate early stage of TGF-β signalling pathway. Among them worth to mention, GTPase RAP2 that suppresses the recycling of activin/nodal receptor, as this way it controls the number of receptors [26]. Upon signalling RAP2 antagonizes the negative response of Smad7, hence positively contributes to the signal propagation. Another regulator called RIN1, promotes endocytosis activity of TGF-β receptor and thus contributes to initiation of signalling [27]. In mammalian cells, PDZ-containing protein erbin (ERBB2IP) avoids association of Smad4 to

Smad2 and Smad3 and sends opposite effects to SARA and endofin [28]. The receptor's endocytosis activity controls the signalling and also regulates availability of ligand on the surface of the cell [29]. The regulatory proteins that are acknowledged above, highlight the connection between the TGF-β signalling and control of receptor internalization. However many of these types of regulatory proteins, yet to be discovered [22].

TGF-β Signalling Activation and Extracellular Factors

Seven factors outside of the cell determine the stimulation to some extent by TGF-β cytokine. The first factor is level of ligand production from the source which could be from different contextual elements [30]. Second factor is the ligands subtypes as they differ in affinity of receptor, such as TGF-β1 or TGF-β2 [31]. Third is numerous ligand-trapping proteins that regulate ligand gradients formation in embryogenesis and stores in adult cells [32, 33]. An example of trap is BMP that traps Noggin, blocks the receptor's contacting domains in ligand [34]. Fourth, ligand's mediators that are released from these traps are proper and their activity could be terminated. The activation of TGF-β1 complex includes conformational motions generated by contacts with integrins in cell surface [35]. Patients with Marfan syndrome carrying mutations in fibrillin 1 gene. This mutation contributes to an inadequate LTBP1 (Latent TGF-β1-Binding Protein 1) anchoring, and attributes faulty release of TGF-β1 in the cell surface of these patients [36]. Fifth factor is type of antagonistic ligands, which are also known as left-right determination factors. These factors suppress nodal-receptor pairing [37], such as inhibin which suppress activin-receptor binding [38]. Sixth, ligands are actually presented to main receptor by accessory receptors. For instance, TGF-β1 is presented to main receptor by β-glycan [38], or nodal-activin receptor binding is by TDGF1 also known as crypto [39]. A disease called haemorrhagic telangiectasia is due to mutation in the gene of accessory receptor endoglin. This mutation similar to its client receptor, ACVRL1 causes this disease [40]. The seventh and the last factor is the combination of all TGF-β1 signalling receptors. These seven factors cooperate together to initiate the TGF-β1 signalling [17].

Characteristics of Smad Protein

MAD is the first intracellular mediator of TGF-β signalling pathway which was found in *Drosophila* [41], and this discovery was followed by finding of orthologs proteins in worm and vertebrates, which were named "Smad" [42]. Smad signalling is the central channel for TGF-β context and responses of the gene and as such, Smad signalling recapitulates the essence of TGF-β context [17]. Smads are divided into three classes, including R-Smad, I-Smad and Co-Smads (Fig. 9.1). Among

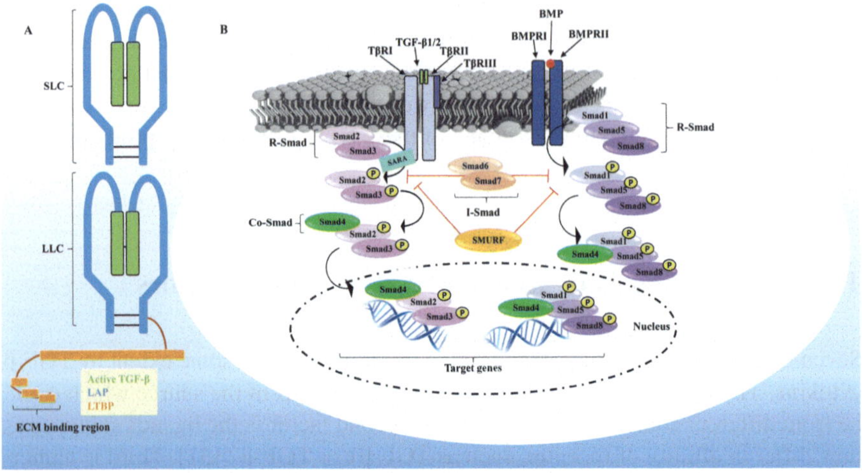

Fig. 9.1 Shows TGF-β in LLC formation and TGF-β/Smad signalling pathway in cytoplasm and nucleus. TGF-β also mediates Smad-independent pathways [43], such as several MAPK pathways, P38, ERK1, ERK2, JNK and PI3K pathway which are reported to be activated as feedback to TGF-β signals [44]. Besides, in some type of cells, TGF-β activates RHO-like GTPase [45]

them, only Smad2 and Smad3 (R-Smads) are phosphorylated and therefore, activated by TβRI. Smads 1, 5 and 8 (another group of R-Smads) respond to BMP subfamily signalling (Fig. 9.1). Smad4, the Co-Smad, along with R-Smads include almost 500 amino-acids with two conserved domain, the C-terminal MH2 and N-terminal MH1 domains. R-Smads has a characteristic SS motif at C-termini. Most of R-Smads and Co-Smad at their MH1 domain appear to contain specific domain for DNA binding activity which probably act in translocation into nucleus. I-Smads including Smad6 and Smad7, in their N-terminal domain has vulnerable sequence homology to the R-Smads' MH1 domain but they do not attach to DNA. MH2 domain is excessively conserved among all types of Smads, possibly because this domain is responsible for interaction with receptor, construction of heteromeric and homomeric Smad complexes and interaction with nuclear pore for nucleo-cytoplasmic transition. Both MH1 and MH2 domains connect to broad range of nuclear proteins, conductive to promote the transcription. The sequences between MH1 and MH2 domains include various phosphorylation sites which allow crosstalk between these domains and other signalling pathways. Among these two domains there is another sequence known as PY motif, which promotes the interaction with Smurfs (Smad ubiquitylation regulatory factor). Smurf1 and Smurf2 are two HECT-domains include E3 ubiquitin ligases that marks Smad-associated TGF-β receptors for degradation via a 26S proteasome (Fig. 9.1). Dephosphorylation of Smads act in TGF-β signalling termination [14].

How Activated TGF-β Receptor Complex Recognizes Smad?

TGF-β receptors are solo serine/theronin kinases which are introduced in human cells. Although why TGF-β facilitates serine/theronin kinases? and why they do not choose Tyr kinases similar to other cytokines with kinase-activating function, remained unclear [17]. TβRI exhibits an augment binding affinity to R-Smad. The question is, how such affinity emerges? The hint originated from a structural comparison between Smad2 and Smad4's MH2 domains. This domain showed the presence of a surface patch with high level of positively charged, located near L3 loop on Smad2 but not on surface of Smad4. This certain surface includes conserved residues in R-Smads, but not in Co-Smads, therefore the mentioned residue on R-Smads was confirmed to be binding site for the phosphorylated GS region of TβRI [46]. A mutation on His331 on basic patch of Smad2 causes reduction of affinity for TβRI and phosphorylation by this receptor [18]. Phosphorylated form of R-Smad in C-terminal sequence, contributes to the specific Smad-receptor interaction. The activated receptor that choose to interact with a particular R-Smad, has the key reason for its selection in the L45 loop of its kinase domain. This location specifies the receptor interaction with R-Smads only. As it is mentioned above, R-Smad also has a corresponding specificity for the receptor which is involved in its L3 loop. Hence the final structural formation and specificity delineation is based on both R-Smad's L3 loop and receptor's L45 loop [14]. Although, several other sequences on R-Smads are also reported to participate in R-Smad-receptor interaction [18].

TGF-β Receptor Stabilization

Tetrameric formation of TGF-β receptor is induced by ligand, which includes two TβRI and two TβRII units. Once these receptors are activated, they are internalized by clathrin-coated pits or by clathrin-independent caveolin-1-positive lipid rafts. Both routes are representative of two distinct results [47, 48]. Clathrin mediates endocytosis and transfers the receptors to the early endosomes, acquire receptors recycling and transfer them back to cell surface, and improve the sustained signalling. Caveolae the member of second route, transfers the receptors into ubiquitination system and to lysosome. The stimulation of TGF-β does not participate in any of these two endocytic routes-internalization patterns. Negative feedback loop of TGF-β signalling is defined by lipid rafts which are part of cell membrane domain. These rafts are defined upon generation of signals by TGF-β-induced I-Smads and E3 Ub ligases which govern a particular degradation of receptor [49, 50]. TGF-β1 induces either Smurf1 or Smurf2, Smad7. BMP signalling induces Smad6 and the HECT (homologous to E6-AP carboxylterminus)-domain E3 Ub ligase Smurf1 [51, 52]. The goal of TGF-β by inducing these proteins is to sustain and increase these proteins cellular levels so consequently their suppressive action improves and this improvement serves signalling better when the signals progress over the threshold. Additionally, induction of I-Smad by TGF-β, causes increase of TGF-β protein levels after they are degraded, so this action serves for initiation of next signalling. The interaction of Smurfs and I-Smads inside the nucleus are via WW domains of Smurfs and proline-tyrosine (PY) motif in I-Smads [53, 54]. The ubiquitination activity is blocked when C2 domain of Smurf2 binds beside catalytic cysteine on HECT domain of its own which leads to interruption of its own Ub thioester, further it causes self-inhibition [55]. Smurf2 diminish its own self-inhibition by binding to Smad7. Smurfs by their C2 domain, tether I-Smad/Smurf complex to part of cell membrane with lipid rafts where receptors bind and are ubiquitinated [56]. Within this activity, Smurf2 facilitates N-terminal sequence of Smad7 by participating in recruitment of E2 conjugating UbcH7 enzyme [57]. Heat shock protein 90 (Hsp90) adjusts Smurf2-mediated receptor ubiquitination, as suppressing Hsp90 caused elevated receptor ubiquitination [58]. The adaptor serine-threonine kinase receptor-associated protein (STRAP) stabilizes and recruits Smad7 to activated TβRI and TβRII receptors [59]. STRAP has WD40 associates with TβRI, TβRII and Smad7, leads to cause an inhibitory effect on this I-Smad. Moreover, two C2-WWHECT domain E3 Ub ligases target Smad7 and modify degradation of receptor [60]. Both of these proteins have similar structure to Smurfs and their functions are similar in regards of promoting receptor down-regulation, which finally causing reduction of R-Smads phosphorylation and also negatively effect on regulation of categories of transcription which initially are mediated by TGF-β. Atrophin1-interacting protein 4 (AIP4/Itch) is another HECT domain E3 ligase implicated in down-regulation of TGF-β receptor. AIP4/Itch attaches to Smad7, consequently leads to recruitment of Smad7 and TβRI [61]. In epithelial cells, if TGF-β is overexpressed, AIP4/Itch causes Smad7 ubiquitination and degradation. Therefore, during early stages of TGF-β receptor internalization, AIP4/Itch coordinates different activities. Degradation of I-Smad/Smurf can be controlled by supplementary regulators. Salt-inducible kinase (SIK) participates in negative feedback loop of TGF-β and regulates degradation of receptor [62]. This activity is accompanied with Smad7, therefore SIK enhances the turnover of TβRI-Smad7 ubiquitination. UCH37 is a deubiquitinating enzyme which binds to Smad7, and this attachment reverses the TβRI ubiquitination, causing TGF-β signalling stabilization [62]. Dapper2 is another regulator of TGF-β signalling components degradation. It causes TβRI's lysosomal degradation [63]. Certainly we can conclude that regulation of degradation is a vital factor to assure an appropriate signalling and better outcome. TGF-β receptors are ubiquitinated and degraded by different co-regulating proteins that contain E3 ligases. The master regulator is I-Smads which controls both ubiquitination and de-ubiquitination. The degradation of TGF-β receptor regulates the TGF-β signalling pathway. Both proteasomal and lysosomal activities modify degradation of TGF-β receptor [52, 62].

R-Smad Stabilization

Intracellular R-Smads are phosphorylated by activated TβRI and their protein level and stabilization is controlled by Ub-proteasome system, either during TGF-β signalling or in steady condition. Smurfs at transcriptional level, are induced by TGF-β signalling. Their constant existence may effect steady condition of stability of Smad, because Smurfs attach to non-activated Smads. To be exact, Smurf1 attaches to Smad1 and Smad5 at PY motif and causes their ubiquitination [64]. Moreover, Smurf2 is able to ubiquitinate Smad1 and Smad2 at steady condition [65]. Smad3 on the other hand, is not a target of Smurf and can be stabilized by a proteasome inhibitor at steady condition [66]. Therefore, ubiquitination of Smad3 occurs in non-signalling condition which indicates Smad3's enough protein level at the beginning of TGF-β signalling. U-box containing carboxyl terminus of Hsc70-interacting protein (CHIP) ubiquitinates Smad3 at steady condition and consequently, regulates cellular sensitivity of TGF-β. CHIP also is reported to ubiquitinate Smad1 [67]. Two other proteins, axin and glycogen synthase kinase 3-b (GSK3b) which are scaffolding proteins, also cooperate in Smad3 basal stabilization [68]. These two scaffolding proteins, enhance Smad3 phosphorylation at Thr66 sequence which causes Smad3 for ubiquitination. When the ligand binds to receptor and signalling is initiated, TβRI receptor phosphorylates R-Smads at SXS-motifs of C-terminal. Next step, R-Smad binds to Co-Smad and in complex formation they translocate into nucleus and attach to chromatin [69]. Upon signalling initiation, the level of Smurf elevates, and those Smurfs induced by BMP, such as Smurf1 at its steady condition, is able to ubiquitinate Smad1 and Smad5 [64]. After Smad1 being phosphorylated by BMP type I receptor at its C-terminal motif, sequential phosphorylation of Smad1 continues by MAPK and GSK3β in the linker region [70]. The linker phosphorylation causes poly-ubiquitination of activated Smad1 through Smurf1. Treatment with proteasome inhibitors maintains higher Smad2 protein level that indicates proteasomes degrades phosphorylated Smad2 [71]. By the time signalling is initiated, Smurf2 interacts with Smad2 and ubiquitinates it [72]. Moreover, the HECT domain of E3 ligase AIP4/Itch that facilitates TβRI-Smad7 recruitment in order to negatively regulate TGF-β signalling, is also able positively enhance this pathway by promoting TβRI and Smad2 attachment, TβRI-mediated phosphorylation of Smad2 and ubiquitination [73]. To explain it more simple, AIP4/Itch by promoting the phosphorylation of Smad2's C-terminal by TβRI, ubiquitinates it. That is why mice without AIP4/Itch, shows lack of TGF-β/Smad signalling. Ubiquitination of Smad2 also requires the RING domain E3 ligase CBL-B. In this case C-terminal phosphorylation of Smad2 by TβRI must be done extremely proper. That is why those mice with silenced CBL-B in their T lymphocytes show resistance to differentiation and growth inhibition induced by TGF-β [74]. On the other hand, Smurf does not effect on Smad3 directly, even though Smad3 has PY motifs which spontaneously binds to Smurfs. Instead, Smad3 in phosphorylated and activated form is degraded by SCF/Roc1 E3 Ub ligase [75]. In fact SCF/Roc1 terminates Smad3-induced signals in nucleus and causes Smad3 exportation from nucleus to cytoplasm to be degraded by proteasome. Phosphorylated Smad2 and Smad3 also can be degraded by a nuclear RING domain E3 ligase known as Arkadia [76]. Arkadia-dependant degradation occurs exactly after Smad's transcriptional activities. As a consequence, Smad3 degradation means termination of Smad3-induced signalling. Besides, another responsibility of Arkadia is to promote R-Smads's promoter transcription activity, as silencing Arkadia in emberionic cells caused recruitment of hypoactive phosphorylated Smad2 and Smad3 [76]. Ubiquitination of Smad3 is also requires sumoylation by sumoylation specific E3 ligase protein inhibitor of activated STAT (PIASy) [77].

Co-Smad Stabilization

Upon phosphorylation of R-Smads, it binds to Co-Smad (Smad4) which together they bind a category of trimeric complex formation includes one Smad4 and two R-Smads (Fig. 9.1). This complex intercalates with DNA and recruits co-factors on DNA specific promoters [69]. Smad4 degradation or survival, depends on different E3 ligases. For instance, SCFbTrCP1 causes ubiquitination of Smad4 [78], and Jab1 causes Smad4 proteasomal degradation [79]. Further, CHIP in corporation with Smad1 degrades Smad4 which leads to BMP signalling inhibition [67]. Mutated Smad4 was associated with various human cancer types due to its instability caused by mutation. Among E3 ligases, SCFkp2 has been shown to be a mediator for mutated Smad4 degradation [80]. Mono-ubiquitination of Smad4 at Lys-507 of its own MH2 domain causes increased transcriptional TGF-β signalling [81]. This type of ubiquitination on Smad4 also causes its transportation from nucleus into cytoplasm [82]. The mono-ubiquitination of Smad4 with one of E3 ligases, and poly-ubiquitination of Smad3 by SCF/Roc1 is stimulated by acetyl-transferase p300 and a nuclear co-activator [75, 83]. Sumoylation of Smad4 targets two goals during TGF-β signalling. First, the sumoylation causes accumulation of Smad4 in nucleus and its stabilization, which in turn, it is beneficial to TGF-β signalling [82], and secondly, Smad4 sumoylation causes inhibition of transcriptional activity of Smad4 [84]. The final level of sumoylation effects on TGF-β signalling are mixture of different sumoylations at receptor, Co-Smad and R-Smad steps. Now we know that Co-Smad and R-Smad both are ubiquitinated by various E3 ligases, which each could perform solo or in sequential steps respectively. Either way, E3 ligases activity gives the right amount of Smads to the cell. The correct amount and types of E3 ligases yet to be discovered.

I-Smad Stabilization

Complex of I-Smad, Smad6 and Smad7 are part of negative feedback of TGF-β signalling pathways [49] (Fig. 9.1). I-Smad alone has multiple responsibilities in different stages of pathway including at receptor, Co-Smad, R-Smad and nuclear Smad complex stages. Smad7 is most studied I-Smad could be an example of multiple responsibilities throughout of pathway. Smad7 is a mediator for ubiquitination and degradation of TβRI, TβRII ad Co-Smad [52], transfers phosphatases, and disrupts interaction between TβRI and R-Smad or between R-Smad and Co-Smad [49]. Smad7 is a target for both acetyl-transferases such as p300/CBP and Ub ligase [52]. To be more specific, Smurf1 and Smurf2 bind to Smad7 linker region of PY motif and ubiquitinate Smad7 [85]. Moreover this activity of Smurfs possibly is connected to Arkadia's activity as in poly-ubiquitination and proteasomal degradation of Smad7, as Smurfs attachment to Smad7 may ubiquitinates a specific protein and that causes Smad7-degradation activity of Arkadia [86]. Stabilization of Smad7 occurs by p300/CBP acetylation of a lysine that can be ubiquitinated as well, therefore this acetylation suppresses the degradation and in another word, stabilizes the Smad7 [87]. In contrast, Smad7 in acetylated form could be deacetylated by histon deacetylases (HDACs) and therefore releases Smad7 for ubiquitination [88]. Smad6 relatively is less known compared to Smad7. Smad6 mediates forming bridge between transcription factor Runx2 and Smurf1, and causing ubiquitination and proteasomal degradation of Runx2 [89]. Smad6 mediates repression of transcription and for this activity requires recruitment of HDACs [90]. PRMT1 is a nuclear protein *N*-methyltransferase that dimethylates Smad6 and alters its transcriptional activity and protein ubiquitination [91]. During TGF-β signalling, protein stability is also regulated by R-Smad-dependent ubiquitination and proeasomal degradation of numerous proteins. R-Smad and I-Smad both can transfer Ub ligases such as Arkadia and Smurfs to the protein that is targeted for ubiquitination [92].

There are additional factors participating in limiting the activity of Smad. Smad's C-terminal phosphorylation is removed by phosphatase. In order to binding to DNA, P300 and CBP (cyclin AMP response element-binding protein) are two co-activators that acetylate Smads on Lys residue which is in MH1 domain [93]. Another post-translational modification on Smad is sumoylation. Smad is also involved with a negative feedback by inducing Smad7 expression which causes recruitment of Smurf to BMP and TGF-β receptors for endocytosis-related poly-ubiquitination and degradation [85]. The ubiquitination of receptor is carried out by a deubiquitinase, USP4 (ubiquitin-specific processing protease 4), along with USP11 and USP15. Smad7 causes recruitment of Smurf2 and USF15 to TGF-β receptor, wherein these two enzymes compete with each other [94]. USP15 also is able to deubiquitinate R-Smad [95]. BAMPB1 (BMP and activin membrane-bound inhibitor) is a decoy receptor, and SKIL interferes with Smad at transcriptional complex which produces negative feedback [96].

Termination of TGF-β Receptor/Smad Signalling

At first glance, the members of TGF-β family, looks simple: a ligand binds to serine/threonine kinase transmembrane receptor that causes the activation of intracellular Smad. Subsequent signal continues in nucleus to regulate gene transcription. Increase of knowledge revealed additional information in each step of TGF-β/Smad signalling pathway. All of TGF-β family members are connected and regulated by other signalling pathways. Moreover, Smad signalling is under broad interactions by other proteins and their post-translational modifications. Besides, developing knowledge brought our attention on how TGF-β signalling is terminated in means of controlling this versatile pathway [49].

There are association between trafficking of TGF-β receptors, function and termination of downstream signalling pathways. TGF-β receptors can connect with intracellular events through at least two routes that anticipate whether signalling receptor must be taken seriously to initiate a response or else, it must be degraded [47]. Internalization of TGF-β receptors through clathrin, boosts the signals by mentoring the receptor to a primal endosome. This route is also stimulated with the cytoplasmic form of promyelocytic leukemia protein (cPML). This stimulation ultimately mediates the receptors to transfer back to surface of the cell [25]. Another TGF-β receptors internalization is through lipid-raft-caveolae-1 that includes receptors attached to I-Smad-Smurf which marks the receptor for degradation [47]. Any kind of interfering with one of the routes would switch the internalization to another route, and in contrast, enhancing one of the routes would inhibit internalization of the other route. The impact of the receptor trafficking on final signalling is different among different cell types. For instance, in cells with caveolin deficiency, TGF-β signalling occurs in absence of clathrin-dependent endocytosis [97]. TGF-β is not involved in trafficking of TGF-β receptor. TβRI is marked for poly-ubiquitylation-mediated proteasomal degradation by the Smad7-Smurf E3 ligase complex [98]. Smurf1 and Smurf2 have higher affinity for I-Smads compared to R-Smads via their paired interaction motifs [57]. Formation of I-Smad-Smurf complex occurs in cell's nucleus in the next step they are targeted to lipid raft vesicles by Smurfs. When Smad7-Smurf2 associate with a TGF-β receptor, both TGF-β receptor and Smad7 are destined to be ubiquitinated and go through degradation by proteasomes and lysosomes [52]. Those mice that carrying Smurf1 deficiency, do not exhibit TGF-β signalling improvement, and that is possibly due to Smurf1 and Smurf2 unnecessary function in TGF-β signalling [99]. Moreover, NEDD4-2 and TGIF interacting ubiquitin ligase (Tiul)1/WWP1 associates with Smad7 that causes degradation of TβRI [100]. Smad7 is known to recruit a phosphatase complex of PP1c and GADD34 to activated TβRI receptor, and consequently inactivate the receptor [101]. In endothelial cells another type of TβRI was shown to be inactivated by dephosphorylation of protein phosphate PP1α which apparently is recruited in a Smad7-dependent manner [102].

Genomic Stability by TGF-β/Smad Signalling

Mutations on those genes that encode TGF-β receptor or ligand occurs rarely in human cancer cells. For instance, in a study of 500 breast cancer cases, 92% of cases had a normal TGF-β/Smad pathway activation due to positive nuclear and phosphorylated Smad2 [103]. In fact, most of transcriptional responses that were mediated by TGF-β, are intact and functional in cancer cells that succeeded to escape the proliferation controls. Increased level of TGF-β in cancerous cells is able to promote neoplastic progression in different possible ways. Upon production of TGF-β in malignant cells, it plays in the way to inhibit immune responses, and accordingly it promotes the production of extracellular matrix [104].

One of the least well-recognized consequences of TGF-β reduction, is genomic instability (GIN), as TGF-β1 deletion causes increased level of GIN [105]. By culturing keratinocytes isolated from TGF-β1 null and wild type mice, it is shown that null cells have a 1000-fold increased level of mutant clones which were resistant to a well-known potent inhibitor of pyrimidine nucleotide synthesis, N-phosphonacetyl-L-aspartate (PALA) [106, 107]. This resistance is due to amplification of a gene that encodes dihydrofolate reductase. Other discovered signs of GIN are reported as chromosomal instability, increased level of centrosome aberration, spontaneous DNA damages even in non-cancerous cells. Mammary epithelial cells which carry heterozygous TGF-β1, with 10–30% abundance of TGF-β in comparison with wild type epithelial cells, they cause more GIN [107]. It is reported that radiation promotes TGF-β activity in both normal and cancer cells *in vitro* and *in vivo* [108]. These type of data simplified the importance of TGF-β as regulator of gene expression that are involved in encoding key DNA damage response (DDR) proteins [109]. On the contrary, epithelial cells carrying TGF-β-null, fail to go through cell cycle arrest or apoptosis as reaction to high-dose of radiation [110]. Failure to go through cell death is due to absence of DDR indicates an obvious decrease of ataxia telangiectasis mutated (ATM) activity induced by radiation [111]. ATM as a PI3K-related kinase, regulates a vast DDR signalling pathways such as some of cell cycle checkpoints, DNA repairing signalling pathway and programmed cell death. Besides, ATM is most firmly activated protein in response to signals from DNA double-strand breaks (DSB) [112]. Mutated ATM causes a type of GIN syndrome ataxia telangiectasia which is characterized by extreme sensitivity to toxic effect of radiation [113]. For instance A-T cell-line is in this condition, therefore being exposed to ionizing radiation causes numerous complex or simple chromosomal abnormality and unrepaired DNA damages [112]. DSB elicits induction of a bulk of proteins involved in checkpoint controls or repairing system, such as checkpoint protein CHK2, DSB repair protein RAD51, tumour suppressor p53-binding protein 1 (53BP1) [114]. In response to DSB, ATM is also activated and as a proper response, it phosphorylates various substrates to regulate the DDR and cell cycle checkpoints by controlling its downstream pathways which each of these pathways are targeting several of their own downstream pathways such as CHK1 and CHK2. For example, ATM activates and stabilizes p53 by phosphorylating this protein. P53 activation and stabilization makes it a major player in G1-S cell cycle checkpoint and in p53-induced apoptosis [112].

TGF-β signalling regulates the kinase activity of ATM, as TGF-β depletion or inhibition in human cells impact on ATM auto-phosphorylation and activity. This impact is followed by DNA damage-induced cell cycle checkpoints and increased sensitivity to radiation [111, 115]. On the other hand, suppression of TGF-β receptors by an inhibitor, mimics the consequences of TGF-β deletion [116]. As it is mentioned above, the receptor associated Smads including R-Smad1, 2, 3, 5 and 8 are localized in cytoplasm. Upon activation of TGF-β receptor by a ligand, R-Smad is phosphorylated and with the aid of Co-Smad4 translocates into nucleus and cause retention of R-Smad to promote expression of target gene [117]. Smad7 is able to bind at or near a DSB regions, maybe due to its ability to bind to any DNA elements with minimal Smad-binding elements CAGA box [118]. Phosphorylated Smad2 is not able to binds to DNA directly, therefore targets DSB by interaction with other repair proteins after chromosomes are remodelled. This was observed mainly in G1 cell cycle phase which can be considered as the role of phospho-Smad2 in G1-S checkpoint or as primary role of phospho-Smad2 in the non-homologous end joining (NHEJ) pathway, in addition to Smad2's role as a transcriptional factor. Besides, Smad2 and Smad3 interact with p53 [119, 120]. Applying ATM inhibitor causes blockage of Smad2 phosphorylation which indicates that the activity of ATM kinase is required for Smad2 to be phosphorylated in response to DNA damage. High dosage of radiation or applying DNA damaging agents causes co-localization of Smad3 with DSB proteins [121]. Smad7 acts as a scaffold in phosphorylation of ATM in prostate cancer cells that are stimulated by TGF-β [122]. Further it is reported Smad7 maintains cell survival after application of DNA damaging agents through facilitating and accelerating the ATM-dependent DNA repair [123]. An study showed deletion of Smad7 protein in embryonic fibroblast cells of mouse, reduces ATM activation and suppresses ATM recruitment to DSB by interrupting the interaction between ATM and Nbs1. Overall observation indicates the notion of complete crosstalk between ATM response signalling pathway and TGF-β/Smad signalling [124].

TGF-β/Smads Signalling and Other Cellular Pathways

MAPK

The linker region of Smad is a hotspot for regulatory input's integration. In response to growth factors, this region is phosphorylated by MAPKs (mitogen-activated protein kinases) [125]. Moreover, Ras-Raf-MAPK pathway exhibits oncogenic potential by phosphorylating Smad3 linker domain via MAPK, further preventing Smad's

C-terminal phosphorylation by a TβRI kinase domain leading to suppression of TGF-β cytostatic effects [126]. In response to stress, same region is phosphorylated by CDK4 in embark of cell progression [127]. The key in this regard is GSK3 phosphorylation of MAPK-primed or CDK-primed Smad linker. WNT is able to suppress GSK3, that it initiates a cooperation between itself (WNT) and BMP pathways [70]. Another association between MAPK and TGF-β pathway is through activating extracellular-signal-regulated kinases 1 and 2 (ERK1 and ERK2) which leads to induction of Epithelial to Mesenchymal Transition (EMT) [128].

FOXO

Cooperation with AKT pathway is available through FOXO factors, which are partners for Smad [129]. TGF-β signalling pathway's cytostatic program includes highly controlled $p21^{Cip1}$ central. C-myc-Miz complex inhibits activation of $p21^{Cip1}$, however, TGF-β relieves this barrier by down-regulating the c-myc. TGF-β-mediated Smad-Foxo complex targets $p21^{Cip1}$ to fulfil the activating task. Investigating the underlying molecular mechanism has revealed that Smad3 and Smad4 bind to Foxo3 or Foxo4 and then bind to $p21^{Cip1}$ promoter which has contiguous SBE and FHBE regions. These regions are necessary for Smad and Foxo binding and transactivation of $p21^{Cip1}$ induced by TGF-β. Silencing Foxo causes loss of $p21^{Cip1}$ induction, while the response of other TGF-β genes stay stable. For instance in HaCat cells, silencing Foxo, suppressed activation of 15 out of 118 TGF-β early gene responses [129].

PI3K/AKT

TGF-β is able to induce apoptosis through repression of PI3K/Akt/survivin pathway. Activation of Akt is linked to survivin, an inhibitor of apoptosis protein (IAP), as Akt mediates expression of survivin gene [130]. Survivin suppresses the activation and overexpression of caspase-3, caspase-7 and caspase-9 which is shown in multiple human cancer types, increases the resistance to numerous apoptotic stimuli [131]. In response to nutrient deprivation stress and growth factor, TGF-β represses endogenous activity of Akt. This action indicates that there is a negative crosstalk between TGF-β tumour suppressor potential and PI3K/Akt pathway [132]. Furthermore, the PI3K/Akt pathway activation is required for TGF-β-induced EMT and migration of the cell [133]. TβRI receptor in association with p85 which is PI3K subunit, activates PI3K activity [134].

RAS

TGF-β and RAS signalling pathways both are central of tumourigenesis and in some cases they are reported to exhibit synergism based on stage of tumour development. TGF-β depends on stage of the tumour, induces the increased level of Cyclin-Dependent Kinase Inhibitors (CDKIs) $p15^{Ink4B}$ and $p21^{Cip1}$, and induces c-MYC repression and repression of Inhibitors of Differentiation (IDs). As a result, cells differentiate, cell cycle is blocked and cell is in the edge of death [13]. This type of TGF-β signalling pathway activity, as it is mentioned before, is upon early stages of tumourigenesis. Unlike tumourigenesis, pro-oncogenic potential of TGF-β signalling pathway includes down-regulation of CDKIs, interfering extracellular matrix (ECM) production, mitogens' autocrine regulation, promotion of cell migration and EMT. The switch between anti-oncogenic potential and pro-oncogenic potential is not fully understood, however evidences show RAS signalling is involved in this conversion [135]. Investigations showed that RAS mainly interferes with Smad2 and Smad3 activation and also disrupts Smad-dependent transcription factor complex [136].

COX

Cyclooxygenase (COX) is a rate-limiting enzyme in prostanoid synthesis. COX1 is constantly expressed but the level of COX2 is not stable and depends on stimuli. Although COX2 is overexpressed in many cancer cells. A predominant prostaglandin, prostaglandin E2 (PGE2), exerts its effect via G protein-coupled receptors such as EP1-EP4, and it is involved with cancer cell proliferation and angiogenesis. In Non-Small Cell Lung Cancer (NSCLC) such as A549 cell-line, epidermal growth factor receptor (EGFR) and COX2 are overexpressed. TGF-β1 induces down-regulation of COX2 protein in A549 cell-line, which is associated with growth suppression and facilitation of fibrotic EMT response. This result suggests TGF-β1 alteration of COX2/PGE2 signalling, is a decision-maker in NSCLC A549 cellular processes [137].

As regards to participation of COX2, mammary epithelial cells from immotile condition to a motile condition through EMT, displayed an abnormal expression of COX2, associated with development of breast cancer tumourigenesis. Elevated expression of COX2 contributes to breast cancer invasion and metastasis to the bone and lung [138], while COX2 inhibition or deficiency contributes in decreased tumourigenesis of mammary cells [139]. It is reported that COX2/PGE2/EP receptor signalling needs GSK3β activity to disrupt TGF-β-induced Smad2/Smad3 signalling, as COX2/PGE2/EP receptor signalling causes reduction of overall cellular Smad3 and impairs the ability of Smad3 to translocate into nucleus as a feedback to TGF-β. In another direction, in order to activate GSK3β, EP receptor activates PI3K/AKT, ERK1, ERK2 and 3′5′-cyclic adenosine monophosphate as well. All of these proteins as response to TGF-β, participate in Smad2 and Smad3 regulation. Hence, they have relative contribution with COX2/PGE2 signalling in response to TGF-β in mammary cells [140].

MED15

MED15 is mediator of RNA polymerase II which its deficiency contributes in attenuation of TGF-β-targeted genes expression namely *P21* and *PAI-1*, decreases TGF-β-induced cell growth and participates in TGF-β-induced EMT through eliminating the phosphorylation and nuclear translocation of Smad2 and Smad3. Smad3 contains a binding domain to a region in MED15 which is critical pairing for Smad3-dependent transcriptional activity. MED15 additionally, is overexpressed in many cancer types which consequently causes hyper-activation of TGF-β signalling through hyper phosphorylation of Smad3. Transfection of MED15 silencing RNA reduced TGF-β/Smad signalling and ultimately decreased metastatic activity of metastatic breast cancer cells. In reverse, TGF-β up-regulates MED15 in both RNA and protein levels, while it did not change the level of other mediators such as MED6 and MED16 in breast cancer. A TGF-β inhibitor, A83-01 appeared to inhibit the up-regulation of MED15 and PAI-1 [141].

TGF-β Pathway Components in Cancer

The cellular responses mediated by TGF-β are vital for tissue homeostasis maintenance. In fact dysregulation of TGF-β pathways are associated with various human disorders such as cancer [8]. In some conditions, due to its apoptotic and anti-proliferative properties, TGF-β is considered tumour suppressor. During progression of the tumour, cancer cells are able to shut down the tumour suppressive responses after activation of TGF-β. Even it is reported that in some conditions, cancer cells provide somatic mutations as response to TGF-β/Smad signalling pathway, to block TGF-β anti-proliferative function. Some cells display resistance to anti-proliferative function of TGF-β even though TGF-β/Smad signalling pathway is not effective [142].

TGF-β signalling pathway includes several layers that complexity of negative regulatory mechanisms impact on each layer differently and ultimately provides a secure supervision on all involved molecules from plasma membrane into nucleus. Deregulation or disruption of any of involved proteins, alters a normal embryonic development or causes different human diseases. Ubiquitination as regulatory system in TGF-β signalling pathway, were discovered from studies that investigated Smad4 mutants in human cancer [143]. In this chapter, among human diseases, we only focus on cancer. TGF-β's role in cancer is complicated as it acts as tumour suppressor in epithelial cells and in early benign tumours, while if the cancer is in its late stage, TGF-β acts as a pro-tumourigenic factor that causes metastasis, cancer cell invasion, immune evasion [144]. Therefore it is understandable that the impact of negative regulators of TGF-β signalling pathway depends on the stage of cancer. The activities that made TGF-β to be considered tumour suppressor, are actually the ability of causing cell cycle arrest and its influence on apoptosis induction in normal epithelial cells or in benign tumours [49]. TGF-β contributes in epithelial to mesenchymal transformation and subsequently contributes to tumour metastasis and invasion [145].

TGF-β is potentially an anti-proliferative factor in immune cells and epithelial via down-regulation of c-myc and induction of two CDK inhibitors (p21Cip1 and p15Ink4b) [8]. TβRI, TβRII and Smad2 are mutated in low incidence (<5%) in various cancer types such as gastric, colon, ovarian, pancreatic, glioma and non-small-cell lung cancer. The incidence of Smad4 mutation is 50% in pancreatic cancer and in higher percentage in colon cancer. Tumours that carry TGF-β components mutation, supposed to show low sensitivity to TGF-β, however, mutation of Smad4 is expected to switch the sensitivity level opposite [146].

Mutations occurs regularly in human cancer cells, therefore TGF-β signalling pathway's core components, Smads and receptors, can be mutated and inactivated. For instance, it has been reported that overexpression of a Ub ligase such as Smurf1 and Smurf2, that negatively regulates Smads, causes inactivation of TGF-β signalling pathway in particular tumour [147]. For instance, in pancreatic cancer cells *Smurf1* gene, and in esophageal squamous cells *Smurf2* gene are overexpressed [148, 149]. Two other Ub ligase genes, *WWP1* gene is amplified in breast and prostate cancer, and NEDD4-2 gene is overexpressed in bladder and prostate cancers [150]. Another Ub ligase, STRAP which stabilizes TGF-β receptor and Smad7, is often mis-expressed in lung and colon cancers [151]. Moreover, among Ub ligases, ectodermin/TIF-1g which is an Ub ligase is mis-expressed in breast and colorectal cancers. Ectodermin/TIF-1g targets Smad4 and lack of this Ub ligase suppresses breast or colorectal cancer cells proliferation, therefore lack of this Ub ligase restores suppressing activity of TGF-β signalling pathway. Besides, some cancer cells do not express Smad4, and silencing ectodermin/TIF-1g in respective cells, does not impact on other layers of TGF-β signalling pathway, which indicates ectodermin/TIF-1g specifically targets Smad4 without involving in regulation of other TGF-β pathway components [152]. The last mentioned Ub ligase is an example how aberrant expression of ectodermin/TIF-1g is associated to oncogenic phenotype of TGF-β signalling pathway.

Smurf1 and Smurf2 are reported to be involved in both tumour suppressing and metastatic phenotypes in several cancer cells. Smurf1 is an Ub ligase for GTPase RhoA, an actin dynamic regulator which ignites the migration of the cell. Therefore, by focusing on this particular activity of Smurf1, we can see the picture of its metastasis-inducing potentials [153]. Smurf2, besides its Ub ligase activity, induces the senescence of cancer cells, which its overexpression in most cancer cells, magnifies its senescence-inducing potential [154]. Mutations in N-terminal MH1 domain in Smad2 and Smad4 causes critical changes in folding of this domain, which leads to instability and Ub-mediated proteasomal degradation of these proteins [148]. Here we discuss several of these type of mutations. For example, R133C in Smad2 and G65V, L43S, P130S and R100T in Smad4 are to be mentioned. G65V and R100T in Smad4 are targeted by the SCF Ub ligase that ultimately causes proteasomal degradation of Smad4. In contrast, cells that contain wild type Smad4, did not go through degradation by this enzyme [80]. Mutation in C-terminal MH2 domain in both Smad2 and Smad4, causes instability, poly-ubiquitination and degradation of these two proteins in pancreatic cancer cells [155]. These type of mutations include, C-terminal domain deletion of 38 amino acids in Smad4, L369R and Q407R mutations in Smad2. Existence of these sort of mutations activates Ub ligases, but does not only cause Smads degradation, but also causes degradation of other proteins that interact with Smads in complexes. Many of Ub ligases, on the other hand, do not only regulate Smads, but also regulate the negative regulators of Smads. E3 ligase for instance, degrades both Smad and Smad repressor proteins, which demonstrates both oncogenic and tumour suppressor potentials of E3 ligase. Because Smad is a tumour suppressor protein, therefore its degradation exhibits a oncogenic profile, and in the opposite, degrading Smad repressor, would turns back the existence and stability of Smad which exhibits tumour suppressing profile of E3 ligase [156]. Two most known Smad repressors are SnoN and Ski play dual potentials in cancer development. The oncogenic potential of SnoN and Ski is by repressing transcriptional activity of Smad through binding to nuclear Smad complexes [157]. Several Ub ligases such as, Smurf2, Arkadia and APC poly-ubiquitinate and degrade SnoN and Ski, and this cooperation brings back the tumour suppressive potential of Smad [158, 159]. On the contrary, Arkadia and Smurf2 target Smad2 and Smad4 as well, therefore their activity removes two main elements of tumour suppressive potential of TGF-β signalling pathway [65, 160, 161].

Targeting TGF-β in Cancer Therapy

As it is mentioned above, TGF-β has a role in controlling DDR, therefore TGF-β inhibition in promoting the positive outcome of radiotherapy is promising [162]. Studies on A-T cells with extreme sensitivity to radiation, revealed that TGF-β inhibition or deletion escalates these cells sensitivity to radiation which is connected to decreased ATM kinase activity [163]. Studies on glioblastoma multiforme (GBM), a type of cancer with high resistance to radiation, showed TGF-β inhibition combined with radiotherapy is promising, as this type of therapy elevated to phase 2 trial in glioblastoma patients. Moreover, it is reported that inhibition of TGF-β and irradiation together synergize to reduce the self-renewal of glioma stem-like cells [164]. These type of cells are reported to produce high level of autocrine TGF-β leads to potentiate their DDR-involved molecules, whereas the TGF-β molecules that are induced by irradiation, regulate self-renewal signals [163]. An increasing body of evidence indicating that inhibition of TGF-β can provide benefits for the aim of radiotherapy, particularly by effecting on ATM [116].

MicroRNA and TGF-β Signalling Pathway Components in Cancer

MicroRNAs (miRNAs) are small non-coding RNA molecules and are introduced as a class of regulatory genes. MiRNAs bind to target messenger RNA (mRNA) and this way regulate the target gene [165, 166]. MiRNAs carry critical roles in pathological and physiological processes. Several recent studies revealed magnificent association between TGF-β signalling pathway and miRNAs, providing new insights in the power of miRNAs in induction of both tumour promoting and tumour suppressive phenotypes of TGF-β signalling pathways in cancer cells [167]. Abnormal expression of miRNAs interferes tightly regulated RNAs in cancer cells, which also is associated with cancer initiation, development and metastasis. To date, in human cells 2578 miRNAs were discovered and introduced. One miRNA in particular is able to target several genes [168]. Figure 9.2 shows different roles of miRNAs in TGF-β signalling pathway as described below.

MiR-452

Among known miRNAs, miR-452 targets WWP1 gene in prostate cancer. WWP1 silencing suppresses cancer cell growth and metastasis which indicates oncogenic potential of WWP1. WWP1 not only ubiquitinates Smad4, but also p53 and p63, therefore silencing WWP1 was associated with expression of p53, activation of caspase-3 and finally promotion of apoptosis. Beside these activities, WWP1 also degrades TβRI receptor and inhibits TGF-β signalling. This indicates dual potential

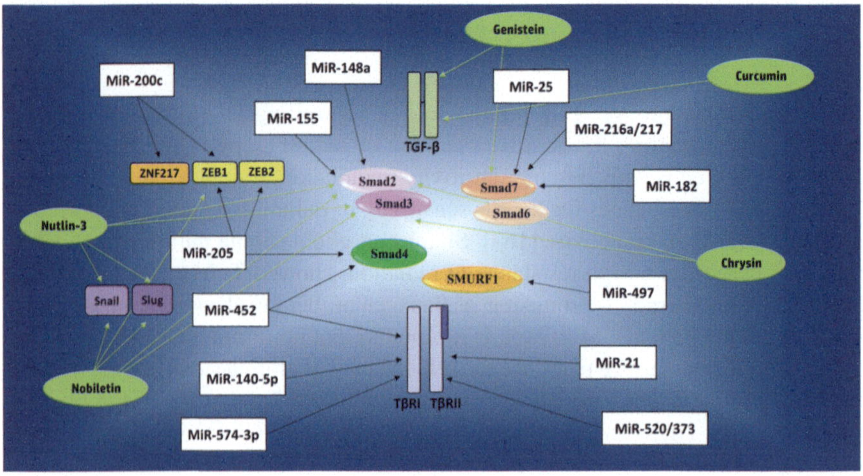

Fig. 9.2 Shows the effect of above mentioned anticancer drugs, in TGF-β signalling pathway

of WWP1. Transfection of miR-452 suppressed prostate cancer cells migrations and metastasis through directly targeting WWP1, as silencing WWP1 did not leave miR-452 any other effect [168].

MiR-497

MiR-497 is down-regulated in ovarian cancer cells, and down-regulation of endogenous miR-497 expression is connected with level of aggression of ovarian cancer cells. In contrary, transfecting exogenous miR-497, inhibits aggression, migration and metastases of ovarian cancer cells. In respective study, Smurf1 was announced as pro-metastatic factor and it was revealed to be exact target of miR-497. Studies showed lowering the level of miR-497 expression was associated with elevated Smurf1 expression and increase of ovarian cancer cell survival. As a result restoring miR-497 expression was offered to be a potential strategy in ovarian cancer treatment [169].

MiR-182

Smad7 is both negative regulator and transcriptional target of TGF-β signalling pathway. TGF-β is reported to activate miR-182 which targets and inhibits Smad7 protein. MiR-182 silencing caused up-regulation of Smad7 and prevented cancer cells invasion. In the contrary, overexpression of miR-182 induced breast cancer cells invasion. Further, the level of miR-182 in cancer cell and protein level of Smad7 are reported to be inversely associated [170].

MiR-21

A study showed miR-21 in coordination with androgen receptor (AR) caused inhibition of TβRII receptor expression in prostate cancer. MiR-21 specifically binds to 3′-UTR of TβRII and AR signalling later potentiates this effect of miR-21 in both transformed and untransformed prostate cancer cells. In early stages of prostate cancer cell development, both miR-21 and AR expressions are increased and this is in parallel with reduction of TβRII expression. In fact, it is reported that miR-21 and AR regulate each other's expression, and their activity causes attenuation of TβRII-mediated Smad2 and Smad3 activation. Down-regulation of TβRII receptor leads to inhibition of tumour suppressive phenotype of TGF-β signalling pathway, which promotes cell migration and invasion. The notion of this study suggests that by targeting miR-21 alone or in combination with targeting AR, may restore the tumour suppressive potential of TGF-β signalling pathway in prostate cancer cells [171].

MiR-216a/217

In hepatocellular carcinoma (HCC cell-line), expression of miR-216a/217 cluster is reported to be up-regulated, and these cells exhibit early tumour recurrence, hepatocarcinogenesis and EMT phenotype. Overexpression of miR-216a/217 caused increased level of EMT, cell migration and invasion. Focused investigation in respective study, revealed that miR-216a/217 target Smad7, along with PTEN which both are down-regulated in HCC cell-line. Since Smad7 is TβRI receptor antagonist, therefore overexpression of miR-216a/217 causes positive feedback regulatory of TGF-β signalling pathway. In addition, TGF-β signalling pathway activation in this cell-line was associated with elevated resistance to sorafenib. In the contrary, suppressing TGF-β signalling pathway could defeat miR-216a/217-induced resistance to sorafenib and avoid HCC cell metastasis [172].

MiR-148a

Glabridin is a flavonoid that inhibits proliferation and metastasis in several hepatocellular carcinoma cell-lines (HepG2, Huh-7, and MHCC97H). The exact mechanism of action of this anti-tumour agent is by miR-148a-mediated suppression of TGF-β/Smad2 signalling pathway in these cells. In fact, glabridin not only inhibits the TGF-β and Smad2 expression and activation, but also promotes the expression of miR-148a. MiR-148a targets *SMAD2*-3′UTR, and decreases expression and activity of Smad2. Silencing miR-148a prohibited the glabridin-mediated inhibition of TGF-β/Smad2 signalling [173].

MiR-200c

Resistance to trantuzumab, an anti-tumour drug, is a causative factor for mortality of HER2-positive breast cancer. A study revealed that HER2-positive breast cancer cells that exhibit resistance to trantuzumab, also exhibit higher level of tumourigenic phenotypes, EMT, elevated TGF-β signalling and invasion in comparison with parental cells. According to this study, in those breast cancer cells that miR-200c was the most down-regulated miRNA, restored sensitivity of the cell to transtuzumab and inhibited the invasion of these cells. The exact mechanism of action of miR-200c is reported to be by targeting a TGF-β transcriptional activator, ZNF217 and ZEB1 a TGF-β mediator. ZNF217 or ZEB1 silencing or restoring the miR-200c, both elevate the sensitivity of the cell to trastuzumab, and inhibits the invasion of breast cancer cells [174].

MiR-140-5p

Another study disclosed that miR-140-5p is decreased in six different hepatocellular carcinoma cell-lines. Increased level of miR-140-5p was connected to survival, cell proliferation and metastasis of these cell-lines. In respective study, miR-140-5p is reported to target TGF-β and MAPK/ERK signalling pathways. The exact molecules of these two pathways to be targeted by miR-140-5p are TβRI and fibroblast growth factor 9 (FGF9) in expression level. TβRI and FGF9 silencing, mimic the phenotype that resemble that phenotype obtained from ectopic miR-140-5p expression. In contrary, TβRI and FGF9 overexpression, narrowed the impact of miR-140-5p in hepatocellular carcinoma cells. Therefore, miR-140-5p exhibits a tumour suppressive feature in hepatocellular cell-lines development [175].

MiR-25

MiR-25 is a member of miR-106b-25 cluster and dys-regulated in most human cancer cells, especially in human colon cancer. When expression of miR-25 was restored, caused suppression of cell growth and cell metastasis. Unlike silencing miR-25 which promoted the same cell's proliferation and invasion, even in xenografts model in *in vivo* study. Underlying molecular mechanism of miR-25 revealed Smad7 is its direct target, by repressing Smad7 to be exact. Therefore miR-25 mimics a tumour suppressor function in colon cancer [176].

MiR-520/373

Profiling the miRNAs in MDA-MB-231, one of breast cancer cell-lines, revealed overexpression of miR-520/373 and its association with down-regulation of TGF-β. MiR-520/373 suppresses the metastasis, attributed by inhibition of *TGFβR2* gene. Silencing the *TGFβR2* mimics the impact of miR-520/373 overexpression on inhibition of Smad-dependent expression of a metastatic promoting genes angiopoietin-like 4, plasminogen activator inhibitor-1 and parathyroid hormone-related protein both *in vitro* and *in vivo*. A negative association between TβRII and miR-520c expressions was discovered in an estrogen receptor negative (ER^-) breast cancer patient but not in (ER^+) patients. Patients with lymph node metastasis of ER^- breast cancers, exhibited decreased of miR-520c expression [177].

MiR-205

Smad4 serves as tumour suppressor by inhibiting epithelial cell growth. Smad4 as Co-Smad in TGF-β/Smad signalling pathway act in TGF-β-induced EMT. MiR-205 targets 3′-UTR of Smad4. Dys-regulation of miR-205 is shown in many human cancer types, such as lung cancer, and it is overexpressed in various non-small cell lung carcinomas, leading to increased cell progression and invasion by targeting tumour suppressor genes, PHLPP2 and PTEN *in vitro* and *in vivo*. Targeting these two tumour suppressor genes, causes activation of Akt/mTOR and Akt/FOXO3a pathways [178]. In breast cancer cells, miR-205 effects EMT by targeting ZEb1/ZEb2, increase of N-caherin and decrease of E-cadherin [179]. Opposed to that, extopic expression of miR-205 in mesenchymal cell-initiated mesenchymal to epithelial transition causes reduced cell migration and up-regulation of E-cadherin. Hence, miR-205 is considered dual functional with both anti-oncogenic and pro-oncogenic properties [180].

MiR-155

Arsenic trioxide (As_2O_3) inhibits the proliferation of prostate cancer cell-lines PC3 and LNCap, by blocking angiogenesis *in vitro* and *in vivo*. Molecular mechanism of angiogenesis blockage, is via miR-155 as this miRNA mediates suppression of TGF-β/Smad signalling pathway. As_2O_3 promotes miR-155 expression through DNA-demethylation. MiR-155 then targets *SMAD2*-3′UTR, wherein reduces the expression and function of Smad2. Silencing miR-155 deleted As_2O_3-induced suppression of TGF-β/Smad2 signalling and ultimately suppressed angiogenesis [181].

MiR-574-3p

TGF-β1 induces up-regulation of miR-574-3p expression, mediated by Smad4 in human gastric cell-line AGS cells, as a TβRI inhibitor, SB431542 suppressed the miR-574-3p expression. This action indicates that miR-574-3p is a tumour suppressor, and its absence due to silencing, caused increase of AGS cells proliferation and migration by TGF-β1. Upon activation by TGF-β1, Smad4 translocates into nucleus, binds to sequence upstream of miR-574-3p gene which ultimately causes up-regulation of miR-574-3p [182]. Previously several other studies have demonstrated other target genes of miR-574-3p, such as EGFR in prostate cancer cells [183], yet, this was the first study reported association of TGF-β/Smad signalling and miR-574-3p [182].

Control of miRNAs by Smads

MicroRNAs as a long pre-miRNA transcript, are transcribed by RNA polymerase II. Similarly as mRNAs, pri-miRNAs are poly-adenylated at 3′ end and at 5′ end have a 7-methyl-guanosine cap. Drosha (RNase III endonuclease) cleaves pri-miRNA with the help of a co-factor (DGCR8), which stabilizes the association of pri-miRNA and Drosha to clarify the exact location of processing.

R-Smads such as Smad2 and Smad3 effect on miRNA by two mechanisms. First, upon activation of Smad2 and Smad3, this complex binds to Smad4, translocate to nucleus and activates mRNA transcription. Second, if Smad2/Smad3 complex does not bind to Smad4, translocates into nucleus and recruits Drosha/DGCR8 complex which leads to promotion of miRNA maturation [184, 185]. Several studies have reported that Smads proteins via binding to R-Smad, regulate transcription of numerous miRNA genes at their promoters [165, 186]. Several miRNAs are reported to be changed in their expression level, after TGF-β treatment [185]. For instance, TGF-β treatment caused elevated expression of miR-216a, miR-155 and miR-217 with their tumour promoting effects [186]. Also, down-regulation of several miRNA by TGF-β was reported. For example, miR-200b and miR-205 both are down-regulated after TGF-β administration. These miRNAs control the expression of ZEB1 and ZEB2 and E-cadherin transcriptional repressors which have effect in metastasis and EMT [187]. TGF-β signalling pathway is also able to regulate the expression of few miRNAs such as miR-145, miR-146a and miR-143 via activation of myocardin-mediated transcription factor (MRTFs) [188]. Smad through its MH1 domain, can associate with pri-miRNA. Smad2 and Smad3 interact with miRNA microprocessor complex known as p68 and Drosha which is facilitated by a 5′-CAGAC-3′ sequence, similar to the Smad binding elements (SBE) located in double-strand sequence of pri-mi-RNA [189]. Further in more recent studies, six more miRNAs have been discovered to increase in their expression level after TGF-β stimulation [190].

TGF-β/Smad in Anticancer Drug Regimes

Cancer treatment includes chemotherapy by using one or several chemotherapeutic anticancer drugs. More efficacious chemotherapy result is expected when combination of synergistic drugs are given to patients than concurrently effect on several cellular signalling pathways [191]. Such as combination of drugs that damage DNA, along with drugs that disrupt DNA repair mechanism components translocate into the nucleus [192]. In this regard drugs that interrupt major cellular signalling pathways are also granted as they promote tumour suppressive potential of cancer cell, or suppress oncogenic potential of cancer cell, by stimulating endogenous molecules such as TGF-β/Smad signalling components. Below several of these anticancer drugs are elaborated.

Genistein

Genistein, a phytoestrogen known for its protective role against estrogen and androgen-mediated carcinogenesis [193]. Genistein exhibits hepatoprotective property against D-Galactosamine (D-GalN)-induced acute liver damage and liver fibrosis, by inhibition of Hepatic Stellate Cells (HSC) activation, accumulating collagen matrix, reducing the expression of alpha Smooth Muscle Actin (a-SMA), increasing the serum alanine transaminase (ALT), and blocking apoptosis and necrosis. In regards to TGF-β/Smad signalling, genistein promotes the elevation of hepatic Smad7 expression followed by blocking the TGF-β in expression level and finally inhibiting the activated TGF-β/Smad signalling [194].

Nutlin-3

Nutlin-3 a series of cis-imidazoline analogue, among various series of synthetic chemicals that were screened to be identified and tested for their possible anticancer potency [195]. Nutlin-3 abolishes E-cadherin down-regulation through TGF-β1 in cancer cells with p53 deficiency. Further, Nutlin-3 exhibited EMT suppression as this drug blocks Smad2 and Smad3 phosphorylation, and subsequently decreased Snail and Slug transcription. Nutlin-3 also promoted the anti-cancer activity of several drugs in EGFR inhibitors class, by disrupting a p53-independent canonical TGF-β1/Smad/Snail/Slug complex [196].

Nobiletin

Nobiletin is a flavonoid extracted from citrus depressa, with anti-cancer properties. In molecular level, nobiletin inactivates TGF-β1-induced EMT in lung adenocarcinoma H1299 and A549 cell-lines. This effect was followed by inhibition of cell invasion and metastasis *in vitro*, continued with attenuation of MMP-2, MMP-9, p-paxillin, p-FAK, p-Src, twist, Snail, Slug and ZEB1 expression. Nobiletin suppressed Smad activity in transcriptional level but without altering the phosphorylation or translocation status of Smad. Smad3 overexpression changed the property of nobiletin in TGF-β1-induced EMT suppression, as for TGF-β1-induced EMT, Smad3 is required. *In vivo* study in the nude mice bearing A549-Luc xenografts, revealed inhibition of metastatic nodules growth, inhibition of tumour growth and finally reversed EMT by nobiletin [197].

Curcumin

Curcumin a major constituent of turmeric, suppresses cancer cell proliferation [193]. As it is reported, curcumin targets breast cancer stem cells, and stimulates the expression of TGF-β in MDA-MB-231 metastatic breast cancer cells. Curcumin plays as TGF-β inhibitor in dose-dependent manner which causes suppression of cell invasion and migration. Besides, curcumin modulates MMP-9 expression by inducing TGF-β in dose- and time-dependent manner [198].

Chrysin

Chrysin (5,7-dihydroxyflavone) is an active component isolated from propolis and honey. Hepatic stellate cells (HSCs) could be activated by TGF-β produced by hepatocytes, endothelial cells and Kupffer cells, which are specialized macrophages located in liver. TGF-β is known as a profibrotic marker, which contributes in liver fibrosis, the first stage of liver scarring that may favour cancer development. TGF-β1 induces transformation of HSCs into myofibroblasts, through promoting collagen synthesis. Chrysin attenuates liver fibrosis through blocking Smad2 and Smad3 signalling [199].

Conclusions and Perspectives

Considering the complex function of TGF- β/Smad signalling pathway in progression or suppression of tumour, any regime to exploit this signalling for therapeutic purposes, face difficulties. Stimulating TGF-β/Smad signalling may inhibit tumour growth, but simultaneously may promote tumour metastasis and invasion. Therapeutic regimes that introduce exogenous administration of TGF-β or induction of TGF-β signalling, would be successful only against tumour cells that are sensitive to anti-oncogenic function of TGF-β.

References

1. Roberts A, Sporn M. The transforming growth factor-βs. In: Peptide growth factors and their receptors I. New York, NY: Springer; 1990. p. 419–72.
2. Coffey R, et al. Selective inhibition of growth-related gene expression in murine keratinocytes by transforming growth factor beta. Mol Cell Biol. 1988;8(8):3088–93.
3. Welch DR, Fabra A, Nakajima M. Transforming growth factor beta stimulates mammary adenocarcinoma cell invasion and metastatic potential. Proc Natl Acad Sci. 1990;87(19):7678–82.
4. Akhurst RJ, Derynck R. TGF-β signaling in cancer a double-edged sword. Trends Cell Biol. 2001;11:S44–51.

5. Huminiecki L, et al. Emergence, development and diversification of the TGF-β signalling pathway within the animal kingdom. Evol Biol. 2009;9(1):28.
6. Kang Y, Chen C-R, Massagué J. A self-enabling TGFβ response coupled to stress signaling: smad engages stress response factor ATF3 for Id1 repression in epithelial cells. Mol Cell. 2003;11(4):915–26.
7. Padua D, et al. TGFβ primes breast tumors for lung metastasis seeding through angiopoietin-like 4. Cell. 2008;133(1):66–77.
8. Siegel PM, Massagué J. Cytostatic and apoptotic actions of TGF-β in homeostasis and cancer. Nat Rev Cancer. 2003;3(11):807–20.
9. Seoane J. The TGFß pathway as a therapeutic target in cancer. Clin Transl Oncol. 2008;10(1):14–9.
10. Annes JP, Munger JS, Rifkin DB. Making sense of latent TGFβ activation. J Cell Sci. 2003;116(2):217–24.
11. Ten Dijke P, Arthur HM. Extracellular control of TGFβ signalling in vascular development and disease. Nat Rev Mol Cell Biol. 2007;8(11):857–69.
12. Lee MK, et al. TGF-β activates Erk MAP kinase signalling through direct phosphorylation of ShcA. EMBO J. 2007;26(17):3957–67.
13. Massagué J. TGFβ in cancer. Cell. 2008;134(2):215–30.
14. Shi Y, Massagué J. Mechanisms of TGF-β signaling from cell membrane to the nucleus. Cell. 2003;113(6):685–700.
15. Massagué J, Seoane J, Wotton D. Smad transcription factors. Genes Dev. 2005;19(23):2783–810.
16. Inman GJ, Hill CS. Stoichiometry of active smad-transcription factor complexes on DNA. J Biol Chem. 2002;277(52):51008–16.
17. Massagué J. TGFβ signalling in context. Nat Rev Mol Cell Biol. 2012;13(10):616–30.
18. Huse M, et al. The TGFβ receptor activation process: an inhibitor-to substrate-binding switch. Mol Cell. 2001;8(3):671–82.
19. Groppe J, et al. Cooperative assembly of TGF-β superfamily signaling complexes is mediated by two disparate mechanisms and distinct modes of receptor binding. Mol Cell. 2008;29(2):157–68.
20. Batut J, et al. Two highly related regulatory subunits of PP2A exert opposite effects on TGF-β/Activin/Nodal signalling. Development. 2008;135(17):2927–37.
21. Kang JS, et al. The type I TGF-β receptor is covalently modified and regulated by sumoylation. Nat Cell Biol. 2008;10(6):654–64.
22. Moustakas A, Heldin C-H. The regulation of TGFβ signal transduction. Development. 2009;136(22):3699–714.
23. Shi W, et al. Endofin acts as a Smad anchor for receptor activation in BMP signaling. J Cell Sci. 2007;120(7):1216–24.
24. Chen Y-G, et al. Endofin, a FYVE domain protein, interacts with Smad4 and facilitates transforming growth factor-β signaling. J Biol Chem. 2007;282(13):9688–95.
25. Lin H-K, Bergmann S, Pandolfi PP. Cytoplasmic PML function in TGF-β signalling. Nature. 2004;431(7005):205–11.
26. Choi S-C, et al. Regulation of activin/nodal signaling by Rap2-directed receptor trafficking. Dev Cell. 2008;15(1):49–61.
27. Hu H, et al. Integration of transforming growth factor β and RAS signaling silences a RAB5 guanine nucleotide exchange factor and enhances growth factor-directed cell migration. Mol Cell Biol. 2008;28(5):1573–83.
28. Dai F, et al. Erbin inhibits transforming growth factor β signaling through a novel Smad-interacting domain. Mol Cell Biol. 2007;27(17):6183–94.
29. Clarke DC, et al. Transforming growth factor β depletion is the primary determinant of Smad signaling kinetics. Mol Cell Biol. 2009;29(9):2443–55.
30. Affolter M, Basler K. The Decapentaplegic morphogen gradient: from pattern formation to growth regulation. Nat Rev Genet. 2007;8(9):663–74.

31. Cheifetz S, et al. The transforming growth factor-β system, a complex pattern of cross-reactive ligands and receptors. Cell. 1987;48(3):409–15.
32. Zakin L, De Robertis E. Extracellular regulation of BMP signaling. Curr Biol. 2010;20(3):R89–92.
33. Zilberberg L, et al. Specificity of latent TGF-β binding protein (LTBP) incorporation into matrix: role of fibrillins and fibronectin. J Cell Physiol. 2012;227(12):3828–36.
34. Groppe J, et al. Structural basis of BMP signalling inhibition by the cystine knot protein Noggin. Nature. 2002;420(6916):636–42.
35. Shi M, et al. Latent TGF-[bgr] structure and activation. Nature. 2011;474(7351):343–9.
36. Lindsay ME, Dietz HC. Lessons on the pathogenesis of aneurysm from heritable conditions. Nature. 2011;473(7347):308–16.
37. Müller P, et al. Differential diffusivity of Nodal and Lefty underlies a reaction-diffusion patterning system. Science. 2012;336(6082):721–4.
38. Wiater E, et al. Identification of distinct inhibin and transforming growth factor β-binding sites on betaglycan functional separation of betaglycan co-receptor actions. J Biol Chem. 2006;281(25):17011–22.
39. Schier AF. Nodal morphogens. Cold Spring Harb Perspect Biol. 2009;1(5):a003459.
40. Marchuk DA. Genetic abnormalities in hereditary hemorrhagic telangiectasia. Curr Opin Hematol. 1998;5(5):332–8.
41. Sekelsky JJ, et al. Genetic characterization and cloning of mothers against dpp, a gene required for decapentaplegic function in Drosophila melanogaster. Genetics. 1995;139(3):1347–58.
42. Derynck R, et al. Nomenclature: vertebrate mediators of TGFβ family signals. Cell. 1996;87(2):173.
43. Derynck R, Zhang YE. Smad-dependent and Smad-independent pathways in TGF-β family signalling. Nature. 2003;425(6958):577–84.
44. Zhang YE. Non-Smad pathways in TGF-β signaling. Cell Res. 2009;19(1):128–39.
45. Edlund S, et al. Transforming growth factor-β–induced mobilization of actin cytoskeleton requires signaling by small GTPases Cdc42 and RhoA. Mol Biol Cell. 2002;13(3):902–14.
46. Wu G, et al. Structural basis of Smad2 recognition by the Smad anchor for receptor activation. Science. 2000;287(5450):92–7.
47. Di Guglielmo GM, et al. Distinct endocytic pathways regulate TGF-β receptor signalling and turnover. Nat Cell Biol. 2003;5(5):410–21.
48. Le Roy C, Wrana JL. Clathrin-and non-clathrin-mediated endocytic regulation of cell signalling. Nat Rev Mol Cell Biol. 2005;6(2):112–26.
49. Itoh S, ten Dijke P. Negative regulation of TGF-β receptor/Smad signal transduction. Curr Opin Cell Biol. 2007;19(2):176–84.
50. Glasgow E, Mishra L. Transforming growth factor-β signaling and ubiquitinators in cancer. Endocr Relat Cancer. 2008;15(1):59–72.
51. Murakami G, et al. Cooperative inhibition of bone morphogenetic protein signaling by Smurf1 and inhibitory Smads. Mol Biol Cell. 2003;14(7):2809–17.
52. Kavsak P, et al. Smad7 binds to Smurf2 to form an E3 ubiquitin ligase that targets the TGFβ receptor for degradation. Mol Cell. 2000;6(6):1365–75.
53. Itoh S, et al. Transforming growth factor β1 induces nuclear export of inhibitory Smad7. J Biol Chem. 1998;273(44):29195–201.
54. Tajima Y, et al. Chromosomal region maintenance 1 (CRM1)-dependent nuclear export of Smad ubiquitin regulatory factor 1 (Smurf1) is essential for negative regulation of transforming growth factor-β signaling by Smad7. J Biol Chem. 2003;278(12):10716–21.
55. Wiesner S, et al. Autoinhibition of the HECT-type ubiquitin ligase Smurf2 through its C2 domain. Cell. 2007;130(4):651–62.
56. Suzuki C, et al. Smurf1 regulates the inhibitory activity of Smad7 by targeting Smad7 to the plasma membrane. J Biol Chem. 2002;277(42):39919–25.
57. Ogunjimi AA, et al. Regulation of Smurf2 ubiquitin ligase activity by anchoring the E2 to the HECT domain. Mol Cell. 2005;19(3):297–308.

58. Wrighton KH, Lin X, Feng X-H. Critical regulation of TGFβ signaling by Hsp90. Proc Natl Acad Sci. 2008;105(27):9244–9.
59. Datta PK, Moses HL. STRAP and Smad7 synergize in the inhibition of transforming growth factor β signaling. Mol Cell Biol. 2000;20(9):3157–67.
60. Komuro A, et al. Negative regulation of transforming growth factor-β (TGF-β) signaling by WW domain-containing protein 1 (WWP1). Oncogene. 2004;23(41):6914–23.
61. Lallemand F, et al. AIP4 restricts transforming growth factor-β signaling through a ubiquitination-independent mechanism. J Biol Chem. 2005;280(30):27645–53.
62. Kowanetz M, et al. TGFβ induces SIK to negatively regulate type I receptor kinase signaling. J Cell Biol. 2008;182(4):655–62.
63. Su Y, et al. The evolutionally conserved activity of Dapper2 in antagonizing TGF-β signaling. Fed Am Soc Exp Biol J. 2007;21(3):682–90.
64. Zhu H, et al. A SMAD ubiquitin ligase targets the BMP pathway and affects embryonic pattern formation. Nature. 1999;400(6745):687–93.
65. Lin X, Liang M, Feng X-H. Smurf2 is a ubiquitin E3 ligase mediating proteasome-dependent degradation of Smad2 in transforming growth factor-β signaling. J Biol Chem. 2000;275(47):36818–22.
66. Inoue Y, et al. Contribution of the constitutive and inducible degradation of Smad3 by the ubiquitin-proteasome pathway to transforming growth factor-β signaling. J Interferon Cytokine Res. 2004;24(1):43–54.
67. Li L, et al. CHIP mediates degradation of Smad proteins and potentially regulates Smad-induced transcription. Mol Cell Biol. 2004;24(2):856–64.
68. Guo X, et al. Axin and GSK3-β control Smad3 protein stability and modulate TGF-β signaling. Genes Dev. 2008;22(1):106–20.
69. Ross S, Hill CS. How the Smads regulate transcription. Int J Biochem Cell Biol. 2008;40(3):383–408.
70. Fuentealba LC, et al. Integrating patterning signals: Wnt/GSK3 regulates the duration of the BMP/Smad1 signal. Cell. 2007;131(5):980–93.
71. Lo RS, Massagué J. Ubiquitin-dependent degradation of TGF-β-activated Smad2. Nat Cell Biol. 1999;1(8):472–8.
72. Bonni S, et al. TGF-β induces assembly of a Smad2–Smurf2 ubiquitin ligase complex that targets SnoN for degradation. Nat Cell Biol. 2001;3(6):587–95.
73. Bai Y, et al. Itch E3 ligase-mediated regulation of TGF-β signaling by modulating smad2 phosphorylation. Mol Cell. 2004;15(5):825–31.
74. Wohlfert EA, et al. Cutting edge: deficiency in the E3 ubiquitin ligase Cbl-b results in a multifunctional defect in T cell TGF-β sensitivity in vitro and in vivo. J Immunol. 2006;176(3):1316–20.
75. Fukuchi M, et al. Ligand-dependent degradation of Smad3 by a ubiquitin ligase complex of ROC1 and associated proteins. Mol Biol Cell. 2001;12(5):1431–43.
76. Mavrakis KJ, et al. Arkadia enhances Nodal/TGF-β signaling by coupling phospho-Smad2/3 activity and turnover. PLoS Biol. 2007;5(3):e67.
77. Imoto S, et al. Regulation of transforming growth factor-β signaling by protein inhibitor of activated STAT, PIASy through Smad3. J Biol Chem. 2003;278(36):34253–8.
78. Wan M, et al. Smad4 protein stability is regulated by ubiquitin ligase SCFβ-TrCP1. J Biol Chem. 2004;279(15):14484–7.
79. Wan M, et al. Jab1 antagonizes TGF-β signaling by inducing Smad4 degradation. EMBO Rep. 2002;3(2):171–6.
80. Liang M, et al. Ubiquitination and proteolysis of cancer-derived Smad4 mutants by SCFSkp2. Mol Cell Biol. 2004;24(17):7524–37.
81. Morén A, et al. Differential ubiquitination defines the functional status of the tumor suppressor Smad4. J Biol Chem. 2003;278(35):33571–82.
82. Lee PS, et al. Sumoylation of Smad4, the common Smad mediator of transforming growth factor-β family signaling. J Biol Chem. 2003;278(30):27853–63.

83. Wang B, Suzuki H, Kato M. Roles of mono-ubiquitinated Smad4 in the formation of Smad transcriptional complexes. Biochem Biophys Res Commun. 2008;376(2):288–92.
84. Jianyin L, et al. Repression of Smad4 transcriptional activity by SUMO modification. Biochem J. 2004;379(1):23–9.
85. Ebisawa T, et al. Smurf1 interacts with transforming growth factor-β type I receptor through Smad7 and induces receptor degradation. J Biol Chem. 2001;276(16):12477–80.
86. Koinuma D, et al. Arkadia amplifies TGF-β superfamily signalling through degradation of Smad7. EMBO J. 2003;22(24):6458–70.
87. Grönroos E, et al. Control of Smad7 stability by competition between acetylation and ubiquitination. Mol Cell. 2002;10(3):483–93.
88. Simonsson M, et al. The balance between acetylation and deacetylation controls Smad7 stability. J Biol Chem. 2005;280(23):21797–803.
89. Shen R, et al. Smad6 interacts with Runx2 and mediates Smad ubiquitin regulatory factor 1-induced Runx2 degradation. J Biol Chem. 2006;281(6):3569–76.
90. Bai S, Cao X. A nuclear antagonistic mechanism of inhibitory Smads in transforming growth factor-β signaling. J Biol Chem. 2002;277(6):4176–82.
91. Inamitsu M, et al. Methylation of Smad6 by protein arginine N-methyltransferase 1. Fed Eur Biochem Soc Lett. 2006;580(28-29):6603–11.
92. Heldin C-H, Moustakas A. A new twist in Smad signaling. Dev Cell. 2006;10(6):685–6.
93. Lönn P, et al. PARP-1 attenuates Smad-mediated transcription. Mol Cell. 2010;40(4):521–32.
94. Eichhorn PJ, et al. USP15 stabilizes TGF-[beta] receptor I and promotes oncogenesis through the activation of TGF-[beta] signaling in glioblastoma. Nat Med. 2012;18(3):429–35.
95. Inui M, et al. USP15 is a deubiquitylating enzyme for receptor-activated SMADs. Nat Cell Biol. 2011;13(11):1368–75.
96. Stroschein SL, et al. Negative feedback regulation of TGF-β signaling by the SnoN oncoprotein. Science. 1999;286(5440):771–4.
97. Lu Z, et al. Transforming growth factor β activates Smad2 in the absence of receptor endocytosis. J Biol Chem. 2002;277(33):29363–8.
98. Zhang L, et al. Determining TGF-β receptor levels in the cell membrane. TGF-β Signal Method Protoc. 2016;1344:35–47.
99. Yamashita M, et al. Ubiquitin ligase Smurf1 controls osteoblast activity and bone homeostasis by targeting MEKK2 for degradation. Cell. 2005;121(1):101–13.
100. Seo SR, et al. The novel E3 ubiquitin ligase Tiul1 associates with TGIF to target Smad2 for degradation. EMBO J. 2004;23(19):3780–92.
101. Shi W, et al. GADD34–PP1c recruited by Smad7 dephosphorylates TGFβ type I receptor. J Cell Biol. 2004;164(2):291–300.
102. Valdimarsdottir G, et al. Smad7 and protein phosphatase 1α are critical determinants in the duration of TGF-β/ALK1 signaling in endothelial cells. Cell Biol. 2006;7(1):16.
103. Xie W, et al. Alterations of smad signaling in human breast carcinoma are associated with poor outcome. Cancer Res. 2002;62(2):497–505.
104. Akhurst RJ. TGF-β antagonists: why suppress a tumor suppressor? J Clin Invest. 2002;109(12):1533–6.
105. Glick AB, et al. Transforming growth factor β1 suppresses genomic instability independent of a G1 arrest, p53, and Rb. Cancer Res. 1996;56(16):3645–50.
106. Agarwal MK, et al. Macrophage inhibitory cytokine 1 mediates a p53-dependent protective arrest in S phase in response to starvation for DNA precursors. Proc Natl Acad Sci. 2006;103(44):16278–83.
107. Maxwell CA, et al. Targeted and nontargeted effects of ionizing radiation that impact genomic instability. Cancer Res. 2008;68(20):8304–11.
108. Tabatabai G, et al. Irradiation and hypoxia promote homing of haematopoietic progenitor cells towards gliomas by TGF-β-dependent HIF-1α-mediated induction of CXCL12. Brain. 2006;129(9):2426–35.

109. Kanamoto T, et al. Functional proteomics of transforming growth factor-β1-stimulated Mv1Lu epithelial cells: Rad51 as a target of TGFβ1-dependent regulation of DNA repair. EMBO J. 2002;21(5):1219–30.
110. Ewan KB, et al. Transforming growth factor-β1 mediates cellular response to DNA damage in situ. Cancer Res. 2002;62(20):5627–31.
111. Kirshner J, et al. Inhibition of transforming growth factor-β1 signaling attenuates ataxia telangiectasia mutated activity in response to genotoxic stress. Cancer Res. 2006;66(22):10861–9.
112. Shiloh Y, Ziv Y. The ATM protein kinase: regulating the cellular response to genotoxic stress, and more. Nat Rev Mol Cell Biol. 2013;14(4):197–210.
113. Savitsky K, et al. A single ataxia telangiectasia gene with a product similar to PI-3 kinase. Science. 1995;268(5218):1749.
114. Whalen MK, et al. Specific ATM-mediated phosphorylation dependent on radiation quality. Radiat Res. 2008;170(3):353–64.
115. Wiegman EM, et al. TGFβ-1 dependent fast stimulation of ATM and p53 phosphorylation following exposure to ionizing radiation does not involve TGFβ-receptor I signalling. Radiother Oncol. 2007;83(3):289–95.
116. Barcellos-Hoff MH, Cucinotta FA. New tricks for an old fox: impact of TGFb on the DNA damage response and genomic stability. Sci Signal. 2014;7:re5.
117. Levy L, Hill CS. Smad4 dependency defines two classes of transforming growth factor β (TGF-β) target genes and distinguishes TGF β-induced epithelial-mesenchymal transition from its antiproliferative and migratory responses. Mol Cell Biol. 2005;25(18):8108–25.
118. Shi X, et al. Study of interaction between Smad7 and DNA by single-molecule force spectroscopy. Biochem Biophys Res Commun. 2008;377(4):1284–7.
119. Cordenonsi M, et al. Links between tumor suppressors: p53 is required for TGF-β gene responses by cooperating with Smads. Cell. 2003;113(3):301–14.
120. Elston R, Inman GJ. Crosstalk between p53 and TGF-β Signalling. J Signal Transduct. 2012;2012:294097.
121. Dubrovska A, et al. TGFβ1/Smad3 counteracts BRCA1-dependent repair of DNA damage. Oncogene. 2005;24(14):2289–97.
122. Zhang S, et al. TGFβ1-induced activation of ATM and p53 mediates apoptosis in a Smad7-dependent manner. Cell Cycle. 2006;5(23):2787–95.
123. Park S, et al. Smad7 enhances ATM activity by facilitating the interaction between ATM and Mre11-Rad50-Nbs1 complex in DNA double-strand break repair. Cell Mol Life Sci. 2015;72(3):583–96.
124. Wang M, Saha J, Cucinotta FA. Smad7 foci are present in micronuclei induced by heavy particle radiation. Mut Res Genet Toxicol Environ Mutagen. 2013;756(1):108–14.
125. Kretzschmar M, Doody J, Massagu J. Opposing BMP and EGF signalling pathways converge on the TGF-β family mediator Smad1. Nature. 1997;389(6651):618–22.
126. Nagata H, et al. Inhibition of c-Jun NH2-terminal kinase switches Smad3 signaling from oncogenesis to tumor-suppression in rat hepatocellular carcinoma. Hepatology. 2009;49(6):1944–53.
127. Matsuura I, et al. Cyclin-dependent kinases regulate the antiproliferative function of Smads. Nature. 2004;430(6996):226–31.
128. Lebrun J-J. The dual role of TGF in human cancer: from tumor suppression to cancer metastasis. Int Scholar Res Notice Mol Biol. 2012;2012:381428.
129. Seoane J, et al. Integration of Smad and forkhead pathways in the control of neuroepithelial and glioblastoma cell proliferation. Cell. 2004;117(2):211–23.
130. Asanuma H, et al. Survivin expression is regulated by coexpression of human epidermal growth factor receptor 2 and epidermal growth factor receptor via phosphatidylinositol 3-kinase/AKT signaling pathway in breast cancer cells. Cancer Res. 2005;65(23):11018–25.
131. Shin S, et al. An anti-apoptotic protein human survivin is a direct inhibitor of caspase-3 and-7. Biochemistry. 2001;40(4):1117–23.
132. Wang J, et al. Transforming growth factor β induces apoptosis through repressing the phosphoinositide 3-kinase/AKT/survivin pathway in colon cancer cells. Cancer Res. 2008;68(9):3152–60.

133. Shin I, et al. Transforming growth factor β enhances epithelial cell survival via Akt-dependent regulation of FKHRL1. Mol Biol Cell. 2001;12(11):3328–39.
134. Yi JY, Shin I, Arteaga CL. Type I transforming growth factor β receptor binds to and activates phosphatidylinositol 3-kinase. J Biol Chem. 2005;280(11):10870–6.
135. Grünert S, Jechlinger M, Beug H. Diverse cellular and molecular mechanisms contribute to epithelial plasticity and metastasis. Nat Rev Mol Cell Biol. 2003;4(8):657–65.
136. Grusch M, et al. The crosstalk of RAS with the TGF-β family during carcinoma progression and its implications for targeted cancer therapy. Curr Cancer Drug Targets. 2010;10(8):849–57.
137. Takai E, Tsukimoto M, Kojima S. TGF-β1 downregulates COX-2 expression leading to decrease of PGE2 production in human lung cancer A549 cells, which is involved in fibrotic response to TGF-β1. Plos One. 2013;8(10):e76346.
138. Minn AJ, et al. Genes that mediate breast cancer metastasis to lung. Nature. 2005;436(7050):518–24.
139. Mazhar D, Ang R, Waxman J. COX inhibitors and breast cancer. Br J Cancer. 2006;94(3):346–50.
140. Neil JR, et al. Cox-2 inactivates Smad signaling and enhances EMT stimulated by TGF-β through a PGE2-dependent mechanisms. Carcinogenesis. 2008;29(11):2227–35.
141. Zhao M, et al. Mediator MED15 modulates transforming growth factor beta (TGFβ)/Smad signaling and breast cancer cell metastasis. J Mol Cell Biol. 2013;5(1):57–60.
142. Wakefield LM, Roberts AB. TGF-β signaling: positive and negative effects on tumorigenesis. Curr Opin Genet Dev. 2002;12(1):22–9.
143. Xu J, Attisano L. Mutations in the tumor suppressors Smad2 and Smad4 inactivate transforming growth factor β signaling by targeting Smads to the ubiquitin–proteasome pathway. Proc Natl Acad Sci. 2000;97(9):4820–5.
144. Pardali K, Moustakas A. Actions of TGF-β as tumor suppressor and pro-metastatic factor in human cancer. Biochim Biophys Acta. 2007;1775(1):21–62.
145. Thiery JP, Chopin D. Epithelial cell plasticity in development and tumor progression. Cancer Metastasis Rev. 1999;18(1):31–42.
146. Levy L, Hill CS. Alterations in components of the TGF-β superfamily signaling pathways in human cancer. Cytokine Growth Factor Rev. 2006;17(1):41–58.
147. Izzi L, Attisano L. Ubiquitin-dependent regulation of TGβ signaling in cancer. Neoplasia. 2006;8(8):677–88.
148. Suzuki A, et al. Identification of SMURF1 as a possible target for 7q21. 3-22.1 amplification detected in a pancreatic cancer cell line by in-house array-based comparative genomic hybridization. Cancer Sci. 2008;99(5):986–94.
149. Fukuchi M, et al. High-level expression of the Smad ubiquitin ligase Smurf2 correlates with poor prognosis in patients with esophageal squamous cell carcinoma. Cancer Res. 2002;62(24):7162–5.
150. Chen C, Matesic LE. The Nedd4-like family of E3 ubiquitin ligases and cancer. Cancer Metastasis Rev. 2007;26(3-4):587.
151. Halder SK, et al. Oncogenic function of a novel WD-domain protein, STRAP, in human carcinogenesis. Cancer Res. 2006;66(12):6156–66.
152. Dupont S, et al. Germ-layer specification and control of cell growth by Ectodermin, a Smad4 ubiquitin ligase. Cell. 2005;121(1):87–99.
153. Sahai E, et al. Smurf1 regulates tumor cell plasticity and motility through degradation of RhoA leading to localized inhibition of contractility. J Cell Biol. 2007;176(1):35–42.
154. Zhang H, Cohen SN. Smurf2 up-regulation activates telomere-dependent senescence. Genes Dev. 2004;18(24):3028–40.
155. Maurice D, et al. Loss of Smad4 Function in pancreatic tumors C-terminal Truncation leads to decreased stability. J Biol Chem. 2001;276(46):43175–81.
156. Lönn P, et al. Regulating the stability of TGFβ receptors and Smads. Cell Res. 2009;19(1):21–35.
157. Sun Y, et al. SnoN and Ski protooncoproteins are rapidly degraded in response to transforming growth factor β signaling. Proc Natl Acad Sci. 1999;96(22):12442–7.

158. Le Scolan E, et al. Transforming growth factor-β suppresses the ability of Ski to inhibit tumor metastasis by inducing its degradation. Cancer Res. 2008;68(9):3277–85.
159. Levy L, et al. Arkadia activates Smad3/Smad4-dependent transcription by triggering signal-induced SnoN degradation. Mol Cell Biol. 2007;27(17):6068–83.
160. Zhang Y, et al. Regulation of Smad degradation and activity by Smurf2, an E3 ubiquitin ligase. Proc Natl Acad Sci. 2001;98(3):974–9.
161. Morén A, et al. Degradation of the tumor suppressor Smad4 by WW and HECT domain ubiquitin ligases. J Biol Chem. 2005;280(23):22115–23.
162. Du S, Barcellos-Hoff MH. Tumors as organs: biologically augmenting radiation therapy by inhibiting transforming growth factor β activity in carcinomas. Semin Radiat Oncol. 2013;23:242.
163. Hardee ME, et al. Resistance of glioblastoma-initiating cells to radiation mediated by the tumor microenvironment can be abolished by inhibiting transforming growth factor-β. Cancer Res. 2012;72(16):4119–29.
164. Zhang M, et al. Blockade of TGF-β signaling by the TGFβR-I kinase inhibitor LY2109761 enhances radiation response and prolongs survival in glioblastoma. Cancer Res. 2011;71(23):7155–67.
165. Ma S, et al. MicroRNA-616 induces androgen-independent growth of prostate cancer cells by suppressing expression of tissue factor pathway inhibitor TFPI-2. Cancer Res. 2011;71(2):583–92.
166. Shi X-B, Tepper CG, deVere White RW. Cancerous miRNAs and their regulation. Cell Cycle. 2008;7(11):1529–38.
167. Guo L, et al. MicroRNAs, TGF-β signaling, and the inflammatory microenvironment in cancer. Tumor Biol. 2016;37(1):115–25.
168. Goto Y, et al. Regulation of E3 ubiquitin ligase-1 (WWP1) by microRNA-452 inhibits cancer cell migration and invasion in prostate cancer. Br J Cancer. 2016;114:1135.
169. Wang W, et al. MicroRNA-497 inhibition of ovarian cancer cell migration and invasion through targeting of SMAD specific E3 ubiquitin protein ligase 1. Biochem Biophys Res Commun. 2014;449(4):432–7.
170. Yu J, et al. MicroRNA-182 targets SMAD7 to potentiate TGFβ-induced epithelial-mesenchymal transition and metastasis of cancer cells. Nat Commun. 2016;7:13884.
171. Mishra S, et al. Androgen receptor and microRNA-21 axis downregulates transforming growth factor beta receptor II (TGFBR2) expression in prostate cancer. Oncogene. 2014;33(31):4097–106.
172. Xia H, Ooi LLP, Hui KM. MicroRNA-216a/217-induced epithelial-mesenchymal transition targets PTEN and SMAD7 to promote drug resistance and recurrence of liver cancer. Hepatology. 2013;58(2):629–41.
173. Jiang F, et al. The repressive effect of miR-148a on TGF beta-SMADs signal pathway is involved in the glabridin-induced inhibition of the cancer stem cells-like properties in hepatocellular carcinoma cells. PloS One. 2014;9(5):e96698.
174. Bai WD, et al. MiR-200c suppresses TGF-β signaling and counteracts trastuzumab resistance and metastasis by targeting ZNF217 and ZEB1 in breast cancer. Int J Canc. 2014;135(6):1356–68.
175. Yang H, et al. MicroRNA-140-5p suppresses tumor growth and metastasis by targeting transforming growth factor β receptor 1 and fibroblast growth factor 9 in hepatocellular carcinoma. Hepatology. 2013;58(1):205–17.
176. Li Q, et al. MicroRNA-25 functions as a potential tumor suppressor in colon cancer by targeting Smad7. Cancer Lett. 2013;335(1):168–74.
177. Keklikoglou I, et al. MicroRNA-520/373 family functions as a tumor suppressor in estrogen receptor negative breast cancer by targeting NF-κB and TGF-β signaling pathways. Oncogene. 2012;31(37):4150–63.
178. Cai J, et al. miR-205 targets PTEN and PHLPP2 to augment AKT signaling and drive malignant phenotypes in non-small cell lung cancer. Cancer Res. 2013;73(17):5402–15.

179. Gregory PA, et al. An autocrine TGF-β/ZEB/miR-200 signaling network regulates establishment and maintenance of epithelialmesenchymal transition. Mol Biol Cell. 2011;22(10):1686–98.
180. Zeng Y, et al. Repression of Smad4 by miR-205 moderates TGF-β-induced epithelial-mesenchymal transition in A549 cell lines. Int J Oncol. 2016;49(2):700–8.
181. Ji H, et al. Inhibition of transforming growth factor beta/SMAD signal by MiR-155 is involved in arsenic trioxide-induced anti-angiogenesis in prostate cancer. Cancer Sci. 2014;105(12):1541–9.
182. Zhang R, et al. Upregulation of microRNA-574-3p in a human gastric cancer cell line AGS by TGF-β1. Gene. 2017;605:63.
183. Chiyomaru T, et al. Genistein up-regulates tumor suppressor microRNA-574-3p in prostate cancer. Plos One. 2013;8(3):e58929.
184. Blahna MT, Hata A. Smad-mediated regulation of microRNA biosynthesis. Fed Eur Biochem Soc Lett. 2012;586(14):1906–12.
185. Butz H, et al. Crosstalk between TGF-β signaling and the microRNA machinery. Trends Pharmacol Sci. 2012;33(7):382–93.
186. McDonald RA, et al. MicroRNA and vascular remodelling in acute vascular injury and pulmonary vascular remodelling. Cardiovasc Res. 2011;93:594.
187. Wang B, et al. miR-200a prevents renal fibrogenesis through repression of TGF-β2 expression. Diabetes. 2011;60(1):280–7.
188. Davis-Dusenbery BN, et al. Down-regulation of Krüppel-like Factor-4 (KLF4) by microRNA-143/145 is critical for modulation of vascular smooth muscle cell phenotype by transforming growth factor-β and bone morphogenetic protein 4. J Biol Chem. 2011;286(32):28097–110.
189. Hata A, Davis BN. Control of microRNA biogenesis by TGFβ signaling pathway—a novel role of Smads in the nucleus. Cytokine Growth Factor Rev. 2009;20(5):517–21.
190. Davis BN, et al. Smad proteins bind a conserved RNA sequence to promote microRNA maturation by Drosha. Mol Cell. 2010;39(3):373–84.
191. Hasanpourghadi M, et al. Targeting of tubulin polymerization and induction of mitotic blockage by Methyl 2-(5-fluoro-2-hydroxyphenyl)-1 Hbenzo [d] imidazole-5-carboxylate (MBIC) in human cervical cancer HeLa cell. J Exp Clin Cancer Res. 2016;35(1):1.
192. Hasanpourghadi M, et al. Mechanisms of the anti-tumor activity of Methyl 2-(-5-fluoro-2-hydroxyphenyl)-1 H-benzo[d]imidazole-5-carboxylate against breast cancer in vitro and in vivo. Oncotarget. 2017;8:28840.
193. Hasanpourghadi M, et al. Phytometabolites targeting warburg effect in cancer cells: a mechanistic review. Curr Drug Targets. 2017;18:1086.
194. Ganai AA, et al. Genistein modulates the expression of NF-κB and MAPK (p-38 and ERK1/2), thereby attenuating d-Galactosamine induced fulminant hepatic failure in Wistar rats. Toxicol Appl Pharmacol. 2015;283(2):139–46.
195. Vassilev LT, et al. In vivo activation of the p53 pathway by small-molecule antagonists of MDM2. Science. 2004;303(5659):844–8.
196. Wu Y, et al. Nutlin-3 inhibits epithelial–mesenchymal transition by interfering with canonical transforming growth factor-β1-Smad-Snail/Slug axis. Cancer Lett. 2014;342(1):82–91.
197. Da C, et al. Nobiletin inhibits epithelial-mesenchymal transition of human non-small cell lung cancer cells by antagonizing the TGF-β1/Smad3 signaling pathway. Oncol Rep. 2016;35(5):2767–74.
198. Mock CD, Jordan BC, Selvam C. Recent advances of curcumin and its analogues in breast cancer prevention and treatment. R Soc Chem Adv. 2015;5(92):75575–88.
199. Balta C, et al. Chrysin attenuates liver fibrosis and hepatic stellate cell activation through TGF-β/Smad signaling pathway. Chem Biol Interact. 2015;240:94–101.

Chapter 10
Natural Agents Mediated Regulation of microRNAs: Do We Need Skilled Archers to Hit the Bullseye

Eun Jung Sohn, Ammad Ahmad Farooqi, and Hwan Tae Park

Abstract MicroRNAs (miRNAs) are short noncoding RNAs, which control gene expression by messenger RNA degradation or translational repression. MicroRNAs are also reportedly involved in regulation of carcinogenesis, angiogenesis, apoptosis, cell proliferation and aging. Recent studies have shown that natural agents exert their anti-cancer effects via regulation of miRNAs. In this work, we summarize the regulation of miRNAs by natural agents as new strategy in cancer treatment.

Keywords MicroRNA · Natural agents · Apoptosis

Introduction

MicroRNAs (miRNAs) are small, ribonucleic acid (RNA) molecules reportedly involved in regulation of gene expression by binding to specific messenger RNAs (mRNAs). Based on the insights gleaned from almost two decades of research, it is now clear that miRNA encoding genes are located within intronic regions and contain their own promoter regions. RNA polymerase II-mediated transcription of long primary transcripts. Drosha, a type-III RNase, along with the co-factor protein

E. J. Sohn (✉)
College of Korean Medicine, Kyung Hee University, Seoul, Republic of Korea

Peripheral Neuropathy Research Center, Department of Physiology, College of Medicine, Dong-A University, Busan, South Korea
e-mail: eunjungs932@hotmail.com

A. A. Farooqi
Institute of Biomedical and Genetic Engineering (IBGE), Islamabad, Pakistan

H. T. Park
Peripheral Neuropathy Research Center, Department of Physiology, College of Medicine, Dong-A University, Busan, South Korea

S. Fayyaz, A. A. Farooqi (eds.), *Recent Trends in Cancer Biology: Spotlight on Signaling Cascades and microRNAs*, https://doi.org/10.1007/978-3-319-71553-7_10

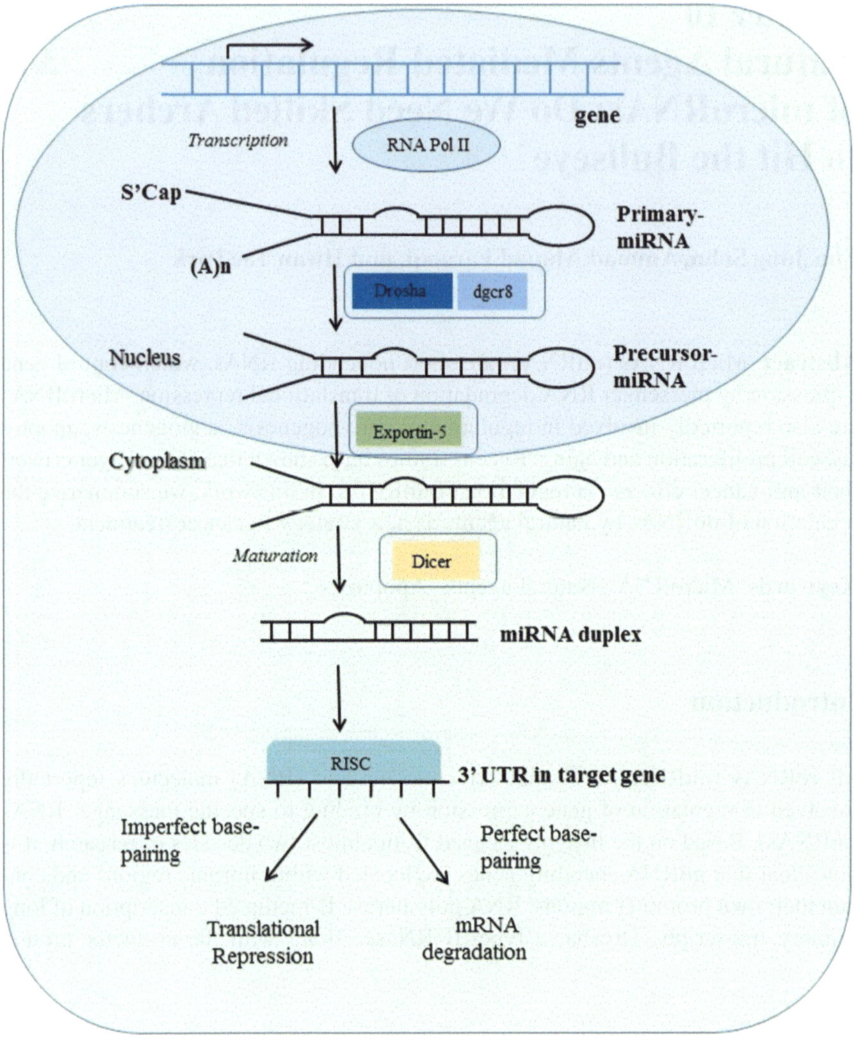

Fig. 10.1 Overview of the biogenesis of microRNAs

DGCR8, structurally associated with primary miRNA (pri-miRNA) transcript. RNase-domains present in Drosha mediated the cleavage of 3′and 5′ strands of pri-miRNAs and generated pre-miRNA. Exportin machinery shipped pre-miRNAs from nucleus to the cytosol. Binding of RNase-III Dicer and TAR RNA binding protein (TRBP) to the pre-miRNAs was noted in cytoplasm to cleave the terminal loop and form a miRNA duplex. miRNA duplex was incorporated into the RNA-induced silencing complex (RISC) (Fig. 10.1). Processing of the miRNA duplex was mediated by argonaute (AGO) proteins and several cofactors such as PACT. After un-winding and strand-selection, mature miRNA is functionally active to recognize its targets. Binding of the mature miRNA to RISC results in the targeting of mRNAs with complementary sites either to translationally repress or degrade mRNAs.

Modulation of microRNAs by Natural Products

Because of off-target effects and toxicities, there is a renewed interest in the identification of chemicals from natural sources to effectively treat cancer (Table 10.1).

Resveratrol

Resveratrol (trans-3,4′,5-trihydroxystilbene), a dietary polyphenolic, non-flavonoid antioxidant, plays an essential role in various cancers and disease [1] and it is currently at the stage of pre-clinical studies for human cancer prevention. Also, Resveratrol inhibited cancer cell growth in pancreatic cancer cells [2] and colon cancer cells [3] and induced apoptosis in several cancer cells such as squamous cancer cell [4]. Resveratrol blocked hypoxia-induced glioma cell migration and invasion via miRNA 34a [5]. Bai et al. showed that Resveratrol exerted anti-tumor mechanism in miRNA 200c transfected cells in lung cancer cells [6]. Also, Resveratrol treatment in bladder and pancreatic cancer cells induced cell death by regulating Bcl-2 of miRNA21 [7, 8]. Tili et al. showed that Resveratrol showed that anti-metastatic or anti-cancer effect via TGF β signaling pathway by miRNA 663 regulation [9].

Curcumin

Curcumin, a natural dietary polyphenol has multi-biological functions such as anti-tumor, anti-inflammatory, antioxidant, and antiangiogenic actions [10, 11]. Ye et al. showed that Curcumin induced apoptosis by activating the p53-miRNA-192-5p/215-XIAP pathway in A549 lung cancer [12]. Also, Curcumin showed a protective effect via miRNA-19/PTEN/AKT/p53 pathway against BPA-associated breast cancer promotion [13]. Yu et al. showed that Curcumin blocked the invasion and proliferation of human osteosarcoma by activating miRNA 138 [14].

Table 10.1 Regulation of miRNAs by natural agents

Phytochemical	Effects on microRNA and pathway	Cells	References
Resveratrol	miRNA 663/TGFβ signaling pathway	SW480 cells	[9]
Curcumin	miRNA 19/PTEN/AKT/p53 axis	MCF-7 breast cancer cell	[13]
Genistein	miRNA 1260/Wnt-signaling	Renal cancer cells	[29]
Quercetin	Let-7c/Notch signaling	Pancreatic cancer	[31]
Tanshinone IIA	miRNA 122/Pyruvate kinase M2 (PKM2)	Esophageal cancer cell	[19]
Ursolic acid	miRNA21/TGFβ/PDCD4 pathway	Glioblastoma cell	[16]

Ursolic Acid

Ursolic acid, a pentacyclic triterpene compound, is known as potent anti-cancer agents in various cancer cells [15]. Ursolic acid repressed proliferation of glioma cells by downregulating of miRNA21 targeting programmed cell death 4 (PDCF4) [16] and induced apoptosis in gastric cancer cells via targeting miRNA 133a [17].

Tanshiones

Tanshiones derived from Danshen, a traditional Chinese herb *Salvia miltiorrhiza* Bunge has considered as potential therapeutic effects in cancer [18]. Tanshinone IIA exerted anti-cancer effect for human esophagus cancer Ec109 cells via upregulating of miRNA122 targeting Pyruvate Kinase M2 (PKM2) expression [19]. Combination treatment with Tanshinone I and TRAIL in prostate cancer induced apoptosis via upregulating of miRNA 145a-3p [20]. Furthermore, Tanshinone I sensitize TRAIL mediated apoptosis via of miR135a-3p mediated Death Receptor 5 in prostate cancer cells [21]. Recent study showed that Tanshinone II A treatment in HepG2 induced apoptosis and cell cycle arrest via miRNA 30b inhibition. Also, inhibition miR30b in Tanshinone IIA induced cells attenuated the level of p53 and Tyrosine-protein phosphatase non-receptor type 11 (PTPN11) in HepG2 cells [21].

Indole-3-Carbinol (I3C)

Indole-3-carbinol (I3C), obtained from cruciferous vegetables such as broccoli, cauliflower, and cabbage has shown anti-proliferation in a variety of human cancer cell lines such as and tumor xenografts model [22]. Wang et al. showed that I3C inhibited the cell growth of hepatocellular carcinoma via decreasing of miRNA21 [23].

Genistein

Genistein, an antimicrobial toxin precursor from the biosynthesis of leguminous is reportedly involved in modulation of wide ranging biological mechanisms [24]. Yang et al. reported that Genistein treatment in A549 human lung cancer cells inhibited the proliferation by activating miRNA-27a and decreasing with MET proto-oncogene signaling [25]. Furthermore, Genistein treatment in human multiple myeloma cells repressed proliferation with an inhibition of nuclear factor-κB and an increase of microRNA-29b [26]. Hirata et al. showed Genistein contributed anti-cancer mechanism in prostate cancer cells, resulting in downregulation of miRNA1260 targeted sFRP1 and Smad4 [27] and in upregulation of miRNA 34a targeted with oncogenic HOTAIR [28]. Also, Genistein repressed the onco-miRNA1260 via alternation of Wnt signaling pathway to induce apoptosis in Renal cancer cells [29].

Quercetin

Quercetin from many fruits and vegetables has shown anti-carcinogenic actions [30]. A recent study showed that Quercetin altered miRNA expression in various cancer cells. Nwaeburu et studied that the let-7c was in the top upregulated miRNAs from the miRNA profiling after Quercetin treatment in Pancreatic cancer cells and targeted NUMB1 which is involved in Notch 1 signaling [31]. Quercetin treatment in ovarian carcinoma cells induced apoptosis by enhancing the miRNA145 expression [32]. Also, Tao et al. showed that Quercetin showed the anti-tumor effect of breast cancer cells via upregulation of miRNA146a *in vitro* and *in vivo* [33].

Mineral Pitch

Mineral Pitch, a dark brown humic matter originating from high altitude rocks has also been tested for efficacy against hepatic cancer cells. miRNA-21 was found to be considerably downregulated after treatment with 100 μg/ml of the Mineral Pitch [34]. There was a 1.67-fold increase in expression of miR-22 upon treatment with Mineral Pitch at a concentration of 50 μg/ml and 1.6-fold increase at a concentration of 100 μg/ml. There was a marked reduction in proliferation of Huh-7 cells reconstituted with miRNA-22. The data clearly suggested that Mineral Pitch exerted its anti-cancer effects via upregulation of miRNA-22 and downregulation of miR-21 [34].

Cantharidin

Cantharidin, a traditional Chinese medicine has been shown to downregulate miRNA-106b-93 in breast cancer cells. Cantharidin dose-dependently inhibited the proliferation of MCF-7 cells. Cantharidin triggered protein expression of PTEN and p21 via downregulation of miR-106b-93 [35].

Grape-Fruit Derived Nanovectors (GNVs)

Grape-fruit derived nanovectors (GNVs) have attracted considerable attention and currently being used as delivering systems for different therapeutic agents. Recently, plant exosome-like nanoparticles (NPs) were isolated from grapefruit juice and NPs were prepared with extracted lipids from grape-fruit exosome-like NPs [36]. To improve the targeting, GNVs were modified and folic acid was used as a targeting molecule. Folic acid had an affiliation for folate receptors and consequently GNVs can be delivered to the target site efficiently. The data clearly suggested that tumor growth was significantly reduced in tumor bearing mice intranasally delivered with miR-17 encapsulated in FA-GNVs [36].

Oridonin

Oridonin, a naturally occurring ent-kaurane diterpenoid obtained from Rabdosia rubescens was found to be effective against pancreatic cancer cells. Results revealed downregulation of miR-103a-3p up to 1.85 times, whereas miRNA-409-3p was upregulated 2.04 times. Moreover, miRNA-107 and miRNA-200b-3p were downregulated 2.13 and 2.22 times respectively in Oridonin treated pancreatic cancer cells [37].

Honokiol

Honokiol, a polyphenol from Magnolia grandiflora, is reportedly involved in upregulating expression level of miRNA-34a. STAT3 (Signal Transducer and Activator of Transcription) is a signaling modulator and noted to transcriptionally downregulate miRNA-34a by occupying the promoter region of miRNA-34a. Honokiol prevented accumulation of STAT3 at the promoter region of miRNA-34a [38].

Ganoderma lucidum

Ganoderma lucidum is a Basidiomycetes mushroom of medicinal importance having known anticancer activity. There was significant downregulation of hsa-miRNA-27a* (4.46-fold) and considerable upregulation of hsa-miRNA-1285 (10.46-fold) [39]. Different miRNAs involved in post-transcriptional regulation of telomerase enzyme, telomerase reverse transcriptase (TERT) were also found to be upregulated. miRNA-1207-5p (upregulation 2.895-fold) and miRNA-3687 (upregulation 2.153-fold) were noted to be changed in treated breast cancer cells [39].

Pien Tze Huang

Pien Tze Huang, a Chinese medicine dose dependently upregulated miRNA-200a, miRNA-200b and miRNA-200c in colorectal carcinoma cells [40].

1,6,7-Trihydroxyxanthone

1,6,7-trihydroxyxanthone (THA), a chemical obtained from *Goodyera oblongifolia* is effective against liver cancer cells. MiRNA-218 and miRNA-128 were noted to be upregulated in THA treated Bel7404 and HepG2 cancer cells [41]. Bmi-1 (B lymphoma mouse Moloney leukemia virus insertion region 1) is an oncogene frequently overexpressed in cancers. MiRNA-218 negatively regulated Bmi-1 and reduced proliferation potential of HepG2 cells [41].

Urolithin A

It is exciting to note that downregulation of β-catenin or miR-21 in colon cancer cells resulted transcriptional repression of T-cell factor/lymphoid enhancer factor (TCF/LEF), whereas miR-21 overexpression resulted in an enhanced TCF/LEF activity [42]. There was a significant increase in the level of β-catenin in miR-21-overexpressing Hepatocellular carcinoma (HCC) cells [43]. Methylated urolithin A repressed miR-21 level and simultaneously enhanced PDCD4 (Programmed Cell Death 4) and PTEN in prostate cancer cells to induce apoptosis. Methylated urolithin A significantly reduced tumor growth in mice xenografted with DU145 prostate cancer cells [44]. Methylated urolithin A also impressively reduced β-catenin level in prostate cancer cells and β-catenin signaling pathway [44].

Andrographolide

Andrographolide isolated from *Andrographis paniculata* effectively reduced cellular proliferation and self-renewal of ALDH1^{+}CD44^{+} oral cancer stem cells (OCSCs). Andrographolide dose-dependently increased miR-218 in OCSCs. Polycomb complex protein (Bmi1) was reduced in the miR-218-overexpressing OCSCs which clearly suggested that miR-218 negatively regulated Bmi1 in OCSCs [45].

Physcion

Physcion isolated from Radix et Rhizoma Rhei dose-dependently suppressed cell viability and colony formation in Cellosaurus cell line CNE-2 cells. Physcion dose-dependently increased the expression of ZBTB10 (Zinc finger and BTB domain containing-10) and simultaneously downregulated miR-27a in CNE-2 cells. Overexpression of Sp1 significantly impaired physcion-mediated autophagy and apoptosis in CNE-2 cells. Data clearly suggested that physcion inhibited miR-27a and upregulated ZBTB10 to reduce levels of Sp1 in CNE-2 cells [46].

1′S-1′-Acetoxychavicol Acetate (ACA)

1′S-1′-acetoxychavicol acetate (ACA) isolated from *Alpinia conchigera* was noted to effectively downregulate miR-210 expression in both Ca Ski and SiHa cells [47]. SMAD4 was directly targeted by miR-210. MiR-210 overexpression reduced SMAD4, while miR-210 inhibition increased SMAD4 levels. ACA mediated effects were considerably pronounced in SMAD4 overexpressing cancer cells. Markedly increased Caspase 3/7 activity was observed in ACA treated SMAD4-overexpressing cells. Data clearly suggested that SMAD4 overexpression augmented apoptosis-inducing effects of ACA [47].

Piperlongumine

Piperlongumine, an alkaloid isolated from *Piper longum* Linn rapidly downregulated cMyc in Panc1 cells [48]. Piperlongumine induced ROS (Reactive Oxygen Species) mediated downregulation of miR-27a and miR-17/20a (part of miR-17-92 cluster) and simultaneous induction of Zinc Finger And BTB Domain Containing proteins (ZBTB4 and ZBTB10) [48].

Conclusion

MicroRNAs have been studied for valid predictive or prognostic biomarker or therapeutic tools in cancer [49]. There has been a renewed interest in identifying pharmacologically active biomolecules from natural sources to treat cancer. Therefore we summarized a list of natural products having proven efficacy against different cancers. These compounds may be beneficial and can be used as templates for design and development of anticancer compounds.

Acknowledgements This work was supported by the National Research Foundation of Korea (NRF) grant funded by (2015R1C1A1A02036842 and 2016R1A5A2007009).

References

1. Shankar S, Singh G, Srivastava RK. Chemoprevention by resveratrol: molecular mechanisms and therapeutic potential. Front Biosci. 2007;12:4839–54.
2. Duan J, Yue W, E J MJ, Lu SE, Gu J, Xu F, Tan XL. In vitro comparative studies of resveratrol and triacetylresveratrol on cell proliferation, apoptosis, and STAT3 and NFkappaB signaling in pancreatic cancer cells. Sci Rep. 2016;6:31672.
3. Feng M, Zhong LX, Zhan ZY, Huang ZH, Xiong JP. Resveratrol treatment inhibits proliferation of and induces apoptosis in human colon cancer cells. Med Sci Monit. 2016;22:1101–8.
4. XD Y, Yang JL, Zhang WL, Liu DX. Resveratrol inhibits oral squamous cell carcinoma through induction of apoptosis and G2/M phase cell cycle arrest. Tumour Biol. 2016;37(3):2871–7.
5. Wang H, Feng H, Zhang Y. Resveratrol inhibits hypoxia-induced glioma cell migration and invasion by the p-STAT3/miR-34a axis. Neoplasma. 2016;63(4):532–9.
6. Bai T, Dong DS, Pei L. Synergistic antitumor activity of resveratrol and miR-200c in human lung cancer. Oncol Rep. 2014;31(5):2293–7.
7. Zhou C, Ding J, Wu Y. Resveratrol induces apoptosis of bladder cancer cells via miR21 regulation of the Akt/Bcl2 signaling pathway. Mol Med Rep. 2014;9(4):1467–73.
8. Liu P, Liang H, Xia Q, Li P, Kong H, Lei P, Wang S, Resveratrol TZ. induces apoptosis of pancreatic cancers cells by inhibiting miR-21 regulation of BCL-2 expression. Clin Transl Oncol. 2013;15(9):741–6.
9. Tili E, Michaille JJ, Alder H, Volinia S, Delmas D, Latruffe N, Croce CM. Resveratrol modulates the levels of microRNAs targeting genes encoding tumor-suppressors and effectors of TGFbeta signaling pathway in SW480 cells. Biochem Pharmacol. 2010;80(12):2057–65.
10. Jurenka JS. Anti-inflammatory properties of curcumin, a major constituent of Curcuma longa: a review of preclinical and clinical research. Altern Med Rev. 2009;14(2):141–53.
11. Mahran RI, Hagras MM, Sun D, Brenner DE. Bringing curcumin to the clinic in cancer prevention: a review of strategies to enhance bioavailability and efficacy. AAPS J. 2017;19:54.
12. Ye M, Zhang J, Zhang J, Miao Q, Yao L, Zhang J. Curcumin promotes apoptosis by activating the p53-miR-192-5p/215-XIAP pathway in non-small cell lung cancer. Cancer Lett. 2015;357(1):196–205.
13. Li X, Xie W, Xie C, Huang C, Zhu J, Liang Z, Deng F, Zhu M, Zhu W, Wu R, et al. Curcumin modulates miR-19/PTEN/AKT/p53 axis to suppress bisphenol A-induced MCF-7 breast cancer cell proliferation. Phytother Res. 2014;28(10):1553–60.
14. Yu D, An F, He X, Cao X. Curcumin inhibits the proliferation and invasion of human osteosarcoma cell line MG-63 by regulating miR-138. Int J Clin Exp Pathol. 2015;8(11):14946–52.
15. Yin R, Li T, Tian J, Xi P, Liu RH. Ursolic acid, a potential anticancer compound for breast cancer therapy. Crit Rev Food Sci Nutr. 2016. https://doi.org/10.1080/10408398.2016.1203755.
16. Wang J, Li Y, Wang X, Jiang C. Ursolic acid inhibits proliferation and induces apoptosis in human glioblastoma cell lines U251 by suppressing TGF-beta1/miR-21/PDCD4 pathway. Basic Clin Pharmacol Toxicol. 2012;111(2):106–12.
17. Xiang F, Pan C, Kong Q, Wu R, Jiang J, Zhan Y, Xu J, Gu X, Kang X. Ursolic acid inhibits the proliferation of gastric cancer cells by targeting miR-133a. Oncol Res. 2014;22(5-6):267–73.
18. Zhang Y, Jiang P, Ye M, Kim SH, Jiang C, Lu J. Tanshinones: sources, pharmacokinetics and anti-cancer activities. Int J Mol Sci. 2012;13(10):13621–66.
19. Zhang HS, Zhang FJ, Li H, Liu Y, Du GY, Huang YH. Tanshinone IIA inhibits human esophageal cancer cell growth through miR-122-mediated PKM2 down-regulation. Arch Biochem Biophys. 2016;598:50–56.

20. Shin EA, Sohn EJ, Won G, Choi JU, Jeong M, Kim B, Kim MJ, Kim SH. Upregulation of microRNA135a-3p and death receptor 5 plays a critical role in Tanshinone I sensitized prostate cancer cells to TRAIL induced apoptosis. Oncotarget. 2014;5(14):5624–36.
21. Ren X, Wang C, Xie B, Hu L, Chai H, Ding L, Tang L, Xia Y, Dou X. Tanshinone IIA induced cell death via miR30b-p53-PTPN11/SHP2 signaling pathway in human hepatocellular carcinoma cells. Eur J Pharmacol. 2017;796:233–41.
22. Weng JR, Tsai CH, Kulp SK, Chen CS. Indole-3-carbinol as a chemopreventive and anticancer agent. Cancer Lett. 2008;262(2):153–63.
23. Wang X, He H, Lu Y, Ren W, Teng KY, Chiang CL, Yang Z, Yu B, Hsu S, Jacob ST, et al. Indole-3-carbinol inhibits tumorigenicity of hepatocellular carcinoma cells via suppression of microRNA-21 and upregulation of phosphatase and tensin homolog. Biochim Biophys Acta. 2015;1853(1):244–53.
24. Rahman Mazumder MA, Hongsprabhas P. Genistein as antioxidant and antibrowning agents in in vivo and in vitro: a review. Biomed Pharmacother. 2016;82:379–92.
25. Yang Y, Zang A, Jia Y, Shang Y, Zhang Z, Ge K, Zhang J, Fan W, Wang B. Genistein inhibits A549 human lung cancer cell proliferation via miR-27a and MET signaling. Oncol Lett. 2016;12(3):2189–93.
26. Xie J, Wang J, Zhu B. Genistein inhibits the proliferation of human multiple myeloma cells through suppression of nuclear factor-kappaB and upregulation of microRNA-29b. Mol Med Rep. 2016;13(2):1627–32.
27. Hirata H, Hinoda Y, Shahryari V, Deng G, Tanaka Y, Tabatabai ZL, Dahiya R. Genistein downregulates onco-miR-1260b and upregulates sFRP1 and Smad4 via demethylation and histone modification in prostate cancer cells. Br J Cancer. 2014;110(6):1645–54.
28. Chiyomaru T, Yamamura S, Fukuhara S, Yoshino H, Kinoshita T, Majid S, Saini S, Chang I, Tanaka Y, Enokida H, et al. Genistein inhibits prostate cancer cell growth by targeting miR-34a and oncogenic HOTAIR. PLoS One. 2013;8(8):e70372.
29. Hirata H, Ueno K, Nakajima K, Tabatabai ZL, Hinoda Y, Ishii N, Dahiya R. Genistein downregulates onco-miR-1260b and inhibits Wnt-signalling in renal cancer cells. Br J Cancer. 2013;108(10):2070–8.
30. Kashyap D, Mittal S, Sak K, Singhal P, Tuli HS. Molecular mechanisms of action of quercetin in cancer: recent advances. Tumour Biol. 2016;37(10):12927–39.
31. Nwaeburu CC, Bauer N, Zhao Z, Abukiwan A, Gladkich J, Benner A, Herr I. Up-regulation of microRNA Let-7c by quercetin inhibits pancreatic cancer progression by activation of Numbl. Oncotarget. 2016;7:58367.
32. Zhou J, Gong J, Ding C, Chen G. Quercetin induces the apoptosis of human ovarian carcinoma cells by upregulating the expression of microRNA-145. Mol Med Rep. 2015;12(2):3127–31.
33. Tao SF, He HF, Chen Q. Quercetin inhibits proliferation and invasion acts by up-regulating miR-146a in human breast cancer cells. Mol Cell Biochem. 2015;402(1-2):93–100.
34. Pant K, Gupta P, Damania P, Yadav AK, Gupta A, Ashraf A, Venugopal SK. Mineral pitch induces apoptosis and inhibits proliferation via modulating reactive oxygen species in hepatic cancer cells. BMC Complement Altern Med. 2016;16:148.
35. Zhang H, Yan X. Cantharidin modulates the E2F1/MCM7-miR-106b-93/p21-PTEN signaling axis in MCF-7 breast cancer cells. Oncol Lett. 2015;10(5):2849–55.
36. Zhuang X, Teng Y, Samykutty A, Mu J, Deng Z, Zhang L, Cao P, Rong Y, Yan J, Miller D, et al. Grapefruit-derived nanovectors delivering therapeutic miR17 through an intranasal route inhibit brain tumor progression. Mol Ther. 2016;24(1):96–105.
37. Gui Z, Li S, Liu X, Xu B, Xu J. Oridonin alters the expression profiles of microRNAs in BxPC-3 human pancreatic cancer cells. BMC Complement Altern Med. 2015;15:117.
38. Avtanski DB, Nagalingam A, Kuppusamy P, Bonner MY, Arbiser JL, Saxena NK, Sharma D. Honokiol abrogates leptin-induced tumor progression by inhibiting Wnt1-MTA1-beta-catenin signaling axis in a microRNA-34a dependent manner. Oncotarget. 2015;6(18):16396–410.

39. Gonul O, Aydin HH, Kalmis E, Kayalar H, Ozkaya AB, Atay S, Ak H. Effects of ganoderma lucidum (higher basidiomycetes) extracts on the miRNA profile and telomerase activity of the MCF-7 breast cancer cell line. Int J Med Mushroom. 2015;17(3):231–9.
40. Shen A, Lin W, Chen Y, Liu L, Chen H, Zhuang Q, Lin J, Sferra TJ, Peng J. Pien Tze Huang inhibits metastasis of human colorectal carcinoma cells via modulation of TGF-beta1/ZEB/miR-200 signaling network. Int J Oncol. 2015;46(2):685–90.
41. WM F, Tang LP, Zhu X, YF L, Zhang YL, Lee WY, Wang H, Yu Y, Liang WC, Ko CH, et al. MiR-218-targeting-Bmi-1 mediates the suppressive effect of 1,6,7-trihydroxyxanthone on liver cancer cells. Apoptosis. 2015;20(1):75–82.
42. Yu Y, Sarkar FH, Majumdar AP. Down-regulation of miR-21 induces differentiation of chemoresistant colon cancer cells and enhances susceptibility to therapeutic regimens. Transl Oncol. 2013;6(2):180–6.
43. Huang YH, Lin YH, Chi HC, Liao CH, Liao CJ, SM W, Chen CY, Tseng YH, Tsai CY, Lin SY, et al. Thyroid hormone regulation of miR-21 enhances migration and invasion of hepatoma. Cancer Res. 2013;73(8):2505–17.
44. Zhou B, Wang J, Zheng G, Qiu Z. Methylated urolithin A, the modified ellagitannin-derived metabolite, suppresses cell viability of DU145 human prostate cancer cells via targeting miR-21. Food Chem Toxicol. 2016;97:375–84.
45. Yang PY, Hsieh PL, Wang TH, CC Y, MY L, Liao YW, Lee TH, Peng CY. Andrographolide impedes cancer stemness and enhances radio-sensitivity in oral carcinomas via miR-218 activation. Oncotarget. 2017;8(3):4196–207.
46. Pang MJ, Yang Z, Zhang XL, Liu ZF, Fan J, Zhang HY. Physcion, a naturally occurring anthraquinone derivative, induces apoptosis and autophagy in human nasopharyngeal carcinoma. Acta Pharmacol Sin. 2016;37(12):1623–40.
47. Phuah NH, Azmi MN, Awang K, Nagoor NH. Down-regulation of microRNA-210 confers sensitivity towards 1′S-1′-acetoxychavicol acetate (ACA) in cervical cancer cells by targeting SMAD4. Mol Cells. 2017;40(4):291–8.
48. Karki K, Hedrick E, Kasiappan R, Jin UH, Safe S. Piperlongumine induces reactive oxygen species (ROS)-dependent downregulation of specificity protein transcription factors. Cancer Prev Res (Phila). 2017;10:467.
49. Iorio MV, Croce CM. MicroRNAs in cancer: small molecules with a huge impact. J Clin Oncol. 2009;27(34):5848–56.

Chapter 11
Sebaceous Carcinoma of the Eyelid

Armando Luis Garcia, Ke Jin, Ravi Doddapaneni, Catherine Jeeyun Choi, Maria Paula Fernandez, Sander Dubovy, David Tse, Daniel Pelaez, and Wensi Tao

Abstract Sebaceous carcinoma (SC) is a malignant neoplasm that frequently affects the eyelids. It is an aggressive malignancy with metastasis rates as high as 25%. Historical studies have shown that mortality rates could be as high as 83%, but with the advent of increased detection, diagnosis, and treatment options, mortality rates have dropped substantially. A major issue with obtaining a proper diagnosis of SC is that it can mimic many other benign and malignant pathologies. Consequently, a delay in the diagnosis of SC can result in a poor prognosis and outcome for patients. The standard of care for SC is surgical excision, however, there is a reoccurrence rate of approximately 4% after excision. In these cases, it is necessary to treat patients with further excision and possibly radiation. Not all patients can have their tumors treated with surgery. In cases where there is diffuse intraepithelial seeding of the

---

Armando Luis Garcia and Ke Jin contributed equally to this work.

A. L. Garcia, M.D. · R. Doddapaneni · C. J. Choi, M.D. · S. Dubovy, M.D.
D. Tse, M.D. · W. Tao, Ph.D. (✉)
Dr. Nasser Al-Rashid Orbital Vision Research Center, Bascom Palmer Eye Institute,
Department of Ophthalmology, University of Miami Miller School of Medicine,
Miami, FL, USA
e-mail: wtao@med.miami.edu

K. Jin, Ph.D.
The Sheila and David Fuente Graduate Program in Cancer Biology, Sylvester Comprehensive Cancer Center, University of Miami Miller School of Medicine, Miami, FL, USA

M. P. Fernandez, M.D.
Department of Ophthalmology, University of Miami School of Medicine,
Miami, FL 33136, United States

D. Pelaez, M.D. (✉)
Dr. Nasser Al-Rashid Orbital Vision Research Center, Bascom Palmer Eye Institute,
Department of Ophthalmology, University of Miami Miller School of Medicine,
Miami, FL, USA

Department of Biomedical Engineering, University of Miami Miller School of Medicine,
Miami, FL, USA
e-mail: dpelaez@med.miami.edu

S. Fayyaz, A. A. Farooqi (eds.), *Recent Trends in Cancer Biology: Spotlight on Signaling Cascades and microRNAs*, https://doi.org/10.1007/978-3-319-71553-7_11

tumor cells, metastasis, or orbital invasion, other methods such as chemotherapy, radiation, and exenteration must be performed. It is essential that we elucidate the cellular mechanisms regarding the etiology of SC to develop better methods of diagnosis and treatment for the future. In this chapter, we will focus on the clinically relevant topics regarding SC, signaling pathways involved in the pathogenesis, and how miRNAs interact with these cascades. Both Wnt and TGF-β signaling have been implicated in the developing sebaceous glands and SC. Several miRNAs have been studied in this disease and downstream targets include: NF-κB, PTEN, c-Myc, and AKT2. These miRNAs may be used as biomarkers for SC or they may be potential therapeutic targets for patients who do not benefit from current interventions.

Keywords Sebaceous carcinoma · Eyelid · Wnt · TGF-β · miRNA

Introduction

Sebaceous carcinoma (SC) is an aggressive, malignant neoplasm that frequently arises from the skin surrounding the periocular region. Historically, it has had a high mortality rate, however, more recent studies have revealed that the survival and mortality of this disease has improved substantially. This is most likely due to a greater clinical awareness of the disease leading to an improvement in diagnosis and treatment options. Many diagnoses are delayed because its clinical presentation may resemble a wide array of benign and malignant clinical entities. This has led to the coining of the term "masquerade syndrome" to describe SC. miRNA usage as biomarkers of this disease may help increase the diagnostic accuracy of this entity. Options for treatment that are currently available include surgical excision, radiation, or chemotherapy. Translational research will benefit patients by providing cellular targets for pharmacologic treatments to improve outcomes where advanced disease is present, and where current interventions have failed or may provide little utility. It was not until recently that the transcription factors determining the cellular fate of the sebaceous glands had been discovered. Many questions are still left to be answered regarding the molecular alterations driving the development of SC. In this chapter, we will discuss most of the relevant clinical information, and how patients can benefit from research elucidating the role that miRNA play in the detection and treatment of SC.

Epidemiology

Five to ten percent of all dermatologic malignancies affect the eyelids, and new cases of eyelid cancer develop in approximately 15.7 out of 100,000 individuals per year in the United States [1, 2]. Sebaceous carcinoma, a malignant neoplasm arising from adnexa of the skin, is usually the second to fourth most common cutaneous malignancy of the eyelid depending on geographical location and study series [2–6]. There are multiple sites on the body where SC can develop; nevertheless, the ocular region is a frequent site of occurrence due to its large quantity of sebaceous glands [5]. Data collected from the Surveillance, Epidemiology, and End Results (SEER) show that 42.8% are located on extraocular head and neck regions, 34.5% are located on the eyelids, and 22.7% are found elsewhere [7]. Cancer of the sebaceous glands is responsible for up to 1–5.5% of periorbital malignancies, and in certain areas of Asia it may account for up to 33% of eyelid cancer [4, 8–11]. Sebaceous carcinoma has been reported to have an incidence of 0.23 per 100,000 person-years increasing at a rate of 3.31% annually [12]. The rising incidence may be due to an increase in detection, but more studies must be performed to determine if the pathogenesis is contributing to an increased incidence [7, 12, 13].

Risk Factors

There are several proposed risk factors that are associated with the development of SC of the eyelid. Early observations have found that women were mainly affected by this disease, but current findings have proven that women have either a slightly lower or equivalent incidence of this pathology compared to men [6, 10, 12–15]. Race may play a role in disease risk due to Asians having a higher incidence of disease risk [4, 13, 16, 17]. In the United States, whites are the most affected race with an overall contribution of 84.23% burden of the disease prevalence [6, 12]. Samples of SC from Japan were found to be positive for human papilloma virus (HPV) as compared to samples from the US [17, 18]. The presence of HPV in the eyelid may be responsible for the different incidences in these regions suggesting environmental differences rather than differences in genetic predisposition. This cancer may occur in younger patients, but it is generally recognized as being more prevalent in older patients. It has a mean age of diagnosis ranging from 50 to 89 years of age (85%) with the largest percentage of patients ranging from 60 to 79 years of age (49%) [6, 12, 13, 19, 20]. When SC occurs in younger patients it is typically due to predisposing factors some of which are discussed further [11, 19]. Long term immunosuppressed states found in solid organ transplant, non-Hodgkin's lymphoma, and acquired immunodeficiency syndrome patients confer an increased susceptibility to this disease [5, 20–25]. Case reports have observed that irradiation to the face increases the chances of developing SC several years after exposure [5, 26–30]. Genetic alterations in p53, Rb, and mismatch repair genes, mainly MSH2 or MLH1, in Muir-Torre syndrome (MTS) are also linked to the occurrence of SC at an earlier age [31–34]. Finally, SC may also arise from a nevus sebaceous [35].

Sebaceous Glands

The skin is a complex, dynamic organ that provides protection from the environment, ultraviolet light, and infectious agents; thermoregulation; sensation; wound repair and regeneration [36]. The sebaceous glands are acinar, holocrine-secreting adnexa that contribute significantly to the barrier of mammalian skin (Fig. 11.1a). They are typically found in the upper portion of the pilosebaceous unit, which is composed of the hair follicle and sebaceous gland, and are filled with sebocytes or

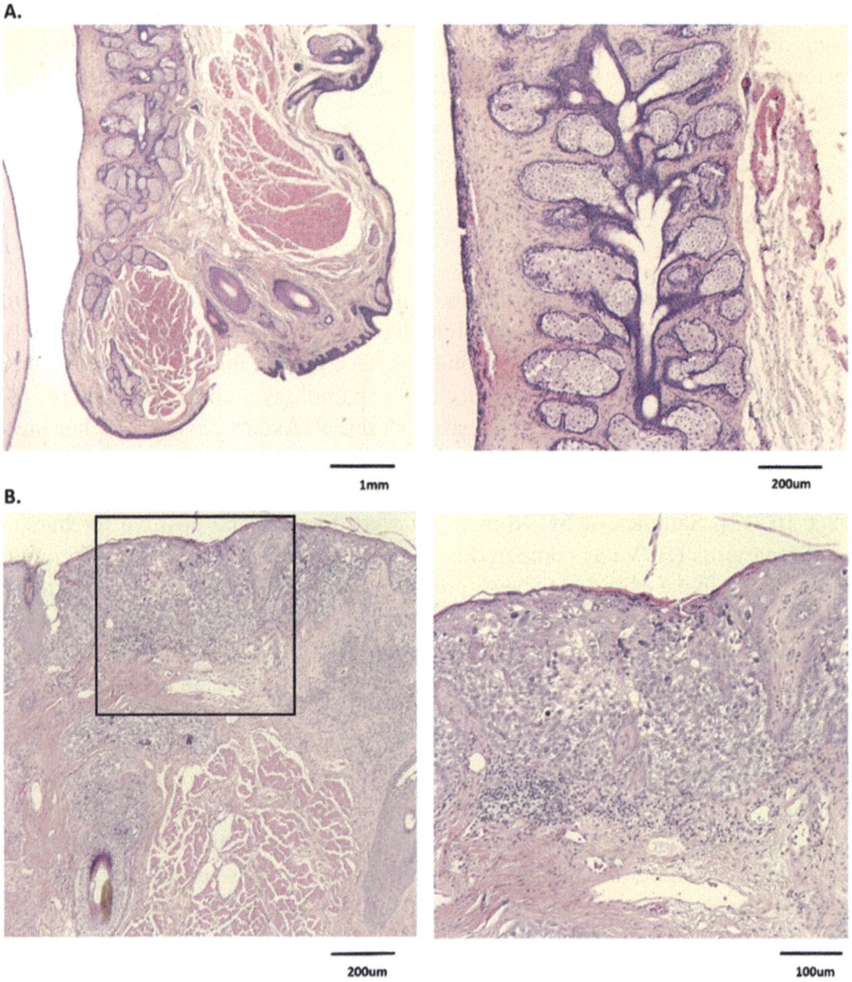

Fig. 11.1 Hematoxylin and eosin stains of sebaceous glands and sebaceous carcinoma. (**a**) The first image is a histologic section of the upper eyelid showing the sebaceous glands, hair follicles, and muscles. The second image is a section of multiple sebaceous glands draining into a single duct. (**b**) Morphology of sebaceous carcinoma with foamy cells, hyperchromatic nuclei, and mitotic figures. (**c**) A section that reveals the progression of pagetoid spread (intraepithelial invasion) of sebaceous carcinoma from deeper to more superficial areas

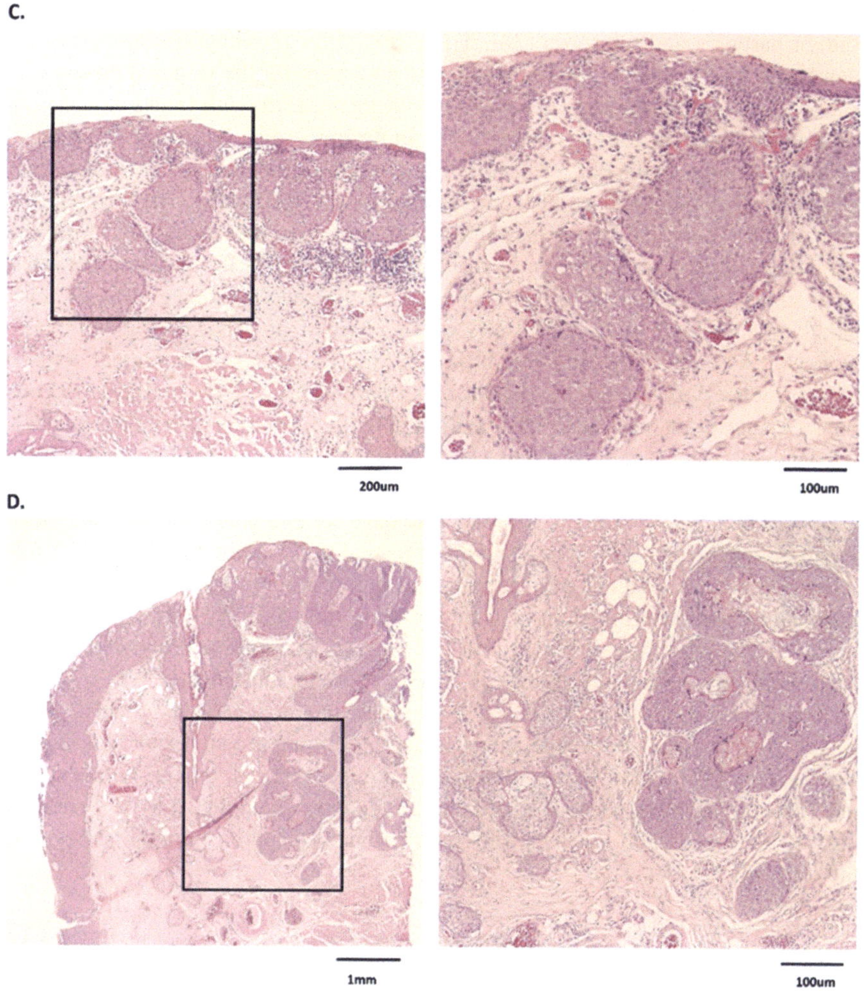

Fig. 11.1 (continued)

meibocytes. In certain parts of the body, such as the Meibomian glands of the eyelid, sebaceous glands may form independently from hair follicles. Lysis of sebocytes or meibocytes results in the excretion of lipids and sebum or meibum via specialized ducts onto the surface of the skin or eye, respectively [16, 37, 38]. Sebum is composed of squalene and wax esters, cholesterol esters, triglycerides, and free cholesterols; and it provides the skin with thermoregulation, reduction of water loss, antioxidants, antimicrobial properties, and pheromones [39–41]. The chemical composition of sebum imparts it with different functions over a wide range of temperatures. It may act as a repellant of rain below 30 °C, an emulsifier of sweat at and above 30 °C, or as a surfactant for eccrine secretions at higher temperatures [42]. Meibum is mainly composed of wax esters, cholesteryl esters, triacyl glycerols, di- and triesters, amphiphilic lipids and intercalated proteins [37]. This specialized lipid is important because it retards the evaporation of the tear film, it is critical to the formation of a smooth optical surface, and it functions as a barrier of foreign particles and microbes [38]. These properties provided by the sebaceous glands and its contents, render their function necessary for the skin and the eye.

Sebaceous glands undergo development during the 13th–16th weeks of gestation from developing hair follicles [39, 41, 43]. In the epidermis, Notch signaling plays an important roles in the normal development, regeneration and cancer formation of its accessory glands, such as sebaceous glands and lacrimal gland. The formation of the hair follicle initiates when progenitor cells grow into mesenchyme to form the cell lineages expressing Sox9 and Lrig1 that will ultimately develop into the sebaceous glands [44–47]. Experiments have traced the genetic lineage of Sox9 cells, and the knockdown of Sox9 results in an absence of sebaceous glands [48]. However, Sox9 and Lrig1 are responsible for the formation of individual cellular compartments. Sox9+ cells are confined to a portion just inferior to the sebaceous gland compartment, and these cells contribute to the formation of both the hair follicle and sebaceous glands. Conversely, cells expressing Lrig1 are most likely responsible for the generation of sebocytes directly because they are situated in the sebaceous gland compartment, and they also have cellular characteristics similar to mature sebocytes [49–51]. Wnt signaling is pivotal in the differentiation of hair follicle stem cells. TCF3/Lef1 are the downstream mediators of Wnt signaling that are responsible for activating stem cells and guiding their fate [52–54]. More importantly, it is the inhibition of Wnt signaling that guide hair follicle progenitor cells to differentiate into sebaceous glands [53, 55]. In addition, inhibition of Wnt signaling via Smad7 overexpression results in hyperplasia of the sebaceous glands, which suggests that crosstalk between the Wnt and transforming growth factor beta signaling cascades [56]. Other factors that contribute to the development, homeostasis, and function of the sebaceous glands are: hedgehog signaling, c-Myc signaling, Blimp1, androgens, peroxisome proliferator-activated receptor, cytokines, enzymes, and retinoids [16, 39, 41].

Clinical Presentation

The term "masquerade syndrome" has been used throughout literature to describe SC [10, 57–60]. SC, the most frequently missed differential diagnosis of a chalazion, may mimic a wide variety of clinical conditions, and misdiagnosis will delay lifesaving treatment [61]. Sebaceous carcinoma can arise from the Meibomian glands, Zeis glands, caruncle, eyebrows, and it may spread to the conjunctiva [10, 13, 19]. The majority of these tumors are found in the upper eyelid due to a greater concentration of sebaceous glands, followed by the lower eyelid, and lastly, diffuse involvement of both eyelids [5, 10]. SC most commonly presents itself as a painless, firm, sessile to round, subcutaneous nodule in the eyelid. Unlike most benign conditions, SC will cause a loss of hair or madarosis, which is a characteristic present in malignant eyelid tumors [13, 62]. Less common presentations include: a diffuse pseudoinflammatory pattern; a pedunculated lesion; a caruncular mass; eyebrow mass; lacrimal gland mass; or widespread involvement of the conjunctiva, eyelids, and orbital tissues [5, 13]. The use of diagnostic clues such as madarosis and familiarization with the clinical variants of SC, will increase the chances of making a proper clinical diagnosis.

Histopathology

As mentioned previously, it is essential to recognize SC as early as possible, and a proper histopathologic diagnosis facilitates this process. Histopathology of this disease carries its own problems including: small or limited specimen, incorrect diagnosis, and improper interpretation of disease margins [9, 13]. SC classically presents as an unencapsulated, lobular collection of sebaceous glands with lipid granules (Fig. 11.1b) that stain positive for oil red O and Sudan black [59, 63–65]. There are four recognized morphological patterns of SC: lobular, comedocarcinoma or trabecular, papillary, and mixed [14, 64, 66]. The cells that compose SC are fundamentally three cell types: basaloid, sebaceous, and intermediate forms [9]. The quantity of basaloid cells is inversely proportional to its level of differentiation, and SC can be classified as: well differentiated, moderately differentiated, or poorly differentiated with the least differentiated tumors having the highest mortality [14]. According to the literature, SC has the capacity to spread into neighboring epithelial tissues [58, 67–69]. Pagetoid invasion (Fig. 11.1c), which is a term that characterizes this phenomenon, can be used to describe any case of SC that involves the conjunctival epithelium [13]. Arriving at an accurate prognosis requires an assessment of the tumor size and its margins, both of which are affected by pagetoid invasion [70]. This complicates matters further because an inaccurate appraisal of the tumor size and margins can lead to an erroneous prognosis.

Immunohistochemistry (IHC) is an important tool that can help differentiate sebaceous carcinoma from other cutaneous malignancies. Nevertheless, the rates of sensitivity and specificity for IHC vary throughout the literature. Thus, it is fundamental that IHC is used in conjunction with a proper morphological appraisal of the tumor. There are several markers used in the diagnosis of SC, yet the markers with the greatest diagnostic utility are: epithelial membrane antigen (EMA) and Ber-Ep4 [66, 71]. EMA is a 75-kDa transmembrane glycoprotein that is expressed in glandular structures, like sebocytes, and squamous structures [72]. It is useful in distinguishing basal cell carcinoma (BCC) from SC because it is positive in most cases of SC and squamous cell carcinoma (SCC), but it is frequently negative in BCC [66, 71]. Ber-Ep4 is a monoclonal antibody against an epithelial cell adhesion molecule that is normally expressed on the basolateral membrane of most epithelium except in adult squamous epithelium, in the secretory portion of eccrine glands, and in the follicular germinative cells in normal skin and eyelids [73–75]. BCC is frequently positive for Ber-Ep4, while SC and SCC are negative in most of the cases [71, 75]. ZEB2, a transcription factor that is involved in epithelial to mesenchymal transition, is another epithelial marker that is regulated by miRNA-200c and miRNA-141, and is upregulated in this disease [76, 77]. The homeostasis and development of the sebaceous glands are partially regulated by androgen receptors (AR), and they are usually positive in SC while being negative in other cutaneous tumors [75, 78]. Lipid stains such as Oil red O and Sudan black IV had been used in the diagnosis of SC, but these stains require frozen tissue samples, they have poor sensitivity rates, and the stains also fade with time [79]. The newer lipid markers, perilipin, adipophilin, and TIP 47 group, offer greater rates of sensitivity and specificity [75, 79–81]. Some stains have been used to detect genetic mutations such as: p53, mismatch repair proteins, and ERRB2 [31, 66, 82–86]. In one study, the stem cell marker Lrig1, which is a known tumor suppressor that is involved in the development of the sebaceous glands, has been linked to a poor prognosis [87]. Of these markers, ERRB2 and AR are very interesting because they may provide routes for pharmacologic treatment by the usage of trastuzumab and thalidomide, respectively [88, 89]. In addition, the utilization of miRNAs, specifically 200-c and 141, as biomarkers may prove helpful in obtaining or confirming a diagnosis of SC [90, 91].

Pathophysiology

Currently, the molecular cues that guide the formation of the stem cell compartments of the sebaceous glands, and the cells that form their tumors are unknown [16]. Sebaceous adenoma, sebaceous hyperplasia, or sebaceous nevus, typically do not lead to the formation of SC. Although most SC arise de novo there are certain molecular alterations that are implicated in its pathogenesis [13]. MTS is an autosomal dominantly inherited defect in the DNA mismatch repair genes. These patients suffer from a multitude of cancers, mainly colorectal cancer, because of their inability to repair genetic errors that occur during DNA replication, especially in areas of multiple nucleotide repeats also known as microsatellites [92, 93]. Sebaceous adenomas and other benign lesions are more tightly linked to MTS, nonetheless, sebaceous carcinomas in MTS patients have decreased levels of mismatch repair proteins [93, 94]. Studies indicate that the loss of mismatch repair genes are only present in SCs from patients with MTS [86, 95]. These findings suggest that the pathogenesis of sebaceous carcinoma in MTS and non-MTS patients may occur via different mechanisms [96]. Signaling of transmembrane receptor tyrosine-protein kinase ERBB-2 promotes cellular proliferation and the inhibition of apoptosis by activating the mitogen-activated protein kinase (MAPK) and the signal transducer and activator of transcription (STAT) pathways [86]. The presence of ERBB-2 overexpression in SC, but not in other eyelid carcinomas or benign lesions was confirmed in two studies [86, 97]. TP53 is a tumor suppressor gene that encodes the protein p53. This protein prevents carcinogenesis by stopping the cell cycle so that DNA can undergo repair or by initiating apoptosis in heavily damaged cells [98]. Oncoproteins in HPV have been shown to promote the degradation of p53 [99]. Moreover, the degradation of p53 by HPV resulted in a more aggressive disease and poorer prognosis [17, 18]. Amino acid substitutions in the N-terminus of Lef1 impair the binding of Lef1 to β-catenin, and result in the inhibition of Wnt target genes. These substitutions have been found in a large percentage of sebaceous adenomas [100]. It was shown that these mutations lead to a decrease in the activity of p53, and the inhibition of Wnt leads to sebaceous differentiation of the tumor [101]. Another important regulator of stem cells and tumor cells is c-Myc. An overexpression of c-Myc in mice lead to an increase in the size and number of their sebaceous glands, and they developed sebaceous adenomas when exposed to carcinogens [102, 103].

Evaluation and Management

According to Shields, the first step in management is to establish a diagnosis and determine the extent of disease. A comprehensive clinical examination of the ocular and periocular region accomplishes this. SC most commonly metastasizes to regional lymph nodes [5, 7, 8, 13, 104–106]. For this reason, palpation of the regional lymph nodes should be performed to check for lymph node metastasis. After a clinical exam, ancillary studies or imaging tests may be useful, but they are not usually performed due to a low rate of distant metastasis [8, 13]. In the case of a high risk patient, a proposed follow-up plan includes a chest X-ray, liver function tests, ultrasound of the regional lymph nodes, and a bone scan every 3 months after diagnosis for 1 year, followed by every 6 months for another year, and then once a year [107]. Sentinel lymph node biopsies (SLNB) are recommended to evaluate for metastasis in periocular SC, but there are not enough studies to support its use [108–111]. Currently, there is an ongoing clinical trial at M.D. Anderson Cancer Center that is evaluating the usage of SLNB in sebaceous carcinoma of the eyelid. In MTS, the patient's first presenting sign is usually SC. Since these patients are at an increased risk for multiple types of cancer, it is important to screen them for other malignancies [21]. There is no consensus for staging, but the American Joint Committee on Cancer (AJCC) seventh edition Tumor, Node, Metastasis (TNM) staging guidelines can be applied to stage SC [8, 112]. Until now, only two studies have correlated the AJCC staging guidelines for SC. They have found that the T category was the strongest predictor of regional lymph node metastasis; and patients with a T category of T2b or worse, a size greater than 10 mm in its greatest dimension, and pagetoid spread all correlate positively with lymph node metastasis [70, 106]. At the time of diagnosis, this information will be used to determine the prognosis, treatment, and surveillance plans for the patient.

The goals of management should include: tumor control, preservation of the globe, maintenance of vision, patient comfort, and cosmetic appearance [13]. The most widely accepted form of treatment for SC of the periocular region is surgical excision [11, 14, 63]. Since the 1950s, wide local excision (WLE) has been the preferred method of surgical treatment for SC [63]. A primary incisional biopsy or Mohs microsurgery may also be employed if the tumor is advanced or if the surgeon wishes to check the tumor margins, respectively [113, 114]. Mohs microsurgery produces excellent results as a method of primary treatment, and may provide lower rates of reoccurrences than other methods [8, 115]. SC is known to reoccur after surgical incision at a rate of 4% [116]. Surgery is the treatment of choice for reoccurrences, and radiation may also be used as an adjuvant [116]. When performing surgical procedures, it is important to preserve the patient's cosmetic appearance. The techniques that have been used incorporate: rotational tarsal flaps, periosteal tissue, amniotic membrane, and buccal tissue in order to reconstruct the eyelid and conjunctiva [117–120]. Radiation may also be used in cases where patients may be unable to tolerate surgery or in cases of diffuse intraepithelial pagetoid spread [121]. Map biopsies are typically performed because SC can invade the conjunctiva diffusely. This technique consists of several biopsies around different regions of the conjunctiva to evaluate if there is epithelial involvement [68]. Subsequently, this information is then used to evaluate the extent of the disease and an appropriate plan for management. Although radiation is associated to worse outcomes than surgery, it has been used as adjuvant therapy to surgery or as an option in poor surgical candidates [122–124]. The use of chemotherapy in pagetoid invasion and metastasis has been documented in case reports, but more studies are required to determine the safety and efficacy of these treatments [125–127]. If orbital invasion or extensive conjunctival involvement is present, the globe cannot be spared and orbital exenteration is performed [13, 14].

Prognosis

This cancer is quite aggressive, and according to a review in 1995, rates of metastasis have been reported to be as high as 14% for locoregional spread to 25% for distant spread [64]. Metastasis rates have been on the decline as revealed by a more recent study where the overall rate of metastasis was shown to be at 2.4%, and 4.4% or 1.4%, for ocular and extraocular sites respectively [7]. In 1982, a study by Rao et al. reported that mortality for SC ranged from 14% to 83%. The rate of mortality would increase with the presence of characteristics that were associated with a poor prognosis such as: vascular, lymphatic, and orbital invasion; involvement of both upper and lower eyelids; poor differentiation; multicentric origin; duration of symptoms >6 months; tumor diameter exceeding 10 mm; a highly infiltrative pattern; and pagetoid invasion of the overlying epithelia of the eyelids [14]. However, a newer case series of 60 patients in 2004 concluded that the mortality from metastasis was 6%, which indicates a decrease in mortality by metastasis [5]. SC has an overall survival rate of 71.1% at 5 years and 45.9% at 10 years [6]. Additionally, data from 2004 to 2012 SEER studies show that the relative population and aged match survival rose from 91.9% to 92.72% for 5 years, and from 79.2% to 86.98% for 10 years [6, 12]. It is important to recognize that the decreased rate of metastasis, the decrease in mortality caused by metastasis, and the increase in survival rate is most likely due to an increase in detection, diagnosis, and treatment options [5, 12].

Molecular Pathways Involved in Carcinogenesis and Associated miRNAs

Since hair follicles in the skin are continuously self-renewing, multipotent stem cells (Fig. 11.2a) that regenerate the cells in the sebaceous glands are required for their normal function. Meanwhile, during malignant transformation, these stem cells are also responsible for the formation of sebaceous carcinoma and other types of eyelid cancers. Genetic profiling of these stem cell population indicates that the Wnt/β-catenin and TGF-β/BMP pathways are critical for both normal stem cell functions in normal sebaceous glands and tumorigenesis in sebaceous carcinoma. Here, we review the important components of these two signaling pathways, and how miRNAs regulate these pathways.

Wnt Signaling

Wnt signaling (Fig. 11.2b) is a highly conserved signaling pathway that controls embryonic development and adult homeostasis through its target gene transcription [128]. Canonical Wnt signaling is focused on regulating the amount of β-catenin. In the absence of Wnt ligands, β-catenin is constantly degraded by the Axin complex in the cytoplasm. Axin functions as a scaffold protein, forming a complex with adenomatous polyposis coli (APC), casein kinase 1 (CK1), and glycogen synthase kinase 3 (GSK3). The Axin complex induces phosphorylation of β-catenin on serine 45 by CK1α and then at threonine 41, serine 37 and serine 33 by GSK3. Phosphorylation of β-catenin creates a binding site for the E3 ubiquitin ligase β-Trcp. β-Trcp leads to the ubiquitination and proteasomal degradation of β-catenin. This constant degradation eliminates β-catenin and prevents β-catenin from translocating to the nucleus where it activates targets of gene transcription. However, when Wnt ligands interact with the seven-pass transmembrane Frizzle (Fz) receptor and the low-density lipoprotein receptor related protein 6 (LRP6) (or LRP5), they form the Wnt-Fz-LRP6 complex that recruits Dishevelled (Dvl) proteins. Dvl induces the phosphorylation of LRP6 that recruits Axin complex. This release β-catenin from the Axin complex, results in the stabilization of β-catenin. Subsequently, β-catenin translocates into the nucleus and forms a transcriptional complex with T cell factor/lymphoid enhancer factor (TCF/LEF) and initiates target gene expression [129–132].

Abnormal activation and dysregulation of Wnt signaling is closely linked to the development of many human cancers [133–135]. Overexpression of β-catenin has been demonstrated in SC [136]. Additionally, patients with Wnt and LRP expression develop metastasis more commonly than patients without Wnt or LRP expression [137]. Moreover, reduced membranous expression of b-catenin at the site of invasion has been showed to be correlated to the invasion and metastasis [138]. These studies suggest a critical role of Wnt signaling in SC.

TGF-β Signaling

The transforming growth factor beta (TGF-β) signaling pathway (Fig. 11.2c) is involved in many cellular processes in both the adult organism and the developing embryo. The canonical TGF-β signaling is initiated by ligand-receptor interaction. Once TGF-β ligands binds to TGF-β receptor 2 (TGFR2), TGFR2 then recruits and activates TGFR1. Subsequently, the active TGFR1 phosphorylates Smad2 and Smad3, which form a complex with Smad4. The Smad2/3/4 complex translocates to the nucleus and binds directly to promoters through Smad3 to induce target gene transcription [139]. Besides the canonical pathway, many signaling

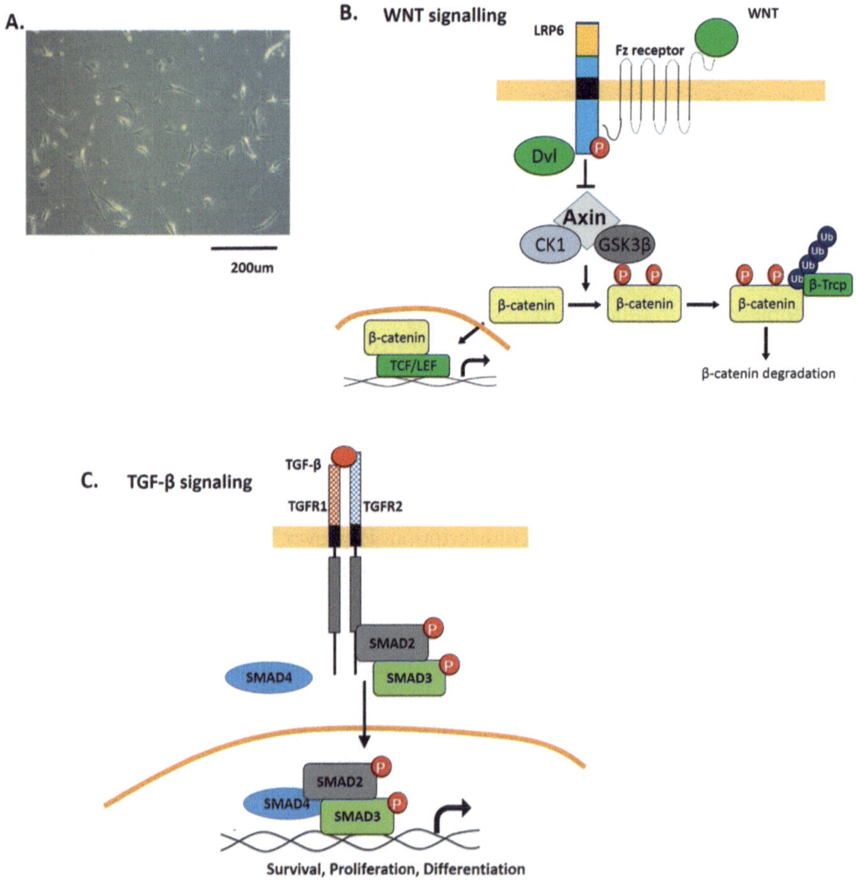

Fig. 11.2 *In vitro* study of molecular pathways in carcinogenesis of sebaceous carcinoma. (**a**) Sebaceous carcinoma cell line was established from a primary tumor. (**b**) The transforming growth factor beta (TGF-β) signaling pathway is involved in carcinogenesis and metastasis of sebaceous carcinoma. (**c**) Wnt/β-catenin signaling pathway is critical for stem cell functions in normal sebaceous glands and tumorigenesis in sebaceous carcinoma

pathways engage in crosstalk with TGF-β signaling. TGF-β activates the MAPKs in a Smad-independent manner, and MAPKs phosphorylate Smad proteins to modulate their transcriptional activity [140–143]. After Wnt signaling cascade activation, β-catenin forms complexes with Smad proteins to enhance gene transcription [144]. TGF-β has also been reported to activate the epidermal growth factor receptor (EGFR) in a ROS-dependent manner [145].

In normal human skin, TGF-β2 was detected predominantly in sebaceous glands [146]. Although the protein levels of TGF-β signaling components have not been reported in SC, a microRNA study released that miR-211, which targets TGF-β signaling, is downregulated in SC compared with sebaceous adenoma [147]. miR-211 always functions as a tumor suppressor by directly targeting TGFR2 and the downregulation of miR-211 in SC. These findings indicate a pro-tumorigenesis role of TGF-β signaling in SC [148–150].

MicroRNAs

MicroRNAs are small, non-coding RNA molecules that function during post-transcriptional regulation of gene expression by targeting mRNA. The abnormal expression of microRNAs has been observed in human cancer where they may function as malignant mechanisms or as biomarkers [151–156]. Studies reveal the abnormal expression of microRNAs in SC. miR-486-5p and miR-184 are overexpressed in SC [147]. miR-486-5p has been reported to be oncogenic, and it targets both NF-κB signaling and PTEN [157, 158]. Conversely, miR-184 suppresses c-Myc, and is suggested to be a tumor repressor through inhibition of AKT2 [159]. Meanwhile, miR-211 and miR-518d are shown to be downregulated in SC [147]. miR-211 is considered as a tumor repressor through targeting TGFR2, which results in an increased expression of c-Myc [148–150]. Dr. Bajaj's group also revealed that low expression levels of miRNA-200c and miRNA-141 were seen in most of SC patients and correlated with shorter disease-free survival [91]. Both miRNA-200c and miRNA-141 are associated with E-cadherin and ZEB1/2 expression [77, 160, 161].

Conclusion

Much of the literature that is available suggests that our understanding of the molecular mechanisms that drive the pathogenesis of SC is limited. A pubmed search with the keywords "sebaceous carcinoma" and "miRNA" only yields two relevant journal articles. In-vitro studies using stem cells with the goal of developing sebaceous carcinoma cells may provide us with the information that is needed to understand the pathogenesis of SC. With this data, it might be possible to understand how miRNAs regulate and interact with pathways like Wnt, TGF-β, among others, which are crucial to the development of the sebaceous glands. These findings may revolutionize the way that we diagnose and treat diseases. In the future, one might be able to perform a simple blood test searching for exosomal miRNAs. These miRNA can provide feedback regarding the availability of pharmacologic targets of different signaling pathways in the tumor cells, or even a patient's response to cancer therapy [162]. The first hurdle that must be addressed to reach this point, is to uncover the mechanisms behind sebaceous gland development, and pathogenesis of sebaceous carcinoma.

References

1. Yin VT, Merritt HA, Sniegowski M, Esmaeli B. Eyelid and ocular surface carcinoma: diagnosis and management. Clin Dermatol. 2015;33:159–69. https://doi.org/10.1016/j.clindermatol.2014.10.008.
2. Cook BE, Bartley GB. Epidemiologic characteristics and clinical course of patients with malignant eyelid tumors in an incidence cohort in Olmsted County, Minnesota. Ophthalmology. 1999;106:746–50. https://doi.org/10.1016/S0161-6420(99)90161-6.
3. Deprez M, Uffer S. Clinicopathological features of eyelid skin tumors. A retrospective study of 5504 cases and review of literature. Am J Dermatopathol. 2009;31:256. https://doi.org/10.1097/DAD.0b013e3181961861.
4. Lin HY, Cheng CY, Hsu WM, Kao WHL, Chou P. Incidence of eyelid cancers in taiwan. A 21-year review. Ophthalmology. 2006;113:2101–7. https://doi.org/10.1016/j.ophtha.2006.06.001.
5. Shields JA, Demirci H, Marr BP, Eagle RC, Shields CL. Sebaceous carcinoma of the eyelids: personal experience with 60 cases. Ophthalmology. 2004;111:2151–7. https://doi.org/10.1016/j.ophtha.2004.07.031.
6. Dasgupta T, Wilson LD, Yu JB. A retrospective review of 1349 cases of sebaceous carcinoma. Cancer. 2009;115:158–65. https://doi.org/10.1002/cncr.23952.
7. Tryggvason G, Bayon R, Pagedar NA. Epidemiology of sebaceous carcinoma of the head and neck: implications for lymph node management. Head Neck. 2012;34:1765–8. https://doi.org/10.1002/hed.22009.
8. Kyllo RL, Brady KL, Hurst EA. Sebaceous carcinoma. Dermatol Surg. 2015;41:1–15. https://doi.org/10.1097/DSS.0000000000000152.
9. Prieto-Granada C, Rodriguez-Waitkus P. Sebaceous carcinoma of the eyelid. Cancer Control. 2016;23:126.
10. Kass LG, Hornblass A. Sebaceous carcinoma of the ocular adnexa. Surv Ophthalmol. 1989;33:477.
11. Ni C, Searl SS, Kuo PK, Chu FR, Chong CS, Albert DM. Sebaceous cell carcinomas of the ocular adnexa. Int Ophthalmol Clin. 1982;22:23–61.
12. Tripathi R, Chen Z, Li L, Bordeaux JS. Incidence and survival of sebaceous carcinoma in the United States. J Am Acad Dermatol. 2016;75:1210–5. https://doi.org/10.1016/j.jaad.2016.07.046.
13. Shields JA, Demirci H, Marr BP, Eagle RC, Shields CL. Sebaceous carcinoma of the ocular region: a review. Surv Ophthalmol. 2005;50:103–22. https://doi.org/10.1016/j.survophthal.2004.12.008.
14. Rao NA, Hidayat LCAA, McLean LCIW, Zimmerman LE. Sebaceous carcinomas of the ocular adnexa: a clinicopathologic study of 104 cases, with five-year follow-up data. Hum Pathol. 1982;13:113–22. https://doi.org/10.1016/S0046-8177(82)80115-9.
15. Yeatts RP, Waller RR. Sebaceous carcinoma of the eyelid: pitfalls in diagnosis. Ophthal Plast Reconstr Surg. 1985;1:35–42.
16. Niemann C, Horsley V. Development and homeostasis of the sebaceous gland. Semin Cell Dev Biol. 2012;23:928–36. https://doi.org/10.1016/j.semcdb.2012.08.010.

17. Hayashi N, Ueno H, Furihata M, Ohtsuki Y. Search for accumulation of p53 protein and detection of human papillomavirus genomes in sebaceous gland carcinoma of the eyelid. Virchows Arch. 1994;424:503–9. https://doi.org/10.1007/BF00191436.
18. Gonzalez-Fernandez F, Kaltreider SA, Patnaik BD, Retief JD, Bao Y, Newman S, Stoler MH, Levine PA. Sebaceous carcinoma: tumor progression through mutational inactivation of p53. Ophthalmology. 1998;105:497–506. https://doi.org/10.1016/S0161-6420(98)93034-2.
19. Sung D, Kaltreider SA, Gonzalez-Fernandez F. Early onset sebaceous carcinoma. Diagn Pathol. 2011;6:81. https://doi.org/10.1186/1746-1596-6-81.
20. Yen MT, Tse DT. Sebaceous cell carcinoma of the eyelid and the human immunodeficiency virus. Ophthal Plast Reconstr Surg. 2000;16:206–10.
21. Imko-Walczuk B, Kryś A, Lizakowski S, Dębska-Ślizień A, Rutkowski B, Biernat W, Wojnarowska F. Sebaceous carcinoma in patients receiving long-term immunosuppresive treatment: case report and literature review. Transplant Proc. 2014;46:2903–7. https://doi.org/10.1016/j.transproceed.2014.09.064.
22. Kaminska ECN, Iyengar V, Tsoukas M, Shea CR. Borderline sebaceous neoplasm in a renal transplant patient without Muir-Torre syndrome. J Cutan Pathol. 2013;40:336–40. https://doi.org/10.1111/cup.12046.
23. Chang TW, Weaver AL, Brewer JD. Sebaceous carcinoma in the clinical setting of non-Hodgkin lymphoma: the Mayo Clinic experience. Int J Dermatol. 2013;52:1210–4. https://doi.org/10.1111/j.1365-4632.2012.05627.x.
24. Seo BF, Jung HW, Choi IK, Rhie JW. Sebaceous carcinoma of the suprapubic area in a liver transplant recipient. Ann Dermatol. 2014;26:395–8. https://doi.org/10.5021/ad.2014.26.3.395.
25. Pande R, Joshi M. Study of surface ocular tumours and its association with human immunodeficiency virus infection. J Clin Ophthalmol Res. 2017;5:35–8.
26. Rumelt S, Hogan NR, Rubin PA, Jakobiec FA. Four-eyelid sebaceous cell carcinoma following irradiation. Arch Ophthalmol. 1998;116:1670. https://doi.org/10.1001/archopht.116.12.1670.
27. Lemos LB, Santa Cruz DJ, Baba N. Sebaceous carcinoma of the eyelid following radiation therapy. Am J Surg Pathol. 1978;2:305–11.
28. Howrey RP, Lipham WJ, Schultz WH, Buckley EG, Dutton JJ, Klintworth GK, Rosoff PM. Sebaceous gland carcinoma: a subtle second malignancy following radiation therapy in patients with bilateral retinoblastoma. Cancer. 1998;83:767–71. https://doi.org/10.1002/(SICI)1097-0142(19980815)83:4<767::AID-CNCR20>3.0.CO;2-P.
29. Hood IC, Qizilbash AH, Salama SS, Young JE, Archibald SD. Sebaceous carcinoma of the face following irradiation. Am J Dermatopathol. 1986;8:505–8.
30. Rundle P, Shields JA, Shields CL, Eagle RC, Singh AD. Sebaceous gland carcinoma of the eyelid seventeen years after irradiation for bilateral retinoblastoma. Eye. 1999;13:109–10. https://doi.org/10.1038/eye.1999.20.
31. Hussain RM, Matthews JL, Dubovy SR, Thompson JM, Wang G. UV-independent p53 mutations in sebaceous carcinoma of the eyelid. Ophthal Plast Reconstr Surg. 2014;30:392–5. https://doi.org/10.1097/IOP.0000000000000121.
32. Draper GJ, Sanders BM, Kingston JE. Second primary neoplasms in patients with retinoblastoma. Br J Cancer. 1986;53:661–71.
33. Gauthier A-S, Campolmi N, Tumahai P, Kantelip B, Delbosc B. Sebaceous carcinoma of the eyelid and muir-torre syndrome. JAMA Ophthalmol. 2014;132:1025. https://doi.org/10.1001/jamaophthalmol.2014.1026.
34. Demirci H, Nelson CC, Shields CL, Eagle RC, Shields JA. Eyelid sebaceous carcinoma associated with muir-torre syndrome in two cases. Ophthal Plast Reconstr Surg. 2007;23:77–9. https://doi.org/10.1097/IOP.0b013e31802c7e45.
35. Izumi M, Tang X, Chiu C-S, Nagai T, Matsubayashi J, Iwaya K, Umemora S, Tsuboi R, Mukai K. Ten cases of sebaceous carcinoma arising in nevus sebaceus. J Dermatol. 2008;35:704–11. https://doi.org/10.1111/j.1346-8138.2008.00550.x.
36. Goldsmith LA, Katz SI, Gilchrest BA, Paller AS, Leffell DJ, Wolff K. Fitzpatrick's dermatology in general medicine, vol. 150. 8th ed. New York, NY: McGraw-Hill; 2012. p. 22. https://doi.org/10.1017/CBO9781107415324.004.

37. Butovich IA. The Meibomian puzzle: combining pieces together. Prog Retin Eye Res. 2009;28:483–98. https://doi.org/10.1016/j.preteyeres.2009.07.002.
38. Bron AJ, Tiffany JM, Gouveia SM, Yokoi N, Voon LW. Functional aspects of the tear film lipid layer. Exp Eye Res. 2004;78:347–60. https://doi.org/10.1016/j.exer.2003.09.019.
39. Smith KR, Thiboutot DM. Thematic review series: skin lipids. Sebaceous gland lipids: friend or foe? J Lipid Res. 2007;49:271–81. https://doi.org/10.1194/jlr.R700015-JLR200.
40. Porter AMW. Why do we have apocrine and sebaceous glands? R Soc Med. 2001;94:236.
41. Zouboulis CC. Acne and sebaceous gland function. Clin Dermatol. 2004;22:360–6. https://doi.org/10.1016/j.clindermatol.2004.03.004.
42. Lupi O. Ancient adaptations of human skin: why do we retain sebaceous and apocrine glands? Int J Dermatol. 2008;47:651–4. https://doi.org/10.1111/j.1365-4632.2008.03765.x.
43. Serri F, Montagna W, Huber WM. Studies of skin of fetus and the child the distribution of alkaline phosphatase in the skin of the fetus. Arch Dermatol. 1963;87:234. https://doi.org/10.1001/archderm.1963.01590140096016.
44. Millar SE. Molecular mechanisms regulating hair follicle development. J Invest Dermatol. 2002;118:216–25. https://doi.org/10.1046/j.0022-202x.2001.01670.x.
45. Schmidt-Ullrich R, Paus R. Molecular principles of hair follicle induction and morphogenesis. Bioessays. 2005;27:247–61. https://doi.org/10.1002/bies.20184.
46. Paus R, Müller-Röver S, Van Der Veen C, Maurer M, Eichmüller S, Ling G, Hofmann U, Foitzik K, Mecklenburg L, Handjiski B. A comprehensive guide for the recognition and classification of distinct stages of hair follicle morphogenesis. J Invest Dermatol. 1999;113:523–32. https://doi.org/10.1046/j.1523-1747.1999.00740.x.
47. Müller-Röver S, Handjiski B, van der Veen C, Eichmüller S, Foitzik K, McKay IA, Stenn KS, Paus R. A comprehensive guide for the accurate classification of murine hair follicles in distinct hair cycle stages. J Invest Dermatol. 2001;117:3–15. https://doi.org/10.1046/j.0022-202x.2001.01377.x.
48. Nowak JA, Polak L, Pasolli HA, Fuchs E. Hair follicle stem cells are specified and function in early skin morphogenesis. Cell Stem Cell. 2008;3:33–43. https://doi.org/10.1016/j.stem.2008.05.009.
49. Frances D, Niemann C. Stem cell dynamics in sebaceous gland morphogenesis in mouse skin. Dev Biol. 2012;363:138–46. https://doi.org/10.1016/j.ydbio.2011.12.028.
50. Jensen KB, Collins CA, Nascimento E, Tan DW, Frye M, Itami S, Watt FM. Lrig1 expression defines a distinct multipotent stem cell population in mammalian epidermis. Cell Stem Cell. 2009;4:427–39. https://doi.org/10.1016/j.stem.2009.04.014.
51. Petersson M, Brylka H, Kraus A, John S, Rappl G, Schettina P, Niemann C. TCF/Lef1 activity controls establishment of diverse stem and progenitor cell compartments in mouse epidermis. EMBO J. 2011;30:3004–18. https://doi.org/10.1038/emboj.2011.199.
52. Lowry WE, Blanpain C, Nowak JA, Guasch G, Lewis L, Fuchs E. Defining the impact of beta-catenin/Tcf transactivation on epithelial stem cells. Genes Dev. 2005;19:1596–611. https://doi.org/10.1101/gad.1324905.
53. Niemann C, Owens DM, Hülsken J, Birchmeier W, Watt FM. Expression of DeltaNLef1 in mouse epidermis results in differentiation of hair follicles into squamous epidermal cysts and formation of skin tumours. Development. 2002;129:95–109.
54. Haegebarth A, Clevers H. Wnt signaling, Lgr5, and stem cells in the intestine and skin. Am J Pathol. 2009;174:715–21. https://doi.org/10.2353/ajpath.2009.080758.
55. Merrill BJ, Gat U, DasGupta R, Fuchs E. Tcf3 and Lef1 regulate lineage differentiation of multipotent stem cells in skin. Genes Dev. 2001;15:1688–705. https://doi.org/10.1101/gad.891401.
56. Han G, Li AG, Liang Y-Y, Owens P, He W, Lu S, Yoshimatsu Y, Wang D, Ten Dijke P, Lin X, Wang X-J. Smad7-induced beta-catenin degradation alters epidermal appendage development. Dev Cell. 2006;11:301–12. https://doi.org/10.1016/j.devcel.2006.06.014.
57. Panda BB, Parija S, Pujahari S, Mallick J. Sebaceous gland carcinoma of eyelid - a tarnished masquerade. J Clin Diagn Res. 2016;10:ND03–5. https://doi.org/10.7860/JCDR/2016/16405.7567.

58. Shet T, Kelkar G, Juvekar S, Mistry R, Borges A. Masquerade syndrome: sebaceous carcinoma presenting as an unknown primary with pagetoid spread to the nasal cavity. J Laryngol Otol. 2004;118:307–9. https://doi.org/10.1258/002221504323012094.
59. Buitrago W, Joseph AK. Sebaceous carcinoma: the great masquerader. Dermatol Ther. 2008;21:459–66. https://doi.org/10.1111/j.1529-8019.2008.00247.x.
60. Brownstein S, Codere F, Jackson WB. Masquerade syndrome. Ophthalmology. 1980;87:259–62.
61. Ozdal PC, Codère F, Callejo S, Caissie AL, Burnier MN. Accuracy of the clinical diagnosis of chalazion. Eye (Lond). 2004;18:135–8. https://doi.org/10.1038/sj.eye.6700603.
62. Kumar A, Karthikeyan K. Madarosis: a marker of many maladies. Int J Trichol. 2012;4:3–18. https://doi.org/10.4103/0974-7753.96079.
63. Straatsma BR. Meibomian gland tumors. AMA Arch Ophthalmol. 1956;56:71–93.
64. Nelson BR, Hamlet KR, Gillard M, Railan D, Johnson TM. Sebaceous carcinoma. J Am Acad Dermatol. 1995;33:1–15. https://doi.org/10.1016/0190-9622(95)90001-2.
65. Wolfe JT, Yeatts RP, Wick MR, Campbell RJ, Waller RR. Sebaceous carcinoma of the eyelid. Errors in clinical and pathologic diagnosis. Am J Surg Pathol. 1984;8:597–606.
66. Jakobiec FA, Mendoza PR. Eyelid sebaceous carcinoma: clinicopathologic and multiparametric immunohistochemical analysis that includes adipophilin. Am J Ophthalmol. 2014;157:186–208.e2. https://doi.org/10.1016/j.ajo.2013.08.015.
67. Lee SC, Roth LM. Sebaceous carcinoma of the eyelid with Pagetoid involvement of the bulbar and palpebral conjunctiva. J Cutan Pathol. 1977;4:134–45.
68. Putterman AM. Conjunctival map biopsy to determine pagetoid spread. Am J Ophthalmol. 1986;102:87–90.
69. Russell WG, Page DL, Hough AJ, Rogers LW. Sebaceous carcinoma of meibomian gland origin. The diagnostic importance of pagetoid spread of neoplastic cells. Am J Clin Pathol. 1980;73:504–11.
70. Choi YJ, Jin HC, Lee MJ, Kim N, Choung H-K, Khwarg SI. Prognostic value of clinical and pathologic T stages defined by the American Joint Committee on Cancer for eyelid sebaceous carcinoma in Korea. Jpn J Ophthalmol. 2014;58:327–33. https://doi.org/10.1007/s10384-014-0321-9.
71. Plaza JA, Mackinnon A, Carrillo L, Prieto VG, Sangueza M, Suster S. Role of immunohistochemistry in the diagnosis of sebaceous carcinoma. Am J Dermatopathol. 2015;37:809–21. https://doi.org/10.1097/DAD.0000000000000255.
72. Ansai S, Arase S, Kawana S, Kimura T. Immunohistochemical findings of sebaceous carcinoma and sebaceoma: retrieval of cytokeratin expression by a panel of anti-cytokeratin monoclonal antibodies. J Dermatol. 2011;38:951–8. https://doi.org/10.1111/j.1346-8138.2011.01274.x.
73. Pai RK, West RB. MOC-31 exhibits superior reactivity compared with Ber-EP4 in invasive lobular and ductal carcinoma of the breast: a tissue microarray study. Appl Immunohistochem Mol Morphol. 2009;17:202–6.
74. Winter MJ, Nagtegaal ID, van Krieken JHJM, Litvinov SV. The epithelial cell adhesion molecule (Ep-CAM) as a morphoregulatory molecule is a tool in surgical pathology. Am J Pathol. 2003;163:2139–48. https://doi.org/10.1016/S0002-9440(10)63570-5.
75. Ansai S, Takeichi H, Arase S, Kawana S, Kimura T. Sebaceous carcinoma: an immunohistochemical reappraisal. Am J Dermatopathol. 2011;33:579–87. https://doi.org/10.1097/DAD.0b013e31820a2027.
76. Bhardwaj M, Sen S, Sharma A, Kashyap S, Chosdol K, Pushker N, Bajaj MS, Bakhshi S. ZEB2/SIP1 as novel prognostic indicator in eyelid sebaceous gland carcinoma. Hum Pathol. 2015;46:1437–42. https://doi.org/10.1016/j.humpath.2015.05.026.
77. Wang X, Chen X, Wang R, Xiao P, Xu Z, Chen L, Hang W, Ruan A, Yang H, Zhang X. microRNA-200c modulates the epithelial-to-mesenchymal transition in human renal cell carcinoma metastasis. Oncol Rep. 2013;30:643–50. https://doi.org/10.3892/or.2013.2530.
78. Asadi-Amoli F, Khoshnevis F, Haeri H, Jahanzad I, Pazira R, Shahsiah R. Comparative examination of androgen receptor reactivity for differential diagnosis of sebaceous carcinoma

from squamous cell and basal cell carcinoma. Am J Clin Pathol. 2010;134:22–6. https://doi.org/10.1309/AJCP89LYTPNVOBAP.
79. Muthusamy K, Halbert G, Roberts F. Immunohistochemical staining for adipophilin, perilipin and TIP47. J Clin Pathol. 2006;59:1166–70. https://doi.org/10.1136/jcp.2005.033381.
80. Boussahmain C, Mochel MC, Hoang MP. Perilipin and adipophilin expression in sebaceous carcinoma and mimics. Hum Pathol. 2013;44:1811–6. https://doi.org/10.1016/j.humpath.2013.02.003.
81. Ostler DA, Prieto VG, Reed JA, Deavers MT, Lazar AJ, Ivan D. Adipophilin expression in sebaceous tumors and other cutaneous lesions with clear cell histology: an immunohistochemical study of 117 cases. Mod Pathol. 2010;23:567–73. https://doi.org/10.1038/modpathol.2010.1.
82. Kiyosaki K, Nakada C, Hijiya N, Tsukamoto Y, Matsuura K, Nakatsuka K, Daa T, Yokoyama S, Imaizumi M, Moriyama M. Analysis of p53 mutations and the expression of p53 and p21WAF1/CIP1 protein in 15 cases of sebaceous carcinoma of the eyelid. Invest Ophthalmol Vis Sci. 2010;51:7–11. https://doi.org/10.1167/iovs.09-4127.
83. Mathiak M, Rütten A, Mangold E, Fischer H-P, Ruzicka T, Friedl W, Propping P, Kruse R. Loss of DNA mismatch repair proteins in skin tumors from patients with Muir-Torre syndrome and MSH2 or MLH1 germline mutations: establishment of immunohistochemical analysis as a screening test. Am J Surg Pathol. 2002;26:338–43.
84. Rajan Kd A, Burris C, Iliff N, Grant M, Eshleman JR, Eberhart CG. DNA mismatch repair defects and microsatellite instability status in periocular sebaceous carcinoma. Am J Ophthalmol. 2014;157:640–7.e1-2. https://doi.org/10.1016/j.ajo.2013.12.002.
85. Plocharczyk EF, Frankel WL, Hampel H, Peters SB. Mismatch repair protein deficiency is common in sebaceous neoplasms and suggests the importance of screening for Lynch syndrome. Am J Dermatopathol. 2013;35:191–5. https://doi.org/10.1097/DAD.0b013e31825f7efe.
86. Kwon MJ, Shin HS, Nam ES, Cho SJ, Lee MJ, Lee S, Park H-R. Comparison of HER2 gene amplification and KRAS alteration in eyelid sebaceous carcinomas with that in other eyelid tumors. Pathol Res Pract. 2015;211:349–55. https://doi.org/10.1016/j.prp.2014.10.006.
87. Pünchera J, Barnes L, Kaya G. Lrig1 expression in human sebaceous gland tumors. Dermatopathology. 2016;3:44–54. https://doi.org/10.1159/000446427.
88. Albanell J, Codony J, Rovira A, Mellado B, Gascón P. Mechanism of action of anti-HER2 monoclonal antibodies: scientific update on trastuzumab and 2C4. Adv Exp Med Biol. 2003;532:253–68.
89. Liu B, Su L, Geng J, Liu J, Zhao G. Developments in nonsteroidal antiandrogens targeting the androgen receptor. ChemMedChem. 2010;5:1651–61. https://doi.org/10.1002/cmdc.201000259.
90. Wang J, Chen J, Sen S. MicroRNA as biomarkers and diagnostics. J Cell Physiol. 2016;231:25–30. https://doi.org/10.1002/jcp.25056.
91. Bhardwaj M, Sen S, Chosdol K, Sharma A, Pushker N, Kashyap S, Bakhshi S, Bajaj MS. miRNA-200c and miRNA-141 as potential prognostic biomarkers and regulators of epithelial-mesenchymal transition in eyelid sebaceous gland carcinoma. Br J Ophthalmol. 2017;101:536–42. https://doi.org/10.1136/bjophthalmol-2016-309460.
92. Larrea AA, Lujan SA, Kunkel TA. SnapShot: DNA mismatch repair. Cell. 2010;141:730–730.e1. https://doi.org/10.1016/j.cell.2010.05.002.
93. Shalin SC, Lyle S, Calonje E, Lazar AJF. Sebaceous neoplasia and the Muir-Torre syndrome: important connections with clinical implications. Histopathology. 2010;56:133–47. https://doi.org/10.1111/j.1365-2559.2009.03454.x.
94. Entius MM, Keller JJ, Drillenburg P, Kuypers KC, Giardiello FM, Offerhaus GJ. Microsatellite instability and expression of hMLH-1 and hMSH-2 in sebaceous gland carcinomas as markers for Muir-Torre syndrome. Clin Cancer Res. 2000;6:1784–9.
95. Gaskin BJ, Fernando BS, Sullivan CA, Whitehead K, Sullivan TJ. The significance of DNA mismatch repair genes in the diagnosis and management of periocular sebaceous cell carcinoma and Muir-Torre syndrome. Br J Ophthalmol. 2011;95:1686–90. https://doi.org/10.1136/bjophthalmol-2011-300612.

96. Stagner AM, Jakobiec FA. Updates on the molecular pathology of selected ocular and ocular adnexal tumors: potential targets for future therapy. Semin Ophthalmol. 2016;31:188–96. https://doi.org/10.3109/08820538.2015.1115257.
97. Cho KJ, Khang SK, Koh JS, Chung JH, Lee SS. Sebaceous carcinoma of the eyelids: frequent expression of c-erbB-2 oncoprotein. J Korean Med Sci. 2000;15:545–50. https://doi.org/10.3346/jkms.2000.15.5.545.
98. Lane DP. p53, guardian of the genome. Nature. 1992;358:15–6. https://doi.org/10.1038/358015a0.
99. Scheffner M, Werness BA, Huibregtse JM, Levine AJ, Howley PM. The E6 oncoprotein encoded by human papillomavirus types 16 and 18 promotes the degradation of p53. Cell. 1990;63:1129–36.
100. Takeda H, Lyle S, Lazar AJF, Zouboulis CC, Smyth I, Watt FM. Human sebaceous tumors harbor inactivating mutations in LEF1. Nat Med. 2006;12:395–7. https://doi.org/10.1038/nm1386.
101. Niemann C, Owens DM, Schettina P, Watt FM. Dual role of inactivating Lef1 mutations in epidermis: tumor promotion and specification of tumor type. Cancer Res. 2007;67:2916–21. https://doi.org/10.1158/0008-5472.CAN-06-3427.
102. Waikel RL, Kawachi Y, Waikel PA, Wang XJ, Roop DR. Deregulated expression of c-Myc depletes epidermal stem cells. Nat Genet. 2001;28:165–8. https://doi.org/10.1038/88889.
103. Honeycutt KA, Waikel RL, Koster MI, Wang X-J, Roop DR. The effect of c-myc on stem cell fate influences skin tumor phenotype. Mol Carcinog. 2010;49:315–9. https://doi.org/10.1002/mc.20617.
104. Song A, Carter KD, Syed NA, Song J, Nerad JA. Sebaceous cell carcinoma of the ocular adnexa: clinical presentations, histopathology, and outcomes. Ophthal Plast Reconstr Surg. 2008;24:194–200. https://doi.org/10.1097/IOP.0b013e31816d925f.
105. Bailet JW, Zimmerman MC, Arnstein DP, Wollman JS, Mickel RA. Sebaceous carcinoma of the head and neck. Case report and literature review. Arch Otolaryngol Head Neck Surg. 1992;118:1245–9.
106. Esmaeli B, Nasser QJ, Cruz H, Fellman M, Warneke CL, Ivan D. American Joint Committee on Cancer T category for eyelid sebaceous carcinoma correlates with nodal metastasis and survival. Ophthalmology. 2012;119:1078–82. https://doi.org/10.1016/j.ophtha.2011.11.006.
107. Pfeiffer ML, Yin VT, Myers J, Esmaeli B. Regional nodal recurrence of sebaceous carcinoma of the caruncle 11 years after primary tumor resection. JAMA Ophthalmol. 2013;131:1091–2. https://doi.org/10.1001/jamaophthalmol.2013.321.
108. Wilson MW, Fleming JC, Fleming RM, Haik BG. Sentinel node biopsy for orbital and ocular adnexal tumors. Ophthal Plast Reconstr Surg. 2001;17:338–44. discussion 44–45.
109. Savar A, Oellers P, Myers J, Prieto VG, Torres-Cabala C, Frank SJ, Ivan D, Esmaeli B. Positive sentinel node in sebaceous carcinoma of the eyelid. Ophthal Plast Reconstr Surg. 2011;27:e4–6. https://doi.org/10.1097/IOP.0b013e3181ef7450.
110. Ho VH, Ross MI, Prieto VG, Khaleeq A, Kim S, Esmaeli B. Sentinel lymph node biopsy for sebaceous cell carcinoma and melanoma of the ocular adnexa. Arch Otolaryngol Head Neck Surg. 2007;133:820–6. https://doi.org/10.1001/archotol.133.8.820.
111. Nijhawan N, Ross MI, Diba R, Ahmadi MA, Esmaeli B. Experience with sentinel lymph node biopsy for eyelid and conjunctival malignancies at a cancer center. Ophthal Plast Reconstr Surg. 2004;20:291–5.
112. Chang AY, Miller CJ, Elenitsas R, Newman JG, Sobanko JF. Management considerations in extraocular sebaceous carcinoma. Dermatol Surg. 2016;42:S57–65. https://doi.org/10.1097/DSS.0000000000000575.
113. Mohs FE. Micrographic surgery for the microscopically controlled excision of eyelid cancers. Arch Ophthalmol. 1986;104:901–9.
114. Leibsohn J, Bullock J, Waller R. Full-thickness eyelid biopsy for presumed carcinoma in situ of the palpebral conjunctiva. Ophthalmic Surg. 1982;13:840–2.

115. Spencer JM, Nossa R, Tse DT, Sequeira M. Sebaceous carcinoma of the eyelid treated with Mohs micrographic surgery. J Am Acad Dermatol. 2001;44:1004–9. https://doi.org/10.1067/mjd.2001.113692.
116. Dowd MB, Kumar RJ, Sharma R, Murali R. Diagnosis and management of sebaceous carcinoma: an Australian experience. ANZ J Surg. 2008;78:158–63. https://doi.org/10.1111/j.1445-2197.2007.04393.x.
117. Kersten RC, Anderson RL, Tse DT, Weinstein GL. Tarsal rotational flap for upper eyelid reconstruction. Arch Ophthalmol. 1986;104:918–22.
118. Weinstein GS, Anderson RL, Tse DT, Kersten RC. The use of a periosteal strip for eyelid reconstruction. Arch Ophthalmol. 1985;103:357–9.
119. Tanaka TS, Demirci H. Cryopreserved ultra-thick human amniotic membrane for conjunctival surface reconstruction after excision of conjunctival tumors. Cornea. 2016;35:445–50. https://doi.org/10.1097/ICO.0000000000000737.
120. Ohshiro T, Tsuchida Y, Harashina T, Wakamatsu K, Yamaguchi S. Reconstruction of the levator function and conjunctiva using an inferiorly based orbital septal flap for subtotal full-thickness defects of the upper eyelid. Ann Plast Surg. 2006;56:336–9. https://doi.org/10.1097/01.sap.0000198729.87563.6f.
121. Yen MT, Tse DT, Wu X, Wolfson AH. Radiation therapy for local control of eyelid sebaceous cell carcinoma: report of two cases and review of the literature. Ophthal Plast Reconstr Surg. 2000;16:211–5.
122. Harvey JT, Anderson RL. The management of meibomian gland carcinoma. Ophthalmic Surg. 1982;13:56–61.
123. Nunery WR, Welsh MG, McCord CD. Recurrence of sebaceous carcinoma of the eyelid after radiation therapy. Am J Ophthalmol. 1983;96:10–5.
124. Hata M, Koike I, Omura M, Maegawa J, Ogino I, Inoue T. Noninvasive and curative radiation therapy for sebaceous carcinoma of the eyelid. Int J Radiat Oncol Biol Phys. 2012;82:605–11. https://doi.org/10.1016/j.ijrobp.2010.12.006.
125. Shields CL, Naseripour M, Shields JA, Eagle RC. Topical mitomycin-C for pagetoid invasion of the conjunctiva by eyelid sebaceous gland carcinoma. Ophthalmology. 2002;109:2129–33.
126. Jung YH, Woo IS, Kim MY, Han CW, Rha EY. Palliative 5-fluorouracil and cisplatin chemotherapy in recurrent metastatic sebaceous carcinoma: case report and literature review. Asia Pac J Clin Oncol. 2016;12:e189–93. https://doi.org/10.1111/ajco.12088.
127. Paschal BR, Bagley CS. Sebaceous gland carcinoma of the eyelid: complete response to sequential combination chemotherapy. N C Med J. 1985;46:473–4.
128. Logan CY, Nusse R. The Wnt signaling pathway in development and disease. Annu Rev Cell Dev Biol. 2004;20:781–810. https://doi.org/10.1146/annurev.cellbio.20.010403.113126.
129. Luu HH, Zhang R, Haydon RC, Rayburn E, Kang Q, Si W, Park JK, Wang H, Peng Y, Jiang W, He T-C. Wnt/beta-catenin signaling pathway as a novel cancer drug target. Curr Cancer Drug Targets. 2004;4:653–71.
130. Kretzschmar K, Clevers H. Wnt/β-catenin signaling in adult mammalian epithelial stem cells. Dev Biol. 2017;428:273. https://doi.org/10.1016/j.ydbio.2017.05.015.
131. Pohl S-G, Brook N, Agostino M, Arfuso F, Kumar AP, Dharmarajan A. Wnt signaling in triple-negative breast cancer. Oncogene. 2017;6:e310. https://doi.org/10.1038/oncsis.2017.14.
132. MacDonald BT, Tamai K, He X. Wnt/β-catenin signaling: components, mechanisms, and diseases. Dev Cell. 2009;17:9–26. https://doi.org/10.1016/j.devcel.2009.06.016.
133. Barker N, Clevers H. Mining the Wnt pathway for cancer therapeutics. Nat Rev Drug Discov. 2006;5:997–1014. https://doi.org/10.1038/nrd2154.
134. Zhou Y, Kipps TJ, Zhang S. Wnt5a signaling in normal and cancer stem cells. Stem Cells Int. 2017;2017:1–6. https://doi.org/10.1155/2017/5295286.
135. Pai SG, Carneiro BA, Mota JM, Costa R, Leite CA, Barroso-Sousa R, Kaplan JB, Chae YK, Giles FJ. Wnt/beta-catenin pathway: modulating anticancer immune response. J Hematol Oncol. 2017;10:101. https://doi.org/10.1186/s13045-017-0471-6.

136. Jayaraj P, Sen S, Kashyap S, Sharma A, Pushker N, Bajaj MS, Ghose S, Chosdol K. Does β-catenin have a role in pathogenesis of sebaceous cell carcinoma of the eyelid? Br J Ophthalmol. 2011;95:284–7. https://doi.org/10.1136/bjo.2009.177204.
137. Kim N, Kim JE, Choung H-K, Lee MJ, Khwarg SI. Expression of Shh and Wnt signaling pathway proteins in eyelid sebaceous gland carcinoma: clinicopathologic study. Invest Ophthalmol Vis Sci. 2013;54:370–7. https://doi.org/10.1167/iovs.12-10671.
138. Jayaraj P, Sen S, Sharma A, Chosdol K, Kashyap S, Rai A, Pushker N, Bajaj MS, Ghose S. Epigenetic inactivation of the E-cadherin gene in eyelid sebaceous gland carcinoma. Br J Dermatol. 2012;167:583–90. https://doi.org/10.1111/j.1365-2133.2012.10968.x.
139. Meng X, Nikolic-Paterson DJ, Lan HY. TGF-β: the master regulator of fibrosis. Nat Rev Nephrol. 2016;12:325–38. https://doi.org/10.1038/nrneph.2016.48.
140. Gruden G, Zonca S, Hayward A, Thomas S, Maestrini S, Gnudi L, Viberti GC. Mechanical stretch-induced fibronectin and transforming growth factor-beta1 production in human mesangial cells is p38 mitogen-activated protein kinase-dependent. Diabetes. 2000;49:655–61.
141. Naito T, Masaki T, Nikolic-Paterson DJ, Tanji C, Yorioka N, Kohno N. Angiotensin II induces thrombospondin-1 production in human mesangial cells via p38 MAPK and JNK: a mechanism for activation of latent TGF-beta1. Am J Physiol Renal Physiol. 2004;286:F278–87. https://doi.org/10.1152/ajprenal.00139.2003.
142. Wang L, Kwak JH, Il KS, He Y, Choi ME. Transforming growth factor-beta1 stimulates vascular endothelial growth factor 164 via mitogen-activated protein kinase kinase 3-p38alpha and p38delta mitogen-activated protein kinase-dependent pathway in murine mesangial cells. J Biol Chem. 2004;279:33213–9. https://doi.org/10.1074/jbc.M403758200.
143. Bakin AV, Rinehart C, Tomlinson AK, Arteaga CL. p38 mitogen-activated protein kinase is required for TGFbeta-mediated fibroblastic transdifferentiation and cell migration. J Cell Sci. 2002;115:3193–206.
144. Kim KK, Wei Y, Szekeres C, Kugler MC, Wolters PJ, Hill ML, Frank JA, Brumwell AN, Wheeler SE, Kreidberg JA, Chapman HA. Epithelial cell alpha3beta1 integrin links beta-catenin and Smad signaling to promote myofibroblast formation and pulmonary fibrosis. J Clin Invest. 2009;119:213–24. https://doi.org/10.1172/JCI36940.
145. Samarakoon R, Dobberfuhl AD, Cooley C, Overstreet JM, Patel S, Goldschmeding R, Meldrum KK, Higgins PJ. Induction of renal fibrotic genes by TGF-β1 requires EGFR activation, p53 and reactive oxygen species. Cell Signal. 2013;25:2198–209. https://doi.org/10.1016/j.cellsig.2013.07.007.
146. Furue M, Kato M, Nakamura K, Nashiro K, Kikuchi K, Okochi H, Miyazono K, Tamaki K. Dysregulated expression of transforming growth factor beta and its type-I and type-II receptors in basal-cell carcinoma. Int J Cancer. 1997;71:505–9.
147. Tetzlaff MT, Curry JL, Yin V, Pattanaprichakul P, Manonukul J, Uiprasertkul M, Manyam GC, Wani KM, Aldape K, Zhang L, Prieto VG, Esmaeli B. Distinct pathways in the pathogenesis of sebaceous carcinomas implicated by differentially expressed microRNAs. JAMA Ophthalmol. 2015;133:1109–16. https://doi.org/10.1001/jamaophthalmol.2015.2310.
148. Xu Y, Brenn T, Brown ERS, Doherty V, Melton DW. Differential expression of microRNAs during melanoma progression: miR-200c, miR-205 and miR-211 are downregulated in melanoma and act as tumour suppressors. Br J Cancer. 2012;106:553–61. https://doi.org/10.1038/bjc.2011.568.
149. Dai X, Rao C, Li H, Chen Y, Fan L, Geng H, Li S, Qu J, Hou L. Regulation of pigmentation by microRNAs: MITF-dependent microRNA-211 targets TGF-β receptor 2. Pigment Cell Melanoma Res. 2015;28:217–22. https://doi.org/10.1111/pcmr.12334.
150. Chu T-H, Yang C-C, Liu C-J, Lui M-T, Lin S-C, Chang K-W. miR-211 promotes the progression of head and neck carcinomas by targeting TGFβRII. Cancer Lett. 2013;337:115–24. https://doi.org/10.1016/j.canlet.2013.05.032.
151. Ambros V. The functions of animal microRNAs. Nature. 2004;431:350–5. https://doi.org/10.1038/nature02871.

152. Bartel DP. MicroRNAs: genomics, biogenesis, mechanism, and function. Cell. 2004;116:281–97.
153. Musilova K, Mraz M. MicroRNAs in B-cell lymphomas: how a complex biology gets more complex. Leukemia. 2015;29:1004–17. https://doi.org/10.1038/leu.2014.351.
154. Schoepp M, Ströse AJ, Haier J. Dysregulation of miRNA expression in cancer associated fibroblasts (CAFs) and its consequences on the tumor microenvironment. Cancers (Basel). 2017;9:54. https://doi.org/10.3390/cancers9060054.
155. Homami A, Ghazi F. MicroRNAs as biomarkers associated with bladder cancer. Med J Islam Repub Iran. 2016;30:475.
156. Inamura K. Major tumor suppressor and oncogenic non-coding RNAs: clinical relevance in lung cancer. Cell. 2017;6:12. https://doi.org/10.3390/cells6020012.
157. Song L, Lin C, Gong H, Wang C, Liu L, Wu J, Tao S, Hu B, Cheng S-Y, Li M, Li J. miR-486 sustains NF-κB activity by disrupting multiple NF-κB-negative feedback loops. Cell Res. 2013;23:274–89. https://doi.org/10.1038/cr.2012.174.
158. Alexander MS, Casar JC, Motohashi N, Vieira NM, Eisenberg I, Marshall JL, Gasperini MJ, Lek A, Myers JA, Estrella EA, Kang PB, Shapiro F, Rahimov F, Kawahara G, Widrick JJ, Kunkel LM. MicroRNA-486-dependent modulation of DOCK3/PTEN/AKT signaling pathways improves muscular dystrophy-associated symptoms. J Clin Invest. 2014;124:2651–67. https://doi.org/10.1172/JCI73579.
159. Foley NH, Bray IM, Tivnan A, Bryan K, Murphy DM, Buckley PG, Ryan J, O'Meara A, O'Sullivan M, Stallings RL. MicroRNA-184 inhibits neuroblastoma cell survival through targeting the serine/threonine kinase AKT2. Mol Cancer. 2010;9:83. https://doi.org/10.1186/1476-4598-9-83.
160. Zhou X, Wang Y, Shan B, Han J, Zhu H, Lv Y, Fan X, Sang M, Liu X-D, Liu W. The downregulation of miR-200c/141 promotes ZEB1/2 expression and gastric cancer progression. Med Oncol. 2015;32:428. https://doi.org/10.1007/s12032-014-0428-3.
161. Kim J, Ryu JK, Lee SH, Kim Y-T. MicroRNA 141 expression is a potential prognostic marker of biliary tract cancers. Gut Liver. 2016;10:836–41. https://doi.org/10.5009/gnl15460.
162. Schwarzenbach H. The clinical relevance of circulating, exosomal miRNAs as biomarkers for cancer. Expert Rev Mol Diagn. 2015;15:1159–69. https://doi.org/10.1586/14737159.2015.1069183.

Chapter 12
Current Knowledge of miRNAs as Biomarkers in Breast Cancer

Maria Luisa Gasparri, Zein Mersini Besharat, Aris Raad Besharat, Ilary Ruscito, Konstantinos Nirgianakis, Ammad Ahmad Farooqi, Andrea Papadia, Elisabetta Ferretti, Pierluigi Benedetti Panici, and Michael David Mueller

Abstract Breast cancer (BC) is the most common malignancy in women worldwide. Breast cancer related mortality has dropped significantly since the widespread adoption of mammographic screening. Approximately, 5% of the patients with BC carry germline mutations that are responsible for their condition. Women carrying a BRCA 1 or 2 mutation have a 57% and 49% lifetime risk of developing breast cancer, respectively. For BRCA1/2 patients reluctant to a prophylactic surgery, the only risk reducing strategy remains an increased imaging and clinical surveillance. Whereas a closer screening mammogram program is helpful in detecting BC at an earlier stage, no laboratory markers exist. New markers would be helpful in identifying BC, once the mutations are identified, and in monitoring the cancer behavior and response to treatment, once the cancer is diagnosed. MicroRNAs (miRNAs) are

M. L. Gasparri (✉)
Department of Gynecology and Obstetrics, "Sapienza" University of Rome, Rome, Italy

Surgical and Medical Department of Translational Medicine, "Sapienza" University of Rome, Rome, Italy

Department of Gynecology and Obstetrics, University Hospital of Bern, Bern, Switzerland
e-mail: marialuisa.gasparri@uniroma1.it

Z. M. Besharat
Department of Experimental Medicine, "Sapienza" University of Rome, Rome, Italy

A. R. Besharat · I. Ruscito · P. Benedetti Panici
Department of Gynecology and Obstetrics, "Sapienza" University of Rome, Rome, Italy

K. Nirgianakis · A. Papadia · M. D. Mueller
Department of Gynecology and Obstetrics, University Hospital of Bern, Bern, Switzerland

A. A. Farooqi
Institute of Biomedical and Genetic Engineering (IBGE), Islamabad, Pakistan

E. Ferretti
Department of Experimental Medicine, "Sapienza" University of Rome, Rome, Italy

Neuromed Institute, Pozzilli IS, Italy

S. Fayyaz, A. A. Farooqi (eds.), *Recent Trends in Cancer Biology: Spotlight on Signaling Cascades and microRNAs*, https://doi.org/10.1007/978-3-319-71553-7_12

key regulatory molecules operating in a post-transcriptional level by regulating gene expression. Aberrant miRNA expression was documented in several pathological conditions, including solid tumors, suggesting their involvement in tumorigenesis. MiRNAs can be detected in human fluids and could be used as biomarkers. A different pattern expression of miRNA in biological fluids of BC patients as compared to healthy control is currently under consideration in many clinical trials. The biologic mechanism of miRNAs, a rationale of its use as biomarker in cancer, and a literature review of the most significant results achieved about miRNAS in BC will be reported and discussed in this chapter.

Keywords Breast cancer · BRCA 1/2 · Biomarkers · miRNA

Breast Cancer

Breast cancer is first in incidence and mortality for women worldwide, with nearly 1.7 million new cases and 521,000 deaths reported in 2012 [1]. Accounting for over 250,000 cases per year and being responsible for more than 40,000 deaths in the United States of America [2]. Five-year survival rates are estimated to be 80% in England and 90% in the United States [3]. Breast cancer mortality rates have been decreasing since the 1970s [4]. This decrease in mortality may be attributed to improved screening and adjuvant therapy in breast cancer.

Indeed, screening with mammographies is currently performed in many countries. However, patients have to consider both its benefits and harms. In a recent systematic review it has been shown that screening mammography reduces the odds of dying of breast cancer by approximately 20% [5]. On the other hand, potential harms of screening programs include false-positive findings that caused associated anxiety as well as overdiagnosis. The latter means cancer types, which are biologically insignificant and would never become clinically evident in the patient's lifetime. These concerns have raised special attention recently [6].

Considering adjuvant therapy in breast cancer is currently based on the identified molecular subtypes. Gene expression profiling analyses have determined the following four molecular subgroups according to hormone receptor expression (HR±) and human epidermal growth factor receptor 2 (HER2): Luminal A (HR+/HER2−), Luminal B (HR+/HER2+), HER2-enriched (HR−/HER2+), and triple-negative (HR−/HER2−). The molecular subtype predicts the prognosis and if a benefit from specific types of therapy should be expected.

Since further improvement in diagnosis and management of breast cancer is pursued, identifying new diagnostic, predictive, or prognostic biomarkers is of great importance. To this direction, several research groups nowadays focus on the potentials of a class of small non-coding RNAs, miRNAs.

Definition and Importance of MicroRNAs

Genome sequencing revealed that only 1.1% of the human genome accounts for protein-coding DNA [7]. Transfer RNA and ribosomal RNA are only a small part of the non-protein-coding DNA genes. The notion that the rest of the DNA was practically "junk" since it did not code for proteins or known types of RNAs was based on the false assumption that it did not contain important information.

MicroRNAs, ~22 nucleotides in length, were previously included in the "junk DNA" since their function was still unknown. The discovery that the microRNA lin-4 downregulated the expression of the gene lin-14 in *Caenorhabditis elegans* changed the field of biology and the way scientists approached these small non-coding RNAs [8]. The downregulation of the expression of lin-14 resulted in modifications in the developmental timing of *Caenorhabditis elegans* [9]. Unfortunately this discovery was at first thought to be unique in the *Caenorhabditis elegans*, since there is no homolog of lin-4 in humans.

The publication of animal microRNA functions 12 years ago [10] brought to the forefront of epigenetic research the post-transcriptional gene regulation. Mature microRNAs bind through a 5′ "seed region" to the 3′ untranslated region (3′ UTR) of target mRNAs driving them to translation repression and/or mRNA degradation [11]. It has been estimated that since microRNAs only need as few as seven nucleotides of complementarity to bind to their target, it has been indicated that microRNA-binding sites can be found in more than 60% of human protein coding genes. Another estimation reported that microRNAs may regulate up to 30% of protein-coding genes in humans [12].

These facts highlight the importance and necessity for microRNA profiling in order to acquire a more complete understanding of microRNAs' identity and role in different biological contests including cancer.

MicroRNA Biogenesis

The mechanism of microRNA biogenesis takes place in different parts of the cell and can be described by four steps, as follows.

In the first step as a result of transcription of microRNA genes by RNA Polymerase II we have the primary microRNAs (pri-miRNAs) in the nucleus. In general, microRNAs can share the regulatory elements and primary transcript of their precursor microRNA host genes and they can reside in their introns. For the remaining microRNA genes transcribed from their own promoters, only a few primary transcripts have been fully identified. In the second step, pri-miRNAs are cleaved in the nucleus by a complex including Drosha and the DGCR8 microprocessor unit, resulting in a precursor microRNA (pre-miRNA), that is ~70 nucleotides in length. In the third step, the precursor microRNAs are then transported to the cytoplasm by exportin-5 (XPO5). There, they are cleaved by a RNase Dicer, AGO2 (Argonaute 2), and TRBP (trans-activation-responsive RNA-binding protein) complex, resulting in the mature microRNAs. In the fourth step, the mature microRNAs are then incorporated to the RNA-induced silencing complex (RISC) and bind to their target mRNAs regulating their expression [13, 14].

The function of the mature microRNAs depends on the complementarity between the microRNA and the mRNA target. If the complementarity is perfect it leads to the degradation of the mRNA target. Otherwise, in case the complementarity is imperfect, it leads to the translational repression of the mRNA target. The fact that microRNAs need as few as seven nucleotides to bind to their mRNA target gives microRNAs the ability to target different mRNAs. This ability to target multiple mRNAs means that the same microRNA can target different levels of the same pathway or different pathways at the same time.

MicroRNA in Cancer: Briefly State of the Art

In the last few years, microRNAs have been the object of many studies in our continued search for explanations regarding diseases, including cancer initiation and progression.

Indeed, 2588 mature microRNAs have been annotated in humans according to miRBase; Release 21 (http://www.mirbase.org/index.shtml).

Through microRNA profiling scientists have reported differences among microRNAs expression in different cell types and even in the same cell in different phases of the cell cycle. Technological advances have permitted us to identify not only single microRNAs that control signalling pathways but also multiple microRNAs that control different/same signalling pathways.

Recent studies have identified microRNA-mRNA networks in different types of cancers [15], including breast cancer. Depending on whether a microRNA targets an oncogene or tumor suppressor gene it can be referred to as "anti-oncomir/tumor suppressor-mir" or "oncomir" respectively and can exert an anti-tumoral or oncogenic function.

Researchers have started identifying microRNA signatures that are associated with the type of cancer, the diagnosis, the response to therapy and survival among others in order to facilitate when possible the prevention of a disease or develop more appropriate treatment regimens. Depending on the function of the microRNA, a mimic or an anti-microRNA can be used to repress/degrade or allow the translation of the mRNA target.

Finally, a significant aspect of microRNAs is the fact that microRNAs are stable in bodily fluids (e.g. plasma, serum, urine, saliva), thus identifying them as circulating biomarker candidates [16–20].

miRNAs in Breast Cancer

Aberrant miRNA tissue expression was firstly correlated with cancer progression in 2002 [21]. A few years later, in 2006, Zhang and colleagues, after an analysis using high-resolution array-based comparative genomic hybridization of 227 human ovarian cancer, breast cancer (BC) and melanoma specimens, showed that miRNA irregularities were present in 72.8% of breast cancers tissues [22]. Later on, several studies evidenced that dysregulation of miRNAs is involved in breast cancer progression by inhibiting apoptosis-related pathways, stimulating cell growth and proliferation and promoting tumor invasion and metastasis [23–27].

Up to date, growing evidences report the aberrant activation of some tissue specific-miRNAs in breast cancer, e.g. miR-21, miR-155, miR-10b, miR-221/222, miR-23b, miR-27b and miR-181 [24, 28–38]. Furthermore, since miRNAs can circulate in body fluids, aberrant miRNAs have been also identified in breast cancer patients' blood, such as let-7a, miR-195, miR-155, miR-145, miR-10b and miR-451 [39–43]. In addition, increasing evidences suggest the value of miRNAs in association with drug-resistance, currently making them particularly attractive as novel biomarkers for breast cancer detection in patients at early stages, prognosis prediction, and patients' monitoring for their response to treatments [44].

Breast Cancer Tissue Specific miRNAs

miR-21

Among the identified dysregulated miRNAs in breast cancer tissue, miR-21 was found to be linked with enhanced tumour cell proliferation and apoptosis inhibition via multiple gene targets, such as tropomyosin alpha-1 chain (TPM1) [28], programmed cell death 4 (PDCD4) [29], p53, phosphatase and tensin homolog (PTEN) and Smad7 [30]. Thanks to tissue microarray expression analysis, upregulation of miR-21 was also found to be correlated to breast cancer advanced clinical stage at diagnosis, patients' poor prognosis and lymph node metastasis [45]. Furthermore, the expression level of miR-21 seems to be closely related to sex hormone receptor expression status and histopathologic breast cancer tumor grades [46].

miR-155

Another miRNA, which has been abundantly proven in breast cancer tissues is miR-155. It has been widely studied during the last decade, indeed, and it has been demonstrated to have a large number of targets among tumor suppressor genes such as FOXO3a, which regulates proapoptotic proteins and cell functions involved in chemosensitivity [31], SOCS1(suppressor of cytokine signaling-1), which regulates STAT3 signalling and is involved in breast cancer cell transformation [32]. miR-155 has also been reported to inhibit apoptosis of the cell line MCF-7 and to promote cell proliferation by downregulation of the tumor protein 53-induced nuclear protein-1 (TP53INP1). TP53INP1 is one of p53 target genes, in detail it is a p53-inducible cell stress response gene [33]. Overexpression of TP53INP1 enhances the p53-mediated apoptosis and induces cell cycle arrest [47]. From a clinical point of view, miR-155 overexpression, as well as miR-21, was also found to be correlated with higher tumor grade, sex hormone receptor expression, advanced tumor stage and lymph node metastasis [46, 48].

miR-10b

While miR-10b was found to be downregulated in most cases of low-stage breast cancers compared to normal breast tissue [24], this miRNA was found overexpressed in approximately 50% of metastatic breast cancers [26]. In vivo miR-10b expression3 attributed invasive properties to non-invasive breast cancer cells; indeed, while control tumors exhibited poor vascularization and no invasive properties, miR-10b overexpressing tumors showed an invasive behavior and were highly vascularized.

miR-221/222, miR-23b and miR-27b

Upregulation of miR-221/222 cluster has been associated with cancer invasion and resistance to tamoxifen, as reported in two different papers published in 2011 [34, 35], whereas miR-23b and miR-27b have been more recently correlated with breast cancer tumorigenesis, progression and metastasis [36, 37]. In particular, Jin et al. firstly reported in 2009 that HER2/neu (Erb-B2), EGF, and TNF-a promote miR-23b/27b expression through the AKT/NF-kB signaling cascade.

miR-181

miR-181 was found to promote breast cancer metastasis and was shown to have high levels of expression in triple negative breast cancer patients with a poor survival [38]. Upregulation of miR-181a by TGF-β and correlation with the breast cancer cell metastatic potential has been reported, while inactivation of miR-181a attenuated TGF-β–mediated epithelial–mesenchymal transition (EMT), invasion, and migration.

Breast Cancer-Specific Circulating miRNAs

Since it has been demonstrated that miRNAs can also circulate in body fluids, such as blood serum, thus being preserved from endogenous RNAse activity, increasing breast cancer-specific circulating miRNAs are being identified and correlated to patients' clinico-pathological and survival characteristics, together with pilot studies on urine samples [49]. It has been reported that miRNAs can circulate in blood serum unbound or packaged into microparticles (such as microvescicles and apoptotic bodies), or connected to Argonaute 2 (Ago2) protein [50].

The importance of studying breast cancer patients' circulating miRNAs profiles resides in their potential clinical usefulness to be early biomarkers of tumor diagnosis and early predictors of treatments response [39].

In 2013, Mar-Anguilar and colleagues carried out miRNA profiling on breast cancer patients' serum, followed by construction of ROC curves to identify the sensitivity and specificity of these biomarkers. Seven miRNAs (miR-10b/21/125b/145/155/191/382) were found to have distinct expression profile between breast cancer patients' serum and controls' serum. In particular, ROC curves highlighted that three serum miRNAs (miR-145, miR-155 and miR-382) could discriminate breast carcinoma from healthy women, thus suggesting their values as new non-invasive biomarkers for breast cancer diagnosis [40].

miR-155

Up to date, several studies have shown a significant higher level of miR-155 detected in breast cancer patients' serum compared to healthy cases [41, 51]. In particular, Sun and coworker, showed that compared to 55 healthy women, the mean fold increase of serum miR-155 in 103 breast cancer patients was 2.94. Interestingly, the same study group observed that miR-155 serum levels decreased after surgery and chemotherapy, thus suggesting the value of this biomarker as a predictor for response to therapy [52].

miR-10b

miR-10b is another miRNA, which was found to be significantly increased in breast cancer patients' serum [40, 53]. In particular, the higher levels of circulating miR-10B were found to be correlated to lymph nodal spread of breast cancer [54].

miR-195, let-7a and Other Promising Circulating miRNAs in Breast Cancer

It has been recently observed that breast cancer patients also showed increased serum levels of miR-195, miR-16, miR-21, mir-451 and miR-145 compared to healthy women. More recently, researchers are trying to define a combination test of circulating miRNAs, which would be able to better discriminate early disease and susceptibility to treatments respect to a single circulating miRNA profile. It has been already identified that both levels of circulating miR-195 and let-7a are reduced in cancer patients after surgery [42]. Furthermore, Enders and coworkers reported increased levels of miR-16, miR-21 and miR-451 in plasma of breast cancer patients'. They also hypothesized that combining plasma levels of miR-451 and miR-145 may potentially act as a specific screening test for breast cancer [43].

On the contrary, some miRNAs seem to show decreased levels in serum of breast cancer patients in comparison to healthy cases. Among them, circulating miR-411, miR-299, miR-215 levels were markedly lower in untreated metastatic breast cancer patients [55].

Conclusion

MiRNAs stability and detectability in peripheral blood and their associated dysregulations in cancer patients, reported to be associated to disease progression and treatment response, make them a new attractive group of biomarkers in regards not only to preclinical diagnosis of cancer but also for treatments success prediction, including the breast cancer setting. Up to now, the results of preliminary studies do not provide definitive conclusions about the usefulness of the clinical application of miRNAs. This can be partially attributed to the not homogeneous samples group employed in different studies (frozen versus paraffin-embedded tissues, serum versus plasma samples), and partially to the different experimental methods applied for the detection of tumor-associated miRNAs (quantitative PCR versus different in situ hybridization techniques or miRNA arrays), together with treatments and tumor heterogeneity analyzed.

Additional studies on larger and more homogeneous cancer populations, together with a better understanding of the best technology able to detect and quantify miRNAs expression in tumor tissue and peripheral blood, are required to clarify the future clinical potential of this growing group of new biomarkers.

References

1. Torre LA, Bray F, Siegel RL, Ferlay J, Lortet-Tieulent J, Jemal A. Global cancer statistics, 2012. CA Cancer J Clin. 2015;65(2):87–108.
2. Siegel RL, Miller KD, Jemal A. Cancer statistics, 2016. CA Cancer J Clin. 2016;66(1):7–30.
3. Cancer survival in England: patients diagnosed between 2010 and 2014 and followed up to 2015. https://www.ons.gov.uk/releases/cancersurvivalforadultsinengland2015.
4. Kohler BA, Sherman RL, Howlader N, Jemal A, Ryerson AB, Henry KA, Boscoe FP, Cronin KA, Lake A, Noone A-M. Annual report to the nation on the status of cancer, 1975–2011, featuring incidence of breast cancer subtypes by race/ethnicity, poverty, and state. J Natl Cancer Instit. 2015;107(6):djv048.
5. Myers ER, Moorman P, Gierisch JM, Havrilesky LJ, Grimm LJ, Ghate S, Davidson B, Mongtomery RC, Crowley MJ, McCrory DC. Benefits and harms of breast cancer screening: a systematic review. JAMA. 2015;314(15):1615–34.
6. Biller-Andorno N, Jüni P. Abolishing mammography screening programs? A view from the Swiss Medical Board. N Engl J Med. 2014;370(21):1965–7.
7. Ponting CP, Hardison RC. What fraction of the human genome is functional? Genome Res. 2011;21(11):1769–76.
8. Lee RC, Feinbaum RL, Ambros V. The C. elegans heterochronic gene lin-4 encodes small RNAs with antisense complementarity to lin-14. Cell. 1993;75(5):843–54.
9. Olsen PH, Ambros V. The lin-4 regulatory RNA controls developmental timing in Caenorhabditis elegans by blocking LIN-14 protein synthesis after the initiation of translation. Dev Biol. 1999;216(2):671–80.
10. Ambros V. The functions of animal microRNAs. Nature. 2004;431(7006):350–5.
11. Bartel DP. MicroRNAs: target recognition and regulatory functions. Cell. 2009;136(2):215–33.
12. MacFarlane L-A, R Murphy P. MicroRNA: biogenesis, function and role in cancer. Curr Genomics. 2010;11(7):537–61.
13. Stefani G, Slack FJ. Small non-coding RNAs in animal development. Nat Rev Mol Cell Biol. 2008;9(3):219–30.
14. He L, Hannon GJ. MicroRNAs: small RNAs with a big role in gene regulation. Nat Rev Genet. 2004;5(7):522–31.
15. Bracken CP, Scott HS, Goodall GJ. A network-biology perspective of microRNA function and dysfunction in cancer. Nat Rev Genet. 2016;17(12):719–32.
16. Chen X, Ba Y, Ma L, Cai X, Yin Y, Wang K, Guo J, Zhang Y, Chen J, Guo X. Characterization of microRNAs in serum: a novel class of biomarkers for diagnosis of cancer and other diseases. Cell Res. 2008;18(10):997–1006.
17. Huang Z, Huang D, Ni S, Peng Z, Sheng W, Du X. Plasma microRNAs are promising novel biomarkers for early detection of colorectal cancer. Int J Cancer. 2010;127(1):118–26.
18. Park NJ, Zhou H, Elashoff D, Henson BS, Kastratovic DA, Abemayor E, Wong DT. Salivary microRNA: discovery, characterization, and clinical utility for oral cancer detection. Clin Cancer Res. 2009;15(17):5473–7.
19. Mitchell PS, Parkin RK, Kroh EM, Fritz BR, Wyman SK, Pogosova-Agadjanyan EL, Peterson A, Noteboom J, O'Briant KC, Allen A. Circulating microRNAs as stable blood-based markers for cancer detection. Proc Natl Acad Sci. 2008;105(30):10513–8.
20. Erbes T, Hirschfeld M, Rücker G, Jaeger M, Boas J, Iborra S, Mayer S, Gitsch G, Stickeler E. Feasibility of urinary microRNA detection in breast cancer patients and its potential as an innovative non-invasive biomarker. BMC Cancer. 2015;15(1):1.
21. Calin GA, Dumitru CD, Shimizu M, Bichi R, Zupo S, Noch E, Aldler H, Rattan S, Keating M, Rai K. Frequent deletions and down-regulation of micro-RNA genes miR15 and miR16 at 13q14 in chronic lymphocytic leukemia. Proc Natl Acad Sci. 2002;99(24):15524–9.
22. Zhang L, Huang J, Yang N, Greshock J, Megraw MS, Giannakakis A, Liang S, Naylor TL, Barchetti A, Ward MR. microRNAs exhibit high frequency genomic alterations in human cancer. Proc Natl Acad Sci. 2006;103(24):9136–41.

23. Findlay VJ, Turner DP, Moussa O, Watson DK. MicroRNA-mediated inhibition of prostate-derived Ets factor messenger RNA translation affects prostate-derived Ets factor regulatory networks in human breast cancer. Cancer Res. 2008;68(20):8499–506.
24. Iorio MV, Ferracin M, Liu C-G, Veronese A, Spizzo R, Sabbioni S, Magri E, Pedriali M, Fabbri M, Campiglio M. MicroRNA gene expression deregulation in human breast cancer. Cancer Res. 2005;65(16):7065–70.
25. Tavazoie SF, Alarcón C, Oskarsson T, Padua D, Wang Q, Bos PD, Gerald WL, Massagué J. Endogenous human microRNAs that suppress breast cancer metastasis. Nature. 2008;451(7175):147–52.
26. Ma L, Teruya-Feldstein J, Weinberg RA. Tumour invasion and metastasis initiated by microRNA-10b in breast cancer. Nature. 2007;449(7163):682–8.
27. Baranwal S, Alahari SK. miRNA control of tumor cell invasion and metastasis. Int J Cancer. 2010;126(6):1283–90.
28. Zhu S, Si M-L, Wu H, Mo Y-Y. MicroRNA-21 targets the tumor suppressor gene tropomyosin 1 (TPM1). J Biol Chem. 2007;282(19):14328–36.
29. Frankel LB, Christoffersen NR, Jacobsen A, Lindow M, Krogh A, Lund AH. Programmed cell death 4 (PDCD4) is an important functional target of the microRNA miR-21 in breast cancer cells. J Biol Chem. 2008;283(2):1026–33.
30. Gong C, Nie Y, Qu S, Liao J-Y, Cui X, Yao H, Zeng Y, Su F, Song E, Liu Q. miR-21 induces myofibroblast differentiation and promotes the malignant progression of breast phyllodes tumors. Cancer Res. 2014;74(16):4341–52.
31. Kong W, He L, Coppola M, Guo J, Esposito NN, Coppola D, Cheng JQ. MicroRNA-155 regulates cell survival, growth, and chemosensitivity by targeting FOXO3a in breast cancer. J Biol Chem. 2010;285(23):17869–79.
32. Jiang S, Zhang H-W, Lu M-H, He X-H, Li Y, Gu H, Liu M-F, Wang E-D. MicroRNA-155 functions as an OncomiR in breast cancer by targeting the suppressor of cytokine signaling 1 gene. Cancer Res. 2010;70(8):3119–27.
33. Zhang C-M, Zhao J, Deng H-Y. MiR-155 promotes proliferation of human breast cancer MCF-7 cells through targeting tumor protein 53-induced nuclear protein 1. J Biomed Sci. 2013;20(1):79.
34. Stinson S, Lackner MR, Adai AT, Yu N, Kim H-J, O'Brien C, Spoerke J, Jhunjhunwala S, Boyd Z, Januario T. miR-221/222 targeting of trichorhinophalangeal 1 (TRPS1) promotes epithelial-to-mesenchymal transition in breast cancer. Sci Signal. 2011;4(186):pt5.
35. Lu Y, Roy S, Nuovo G, Ramaswamy B, Miller T, Shapiro C, Jacob ST, Majumder S. Anti-microRNA-222 (anti-miR-222) and-181B suppress growth of tamoxifen-resistant xenografts in mouse by targeting TIMP3 protein and modulating mitogenic signal. J Biol Chem. 2011;286(49):42292–302.
36. Jin L, Wessely O, Marcusson EG, Ivan C, Calin GA, Alahari SK. Prooncogenic factors miR-23b and miR-27b are regulated by Her2/Neu, EGF, and TNF-α in breast cancer. Cancer Res. 2013;73(9):2884–96.
37. Wang Y, Rathinam R, Walch A, Alahari SK. ST14 (suppression of tumorigenicity 14) gene is a target for miR-27b, and the inhibitory effect of ST14 on cell growth is independent of miR-27b regulation. J Biol Chem. 2009;284(34):23094–106.
38. Taylor MA, Sossey-Alaoui K, Thompson CL, Danielpour D, Schiemann WP. TGF-β upregulates miR-181a expression to promote breast cancer metastasis. J Clin Invest. 2013;123(1):150–63.
39. O'Bryan S, Dong S, Mathis JM, Alahari SK. The roles of oncogenic miRNAs and their therapeutic importance in breast cancer. Eur J Cancer. 2017;72:1–11.
40. Mar-Aguilar F, Mendoza-Ramírez JA, Malagón-Santiago I, Espino-Silva PK, Santuario-Facio SK, Ruiz-Flores P, Rodríguez-Padilla C, Reséndez-Pérez D. Serum circulating microRNA profiling for identification of potential breast cancer biomarkers. Dis Markers. 2013;34(3):163–9.
41. Zhu W, Qin W, Atasoy U, Sauter ER. Circulating microRNAs in breast cancer and healthy subjects. BMC Res Notes. 2009;2(1):89.

42. Heneghan HM, Miller N, Kelly R, Newell J, Kerin MJ. Systemic miRNA-195 differentiates breast cancer from other malignancies and is a potential biomarker for detecting noninvasive and early stage disease. Oncologist. 2010;15(7):673–82.
43. Ng EK, Li R, Shin VY, Jin HC, Leung CP, Ma ES, Pang R, Chua D, Chu K-M, Law W. Circulating microRNAs as specific biomarkers for breast cancer detection. PLoS One. 2013;8(1):e53141.
44. Bahrami A, Aledavoud SA, Anvari K, Hassanian SM, Maftouh M, Yaghobzade A, Salarzaee O, ShahidSales S, Avan A. The prognostic and therapeutic application of microRNAs in breast cancer: tissue and circulating microRNAs. J Cell Physiol. 2017. https://doi.org/10.1002/jcp.25813.
45. Yan L-X, Huang X-F, Shao Q, Huang M-Y, Deng L, Wu Q-L, Zeng Y-X, Shao J-Y. MicroRNA miR-21 overexpression in human breast cancer is associated with advanced clinical stage, lymph node metastasis and patient poor prognosis. RNA. 2008;14(11):2348–60.
46. Wang F, Zheng Z, Guo J, Ding X. Correlation and quantitation of microRNA aberrant expression in tissues and sera from patients with breast tumor. Gynecol Oncol. 2010;119(3):586–93.
47. Tomasini R, Samir AA, Carrier A, Isnardon D, Cecchinelli B, Soddu S, Malissen B, Dagorn J-C, Iovanna JL, Dusetti NJ. TP53INP1s and homeodomain-interacting protein kinase-2 (HIPK2) are partners in regulating p53 activity. J Biol Chem. 2003;278(39):37722–9.
48. Chen J, Wang BC, Tang JH. Clinical significance of MicoRNA-155 expression in human breast cancer. J Surg Oncol. 2012;106(3):260–6.
49. Gasparri ML, Casorelli A, Bardhi E, Besharat AR, Savone D, Ruscito I, Farooqi AA, Papadia A, Mueller MD, Ferretti E, et al. Beyond circulating microRNA biomarkers: Urinary microRNAs in ovarian and breast cancer. Tumor Biol. 2017;39(5):1010428317695525.
50. Arroyo JD, Chevillet JR, Kroh EM, Ruf IK, Pritchard CC, Gibson DF, Mitchell PS, Bennett CF, Pogosova-Agadjanyan EL, Stirewalt DL. Argonaute2 complexes carry a population of circulating microRNAs independent of vesicles in human plasma. Proc Natl Acad Sci. 2011;108(12):5003–8.
51. Roth C, Rack B, Müller V, Janni W, Pantel K, Schwarzenbach H. Circulating microRNAs as blood-based markers for patients with primary and metastatic breast cancer. Breast Cancer Res. 2010;12(6):R90.
52. Sun Y, Wang M, Lin G, Sun S, Li X, Qi J, Li J. Serum microRNA-155 as a potential biomarker to track disease in breast cancer. PLoS One. 2012;7(10):e47003.
53. Hagrass HA, Sharaf S, Pasha HF, Tantawy EA, Mohamed RH, Kassem R. Circulating microRNAs-a new horizon in molecular diagnosis of breast cancer. Genes Cancer. 2015;6(5–6):281.
54. Chen W, Cai F, Zhang B, Barekati Z, Zhong XY. The level of circulating miRNA-10b and miRNA-373 in detecting lymph node metastasis of breast cancer: potential biomarkers. Tumor Biol. 2013;34(1):455–62.
55. van Schooneveld E, Wouters MC, Van der Auwera I, Peeters DJ, Wildiers H, Van Dam PA, Vergote I, Vermeulen PB, Dirix LY, Van Laere SJ. Expression profiling of cancerous and normal breast tissues identifies microRNAs that are differentially expressed in serum from patients with (metastatic) breast cancer and healthy volunteers. Breast Cancer Res. 2012;14(1):R34.

Chapter 13
Signaling Landscape of AML: The Story So Far

Chiara Martinelli

Abstract Acute myeloid leukemia (AML) is the most frequent leukemia in adults and presents a very high incidence all over the world. The most important aberrations involve mutations and large chromosomal translocations in the genes responsible for hematopoiesis, resulting in an abnormal signal transduction activation that boosts survival and proliferation of progenitor cells and a typical accumulation of poorly differentiated myeloid cells. Acute myeloid leukemia is an extremely complex malignancy with considerable genetic, epigenetic, and phenotypic heterogeneity. Most AML genomes present very few mutations, which are responsible for the aberrant phenotypes observed. The possibility to characterize the mutations present in single cells and the studies on hematopoiesis performed both *in vitro* and *in vivo*, make AML an ideal model for investigating the underlying mechanisms of tumorigenesis. In the last years, the signaling proteins identified as specifically mutated in AML have raised huge consideration as attractive therapeutic targets and many efforts are currently ongoing in order to design *ad hoc* strategies to improve prognosis and therapy. Recent advances in the conventional treatments, together with innovative therapies, show significant promises for curing AML patients.

Keywords Acute myeloid leukemia (AML) · Mutations · Signaling pathways · Therapy

List of Abbreviations

ALCL	Anaplastic large cell lymphomas
AML	Acute myeloid leukemia
APL	Acute promyelocitic leukemia
ATRA	All-trans retinoic acid

C. Martinelli (✉)
Department of Biosciences, University of Milan, Milan, Italy
e-mail: chiara.martinelli@protonmail.com

S. Fayyaz, A. A. Farooqi (eds.), *Recent Trends in Cancer Biology: Spotlight on Signaling Cascades and microRNAs*, https://doi.org/10.1007/978-3-319-71553-7_13

CART	Chimeric antigen receptor T-cell
CBF	Core binding factor
CK	Complex karyotype
CN	Cytogenetically normal
CR	Complete remission
DNMT	DNA methyltransferases
FAB	French American British
GO	Gemtuzumab ozogamicin
HDAC	Histone deacetylase
HMA	Hypomethylating agent
ITD	Internal tandem duplication
MEF	Mouse embryonic fibroblast
NCCN	National comprehensive cancer network clinical guidelines in oncology
OS	Overall survival
PTD	Partial tandem duplication
RTK	Receptor tyrosine kinase
scFv	Single chain variable fragment
TKD	Tyrosine kinase domain
WHO	World health organization

Introduction

Acute myeloid leukemia (AML) is a cancer of the myeloid cell line characterized by a peculiar proliferation and accumulation of anomalous white blood cells in the bone marrow, inducing typical symptoms such as a drop in the number of red blood cells, platelets, and normal white blood cells [1–3]. Although many putative risk factors involved in leukemogenesis have been identified, the exact origin of the disease has yet to be clearly determined. In fact, AML can develop both after exposure to genotoxic agents and following an antecedent hematological disorder (e.g., marrow failure syndrome). Usually, AML progression is very rapid and death occurs in a few weeks or months if not promptly treated. Different AML subgroups have been determined in order to facilitate diagnosis and prognosis of patients and each of them presents different ranges of survival and relapse. Approximately 35–40% of patients younger than 60 years of age may obtain long term survival with current forms of therapy. At an early stage, AML patients undergo chemotherapy treatment in order to obtain a remission and later they can receive additional chemotherapy or a hematopoietic stem cell transplant. Recent advances in understanding the molecular features of AML has led to the creation of tools that help clinicians in selecting the best therapy and predicting which will be the potential outcome.

AML is a very complex disease with significant genetic, epigenetic, and phenotypic heterogeneity and the most important prognostic factor remains the cytogenetic status: different cytogenetic abnormalities are associated with different outcomes. This complexity gives rise to the great challenge of targeting altogether these pathways, without inducing toxicity to normal cells. The ideal successful treatment would effectively eradicate the leukemic stem cell and its subclones so that residual disease cannot induce recurrence. Although the promise of a personalized treatment approach has yet to be achieved, clinicians need as much information as possible in order to choose the best approach for each patient. Importantly, thanks to the recent advances in the discovery of new molecular pathways involved in AML pathogenesis, a large number of novel drugs that can target specific subsets of patients are being developed for clinical applications.

In this chapter, we will give the current definition and classification of AML. Then, we will describe its pathophysiology highlighting the most crucial prognostic factors that have to be considered when diagnosing the disease. The main core will be the description of some of the essential molecules involved in the abnormal regulation of intracellular signaling cascades and responsible for the onset of AML. Finally, we will report recent findings concerning the most innovative approaches actually ongoing in AML therapy.

Definition and Classification of AML

Despite of its etiology, AML is characterized by the abnormal proliferation of white blood cells interfering with normal hematopoiesis. Clinically, AML is described as the presence of undifferentiated myeloid cells mainly localized in the bone marrow and peripheral blood. Common symptoms found in the majority of cases are leukocytosis and anemia with thrombocytopenia. An excessive abrupt weight loss, together with severe fatigue, bring the patient to death within months after infection or bleeding if not promptly treated. AML is the most frequent type of leukemia in adults and the second one in children [1–3]. It can develop as a *de novo* disease or it can appear as a consequence of chemo/radiotherapy treatments [4]. Currently, AML is classified according to many diagnostic techniques: morphologic evaluation of bone marrow specimens and blood smears, flow cytometry, cytogenetics and, recently, screening for specific mutations. The French-American-British (FAB) classification system represented the first definition of the eight different subtypes of AML based on their morphological and cytochemical features. In 2001, the "Classification of Tumours of Haematopoietic and Lymphoid Tissues" was introduced by the World Health Organization (WHO) integrating the most recent diagnostic and management advances in AML: it was revised in 2008 [5]. In 2016, the "WHO classification of AML and related neoplasms" was released, that defined six main groups: (1) AML with recurrent genetic abnormalities; (2) AML with myelodysplasia-related features; (3) therapy-related AML; (4) AML not otherwise specified; (5) myeloid sarcoma; and (6) myeloid proliferation related to Down syndrome [6]. In the 2008 revision, mutations in nucleophosmin 1 (NPM1) and

CCAAT/enhancer binding protein α (CEBPA) were recognized as distinct entities [5], while BCR-ABL1 and mutated RUNX1 were added in the 2016 version based on their clinicopathological features [6].

Pathophysiology of AML

It is well known that specific chromosomal translocations are responsible for creating chimeric proteins that affect the normal maturation steps of myeloid precursor cells. However, more than 97% of AML patients present specific mutations [7], even without any large chromosomal aberration [8]. Nowadays, the "two-hit" model of leukemogenesis allows to rationally classify the mutations responsible for the onset of AML. Accordingly, different types of mutations are defined: (1) class I mutations, activating pro-proliferative pathways and (2) class II mutations, prejudicing normal differentiation and contributing to myeloid maturation arrest. The two mutations need to combine in order to develop the pathology [9, 10]. Mutations in Fms Related Tyrosine Kinase 3 (FLT3) (internal tandem duplications, ITD, and tyrosine kinase domain, TKD), K/NRAS, TP53 and c-KIT belong to class I and are found in approximately 28, 12, 8 and 4% of cases, respectively [8, 11]. Abnormal cellular proliferation and survival have been demonstrated to be enhanced also by mutations in Signal Transducer and Activator of Transcription 3 (STAT3) [12, 13]. Approximately 50% of AML cases with worse prognosis display enhanced tyrosine phosphorylation of STAT5 [14] or mutations in FLT3 and JAK2 [15, 16]. NPM1 and CEBPA (approximately 27% and 6% of cases, respectively) belong to class II and usually present a better prognosis [8]. Mutations in DNMT3A, TET2, and IDH-1 and IDH-2, responsible for epigenetic regulation and found in more than 40% of AML patients, have been recently included in class III [7, 8]. According to the current model, the interaction between the different kinds of proteins mutated in the three classes is responsible for the pathogenesis of AML. Frequently, c-KIT mutation has been associated with t(8;21) or inv(16), while NPM1 frequently occurs together with the mutation in FLT3-ITD, DNMT3A and IDH-1 or IDH-2 [7]. AML is commonly diagnosed if 20% or more blasts are detected in the bone marrow or peripheral blood [17]. Further positivity to myeloid markers like CD13, CD33, CD117 and of an extramedullary tissue infiltrate, or chromosomal aberrations in the appropriate clinical setting can convey to AML diagnosis [5].

Prognostic Factors in AML

Evaluation of the prognosis is essential for the right management of AML patients. Their classification has to consider both the risk of developing treatment resistance and the risk of relapse and mortality related to the treatment, and then establish the best strategy in terms of kind and intensity of therapy to be administered. As an example, the diagnostic evaluation of a poor performance status together with increased age is often linked to poor outcomes in terms of complete remission (CR) and overall survival (OS) [18, 19]. However, many model analyses consider also other factors equally important to predict the risks related to mortality after treatment, especially in older patients [20] and a poor prognosis is often linked to a prior hematological malignancy [21]. The strongest prognostic factor remains cytogenetic changes, that conventionally help clinicians in classifying patients in different categories, according to their prognostic risk, as favorable, intermediate or adverse groups. While some chromosomal aberrations are currently correlated to a favorable prognosis, other conditions such as a complex karyotype or monosomy 5 or 7 remain associated with a poor outcome [2, 17, 22]. Thanks to the improved knowledge of AML mutations it has been possible to refine the classification of the risk: for example, patients that present t(8;21) together with c-KIT mutation display a higher risk of relapse [2, 7, 23]. Evaluation of gene mutations has been demonstrated to be particularly relevant also in the prognosis of patients with a cytogenetically normal (CN)-AML (approximately 50% of *de novo* AML cases). Patients that display a CN-AML with mutations in CEBPA or NPM1 in absence of FLT3-ITD show a prognostic risk similar to that of favorable risk patients [17, 24], while older patients that present a CN-AML with FLT3-ITD belong to the adverse prognostic risk group [2, 25] and display a worse prognosis [7, 26]. Mutations in tumor suppressor genes like TP53 are very rare in AML patients [7, 8] and give a very poor prognosis [10] especially when associated with unfavorable cytogenetics and complex karyotype [27]. Genes involved in epigenetic regulation of transcription have been found mutated in AML patients and have been included as important parameters for a correct prognosis of patients. For example, mutated DNMT3A and PTDs in MLL1 correlate with worse prognosis in CN-AML [7, 10, 27, 28], while the involvement of IDH-1/IDH-2 mutations has not been clearly elucidated yet. An important role in prognosis is given to the informations collected after treatment initiation: patiens who achieve a complete remission after induction therapy display a better outcome as compared with treatment resistant patients [29, 30].

Recently, the definition of minimal residual disease has been improved by technological advancements that have led to the introduction of RT-PCR and flow cytometry techniques performed in patients with CR. An important correlation linked to an increased incidence of relapse has also been established between high levels of RUNX1-RUNX1T1 transcripts after induction therapy and t(8;21) [31, 32].

Cytogenetic and Molecular Risk

Nowadays, AML patients are classified in different prognostic risk categories based on their cytogenetic and molecular features. Patients with recurrent cytogenetic aberrations (approximately 45%) have been divided into favorable, intermediate, and poor-risk categories [33–35]. Recently, new methodologies and technical advancements in AML diagnosis have allowed the identification of new molecular biomarkers, like gene mutations, gene and noncoding RNA expression signatures, and DNA methylation profiles associated with distinct cytogenetic groups [36, 37]. NPM1 and CEBPA mutations and FLT3-ITD have been included in a classification of AML patients into four groups: (1) favorable (with NPM1 and CEBPA mutations without FLT3-ITD), (2) intermediate I (CN patients with FLT3-ITD), (3) intermediate II (patients with intermediate cytogenetic risk), and (4) adverse (with a poor cytogenetic risk) [17, 22, 38]. Finally, BAALC (Brain and Acute Leukemia, Cytoplasmic) is implicated in acute myeloid leukemia and its overexpression has been demonstrated to be associated to poor prognosis.

From these data, it is evident that many components of the intracellular signaling networks involved in controlling proliferation and survival of hematopoietic cells are responsible for the onset of AML. In the next sections, we will describe some of the most common mutations in key signaling molecules involved in AML pathogenesis and particularly relevant from a therapeutic point of view: (1) receptor tyrosine kinases, (2) transcription factors, (3) NPM1 mutations and (4) epigenetic modifiers.

Receptor Tyrosine Kinases and Their Signaling Intermediates

These proteins usually work as transducers activated by extracellular growth factors [38]. FLT3 is a member of the class III receptor tyrosine kinases (RTK) involved in hematopoiesis. It has been frequently found mutated in AML and its abnormal activation leads to uncontrolled blast proliferation [39–42] and to constitutive activation of downstream signaling cascades [43] (Fig. 13.1a). Mutations like FLT3-ITD have been found in 25–35% of AML patients, they give proliferative advantage to the hematopoietic progenitor cells and patients often display increased risk of relapse and mortality [22]. Mutations present in the FLT3-TKD [44] have been found in 7% of patients and have a favorable prognostic outcome [45]. FLT3 mutations are responsible for the activation of the AKT and ERK1/2 mediated pathways [46, 47]. It has been demonstrated that also the wild-type FLT3 receptor *per se* can be overexpressed in AML patients, giving rise to its constitutive activation [48, 49] and impaired negative regulation of its downstream effectors [50, 51]. Patients affected by core binding factor (CBF)-AML with t(8;21), inv(16), t(16;16) may also present FLT3 mutations, eventhough the prognostic impact remains to be fully understood [52, 53].

Mutations in c-KIT, a gene encoding another member of the class III RTK have been described in 20% of patients that often present relapse, especially when combined with t(8;21) [22, 54] (Fig. 13.1a). c-KIT is involved in hematopoiesis and acts on PI3K and consequently on RAS/RAF, MEK/ERK downstream pathways [55]. Mutations lead to its permanent activation and 80–90% of AML cases display constitutively phosphorylated wild-type c-KIT.

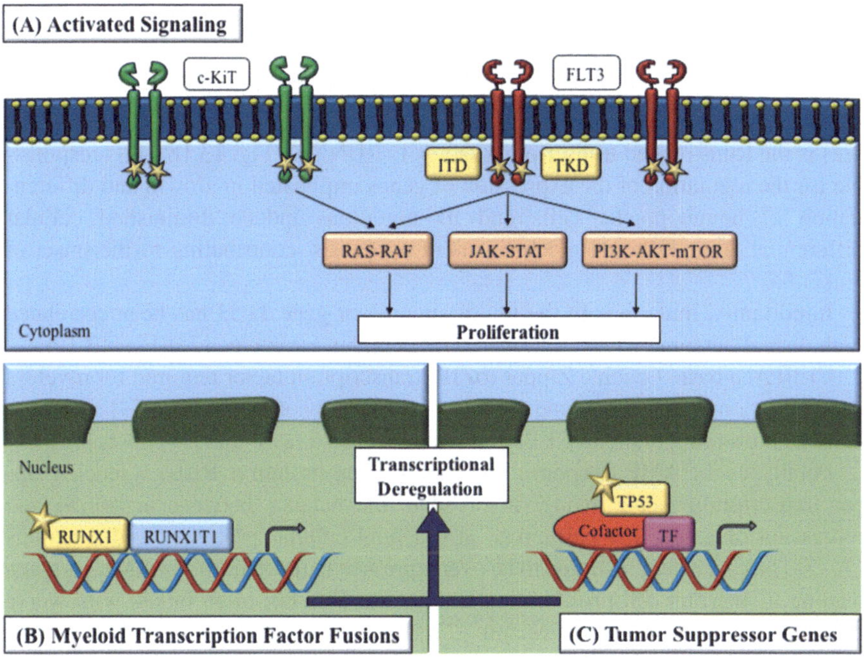

Fig. 13.1 Tyrosine kinase receptors, transcription factors and tumor suppressors are mutated in acute myeloid leukemia. (**a**) Mutations in FLT3 and c-KIT cause their constitutive activation and permanent signaling to downstream pathways RAS-RAF, JAK-STAT, and PI3K-AKT-mTOR contributing to uncontrolled proliferation. (**b**) Myeloid transcription factor fusions (RUNX1-RUNX1T1) cause transcriptional deregulation. (**c**) Mutations in tumor suppressor genes (TP53) cause transcriptional deregulation. Yellow stars indicate mutations

Aberrant activity of RTKs impacts on the regulation of many signaling intermediates. For example, they can influence RAS proteins, that normally act as intermediates, modulating the signal to downstream cellular targets, as previously mentioned [56]. In AML, abnormal activation of upstream receptor tyrosine kinases and also somatic mutations of RAS are able to permanently activate it and all its downstream effectors [57–60]. It has been shown that stable activation of the MEK1/2-ERK1/2 pathways is responsible for regulating the levels of the pro-apoptotic molecule Bim [61].

The constitutive activation of the PI3K/AKT/mammalian target of rapamycin (mTOR) [62–64] together with mitogen-activated protein kinase (MAPK) pathway play a pivotal role in AML cell proliferation and survival [65–68]. In particular, activation of mTOR complex 1 and 2 (mTORC1-2) is responsible for the control of oncogenic protein synthesis, cell cycle progression and proliferation, aberrant regulation of tumor cells metabolism [65, 66].

AML is rarely associated with mutational activation of cytoplasmic tyrosine kinases like JAK/STAT [69]. Interestingly, STAT3 activation has been found in *de novo* AML [15] and the t(8;21) fusion has been shown to stimulate the JAK/STAT pathway [70] together with FLT3-ITD mutations [71].

Transcription Factors

Many transcription factors contribute to leukemogenesis. A transcription factor that has been found mutated in approximately 13% of AML patients with poor outcome [72] is the Runt-related transcription factor 1 (RUNX1) (Fig. 13.1b). It is responsible for the regulation of the expression of genes implicated in growth and differentiation of hematopoietic cells and its mutations induce diminished cellular differentiation and altered mechanisms of apoptosis, contributing to the onset of leukemia.

Importantly, mutations in the tumor suppressor gene TP53 has been correlated with dismal outcome when combined with complex karyotype [73](Fig. 13.1c).

CEBPA, a basic Leucine Zipper (bZIP) transcription factor required for myeloid differentiation, has been found mutated in 10–15% of CN-AML [74] and only patients presenting biallelic CEBPA mutations have a favorable prognosis [75].

Finally, cyclic AMP Response Element Binding Protein (CREB), a leucine zipper transcription factor mainly involved in the balance between activation and repression of genes that participate in essential cellular processes and functions [76–78] has been demonstrated to be overexpressed in the majority of AML patients, leading to aberrant cell proliferation and growth [79, 80] often linked with worse outcome and increased risk of relapse [77]. Interestingly, CREB is negatively regulated by miR34b [81, 82] that is aberrantly silenced in AML, leading to uncontrolled cell proliferation [81].

NPM1 Mutations

One of the most frequent mutations described in *de novo* AML patients with normal karyotype is located in the last exon of nucleophosmin (NPM1) gene [83]. NPM1 is an abundant nuclear cytoplasmic shuttling protein that participates in many different cellular processes such as centrosome duplication, cell cycle progression and stress response [84–87]. It physically interacts with several nuclear proteins, including nucleolin [88], p120 [89], p53 [90] and Mdm2 [91]. Recent works have shown that NPM1 interacts and complexes with the tumor suppressor p19/Arf [92] and it is absolutely necessary for its physiological localization and stabilization in the nucleolus [93]. Mutations in NPM1 are heterozygous and consist in the insertion of short nucleotide stretches, leading to a reading frameshift and to a *de novo* formation of a chromosome region maintenance 1 (CRM1/Exportin 1)-dependent nuclear export signal (NES, [94] Fig. 13.2). The most frequent mutation of NPM1 gene (mutation A) has been described in 75–80% of cases and involves the duplication of the TCTG tetranucleotide [95]. Loss of the two tryptophan residues located in the C-terminal portion of the protein and needed for nucleolar localization [96] and creation of a new sequence of eleven amino acids

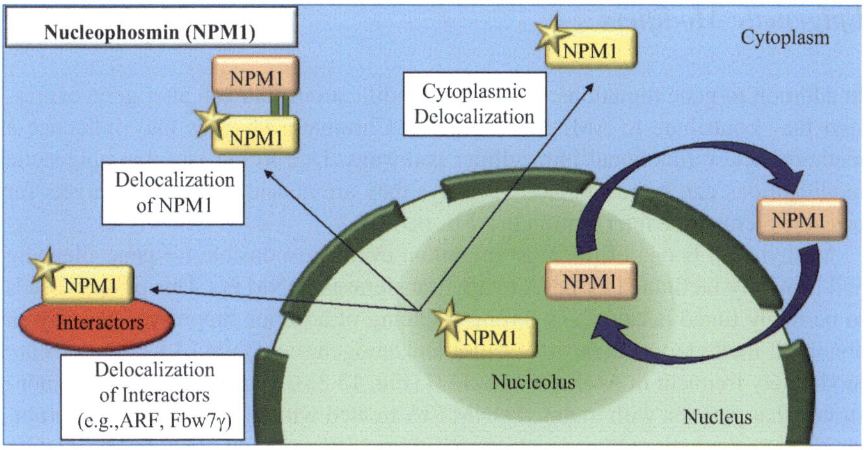

Fig. 13.2 Nucleophosmin is mutated in acute myeloid leukemia. Mutations in nucleophosmin (NPM1) lead to the aberrant cytoplasmic delocalization of the mutated and wild-type NPM1 proteins and of its physiological intracellular interactors. Yellow stars indicate mutations

contribute to the generation of the supplementary novel NES responsible for mutant NPM1 cytoplasmic delocalization. AML with mutated NPM1 usually presents a monocitic morphology, high blood count and many circulating blasts and often displays negativity to CD34 marker. Three chromosomal translocations related to NPM1 have been described so far. The t(2;5) translocation involves the N-terminal portion of NPM1 and leads to a fusion with the catalytic domain of the membrane-associated receptor tyrosine kinase ALK (anaplastic lymphoma kinase), that renders it constitutively active [97]. This aberration is present in about 30% of anaplastic large cell lymphomas (ALCL) [98]. The t(5;17) (q35;q21) is rare and it has been described in acute promyelocitic leukemias (APL). Also in this case, the N-terminal end of NPM1 is fused to the RARα C-terminal portion. This event leads to the binding of co-repressors that interfere with RARα-dependent transcriptional activities [99]. As a consequence, myeloid differentiation is blocked: patients that present this genetic abnormality can be treated with super-physiological doses of all-trans retinoic acid (ATRA) that release the block [100]. The third translocation t(3;5)(q25;q35) is very rare [101] and implicates the fusion of NPM1 to myelodisplasia/myeloid leukemia factor 1 (MLF1), a cytoplasmic protein that has a putative role in normal hematopoietic differentiation and has been shown to be able, together with the oncogene RASV12, of transforming mouse embryonic fibroblasts (MEFs) [102]. Interestingly, some of the most recurrent AML genomic aberrations are mutually exclusive with NPM1 mutations [103]. Due to the stability of this mutation, different diagnostic techniques have been established for the detection of mutated NPM1 [104–113].

Epigenetic Modifiers

In addition to gene mutations, epigenetic modifications that can alter gene expression may contribute to AML pathogenesis. Chromatin changes may influence a plethora of key functional intracellular pathways. Due to the fact that epigenetic modifications can be reversible, nowadays they are considered critical targets for possible therapeutic intervention in AML patients.

Methylation is one of the most important events responsible for gene silencing and is mainly facilitated by DNA methyltransferases (DNMTs). This process needs to be finely tuned because any wrong silencing of a tumor suppressor gene by an abnormal methylation event may cause leukemogenesis. DNMT3A mutations are moderately frequent in AML [7, 114–118] (Fig. 13.3a). Many studies have demonstrated that patients with mutated DNMT3A treated with high doses of daunorubicin displayed a better OS and outcome [7, 118, 119] and with the hypomethylating agent (HMA) decitabine respect to the ones with wild-type DNMT3A [120]. The mechanism by wich DNMT3A is involved in leukemogenesis has not yet been completely delucidated: the current hypothesis is that the mutated protein can have a role as a dominant negative over wild-type DNMT3A. Experiments performed *in vivo* have demonstrated that mutations in DNMT3A are able to induce myeloproliferative diseases and mice that develop AML acquire also mutations in signaling

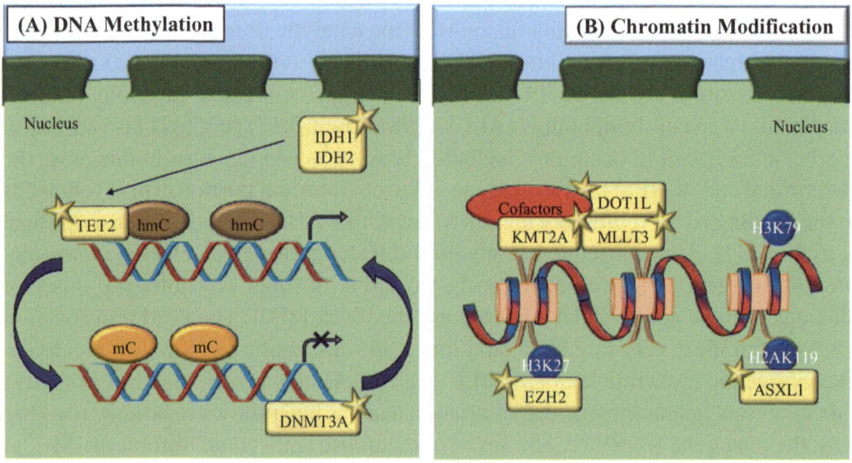

Fig. 13.3 Epigenetic regulators of DNA metabolism are mutated in acute myeloid leukemia. (**a**) Mutations in genes involved in DNA metabolism such as DNMT3A, TET2, IDH1 and IDH2 cause deregulation of DNA methylation. hmC indicates 5-hydroxymethylcytosine, mC indicates 5-methylcytosine. (**b**) Mutations in genes involved in epigenetic regulation such as ASXL1 and EZH2, cause deregulation of chromatin modification (H3K79, H2AK119, H3K27). KMT2A-MLLT3 fusion influences the functionality of other methyltransferases such as DOT1L. Yellow stars indicate mutations

molecules such as c-KIT [121]. This is similar to what happens in AML patients where FLT3-ITD mutations are often found cooperating with DNMT3 mutations.

Missense mutations in the two isoforms of IDH-1 and IDH-2 have been demonstrated to be present in AML patients [122–124]. They can promote tumorigenesis through a mechanism that possibly involves epigenetic changes mediated by inhibition of TET2, that is responsible for the oxidation of 5-methylcytosine (5mC) to 5-hydroxymethylcytosine (5hmC) depending on α-ketoglutarate [125]. Mutations in TET2 cause the persistence of hypermethylation and silencing of genes essential for normal cellular functions. Mutations in TET2 have been detected in 8–27% of AML patients and are linked to reduced levels of 5hmC [125]. IDH proteins catalyze the conversion of isocitrate to α-ketoglutarate and when they are mutated, a particular metabolite is produced, 2-hydroxyglutarate [126] that works as an inhibitor of α-ketoglutarate, and consequently of TET2 (Fig. 13.2a).

The detection of IDH mutations in AML has some positive outcomes in predicting the response to treatment with IDH inhibitors. Interestingly, it has been demonstrated that association between TET2 mutations and NPM1 wild-type/FLT3 wild-type display a poor outcome [7], eventhough its prognostic value is not completely clear [122, 127].

A positive regulator of gene transcription, the mixed-lineage leukemia gene (MLL) has been correlated to AML in up to 10% of patients and often correlates with a relatively unfavorable prognosis [128, 129] (Fig. 13.3b). MLL is a histone methyltransferase that targets H3K4 and leaves marks of transcriptional activation. It is generally involved in translocations at chromosomal band 11q23 and its prognostic value is linked to its fusion partner. AML patients displaying t(9;11) result in many different epigenetic aberrations that promote cell survival [130, 131] and show a better outcome respect to other translocations. Patients with CN-AML and MLL partial tandem duplication (PTD) have been associated with worse outcomes, eventhough intensive consolidation treatments have displayed long term disease-free survival [129].

Another transcriptional repressor methyltransferase is enhancer of zeste 2 polycomb repressive complex 2 subunit (EZH2). In myeloid malignancies, EZH2 mutations lead to loss of function [132, 133]. ASXL1 has an important role in the recruitment of the enzyme at the target loci. Its mutations are responsible for loss of H3K27 methylation and may be associated with adverse prognosis, an effect that seems to be correlated to the presence of RUNX1 mutations [134] (Fig. 13.3b).

Histone acetylation has also been correlated to tumorigenesis. In fact, histone deacetylases (HDACs) may be aberrantly recruited by myeloid oncoproteins, such as EVI1 or PML/RARα [135, 136] leading to chromatin remodeling that can affect transcription of putative target genes [137–139]. Acetylated lysine residues in histones can be recognized by reader proteins containing bromodomains, such as the bromodomain and extra terminal proteins (BET) BRD2, BRD3, and BRD4. BRD4 belongs to a complex including mediator and pTEF-b and is able to connect histone acetylation to transcription [140, 141]. Although BRD4 and other BET proteins are ubiquitously present at gene promoters and enhancers, inhibition of BET proteins results in large changes in expression of specific genes [142–144].

Other Signaling Pathways

Triad1, an E3 ubiquitin ligase encoded by ARIH2 gene is expressed in bone marrow progenitor cells [145]. It has been demonstrated that its interaction with Ubc7 or Ubc13 leads to proteasomal or lysosomal degradation [146, 147]. Clinical works suggest that Triad1 could be considered a leukemia suppressor. Deletions on chromosome 3p21 have been reported in AML patients [148–150] and a decrease in Triad1 mRNA has been associated with translocations of MLL1-gene and MYST4/CREBB and mutations such as FLT3-ITD [151]. Interestingly, HoxA10 is able to activate ARIH2 gene transcription during myelopoiesis [152], but when overexpressed in bone marrow progenitor cells it increases cytokine dependent proliferation *in vitro* and induces the development of myeloproliferative neoplasms degenerating in AML *in vivo*. It has been shown that increased expression of Triad1 can block the detrimental effects due to HoxA10 overexpression by favouring the degradation of proteins involved in the growth of hematopoietic stem and progenitor cells [152]. Taken together, these observations clearly corroborate the hypothesis that many additional mutations and deregulations in different intracellular pathways are needed for leukemogenesis [16, 153–155].

Current Established Therapies

The common therapeutic strategy adopted after AML diagnosis depends on the eligibility of the patient for intensive induction chemotherapy [17]. Usually, the regimen consists in continuous administration of cytarabine with an anthracycline. The efficacy of the treatment in adults who are <60 years of age ranges between 60% and 85% while in older patients the percentage of successes remains lower. Some patients have been shown to benefit by the combination of induction therapy and the administration of gemtuzumab ozogamicin (GO), a humanized anti-CD33 monoclonal antibody conjugated with the cytotoxic agent calicheamicin [156]. After intensive chemotherapy and when remission is complete, a post-remission regimen called "consolidation therapy" is frequently established, consisting in a new intensive chemotherapy treatment and/or an allogeneic hematopoietic stem cell transplantation [17, 158]. Monitoring minimal residual disease is essential for post-remission and pre-emptive salvage therapies. Patients >60 years of age and with intermediate risk genetic factors that present cure rates of only 10–15% display the worst outcomes. In these cases, an experimental treatment that may include new maintenance therapies can be administered. Generally, post-remission therapy consists in an allogeneic hematopoietic stem cell transplantation [157–165]. In some cases, patients can experience a relapse and subsequently undergo a reinduction therapy [166]. The percentage of successes in remissions after reinduction can range between 20% and 30%. The outcome after relapse is determined by many factors, especially by a short duration of remission, adverse genetic factors, prior allogeneic transplantation, older age, and poor general health status. When the relapse is diagnosed the most important factor that has to be considered is the possibility to receive intensive salvage therapy finalized to a complete remission before allogeneic hematopoietic stem cell transplantation [167]. When these protocols are not feasible, patients can udergo low-intensity therapy or best supportive care with the options of declining treatment or of receiving new experimental therapies.

Innovative Therapeutic Approaches

Although conventional therapies can provide a better outcome in AML patients, the overall survival remains less than 30% in adults and approximately 60% in children [3]. Moreover, treatment of AML can induce secondary effects like high morbidity and resistant disease with subsequent relapse. Current studies are therefore focused on finding new strategies in order to inhibit proteins involved in drug resistance and cell survival. Many new compounds under development target different molecules and pathways such as (1) signaling through tyrosine kinases, (2) nuclear export of proteins, (3) epigenetic regulation of DNA and chromatin, and (4) specifically expressed antigens.

The very high frequency of mutations in RTKs has led to the design of targeted inhibitors that unfortunately up to now have given poor results [168], especially due to their relatively high toxicity. Some JAK inhibitors have also been introduced in clinical trials and encouraging results have been obtained so far as they have displayed good anti-tumor effects. Several types of STAT inhibitors are currently under investigation for therapy [169]. In fact, STAT3 is hyperphosphorylated in up to 50% of AML patients giving worse prognosis. Therefore, STAT3 targeting has been shown to be effective in the treatment of chemotherapy resistant AML patients.

MAPK signaling pathways have also been targeted by molecules giving good results [170, 171]. Inhibition of mTORC1 pathways by rapamycin or other rapalogs has been shown effective both *in vitro* and *in vivo* [172–179].

High expression of CRM1, an essential nuclear export protein has been correlated with short survival in AML [180], therefore its inhibition could be a promising strategy: a clinical experimentation is currently undergoing [181]. A recent work demonstrated the feasibility of isolating and producing a functional recombinant scFv intrabody univocally recognizing the mutated form of NPM1 and fused to nuclear localization signal(s) in an attempt of relocalizing the aberrant protein to its correct location. However, the authors underline that the lack of precise information regarding the strength of signal peptides makes the possibility to create molecules able to counteract cellular mislocalization of shuttling NPM1very challenging [182].

Due to the fact that mutations in genes involved in epigenetic regulation have been frequently found in AML, new possibilities have opened for the development of innovative epigenetic therapies [183, 184]. As an example, inhibition IDH-1 and IDH-2 is undergoing a recent interest [185]. Both *in vitro* studies and phase 1 trials have demonstrated the possibility of specifically blocking terminal differentiation of leukemic blasts in IDH mutated AML [186] and to obtain high sensitivity to BCL2 inhibitor venetoclax in IDH-1/-2-mutated cells [187]. Two epigenetic compounds currently approved for clinical use in myeloid malignancies are azacitidine and decitabine, two inhibitors of DNA methyltransferases [188]. Their mechanism of action is probably based on reversing aberrant DNA hypermethylation and thereby restoring expression of critical (tumor suppressor) genes. These compounds are reported to be particularly useful for patients not eligible for intensive chemotherapy [189–192].

HDAC inhibitors were among the first epigenetic drugs developed and several of them have been tested in clinical trials for malignancies, including AML [193]. Recently, combinations of DNMT and HDAC inhibitors have been experimented [194–196]. This is supported by recent findings involving the targeting of transcription elongation factors such as bromodomain proteins using small molecules. Bromodomain proteins, including BET proteins, exert their function by binding acetylated lysine residues in histone proteins, such as H3K27 [183]. BET inhibitors showed therapeutic efficacy both *in vitro* and *in vivo* [142, 197–199] and may be effective in AML associated with MLL translocations [142].

Antibody therapy of AML has recently raised a great interest. The main strategy is currently focused on the development of specific monoclonal antibodies directed against CD33, a marker specific of myeloid cells. Anti-CD33 antibodies can be used either alone or in combination with anti-CD3 antibodies, possibly conjugated to specific cytotoxic molecules [200]. An example of conjugated antibody is gemtuzumab ozogamicin that was approved in 2000 for treating CD33-positive AML patients >60 years of age at first relapse. In 2009, it was removed from the market due to lacking of significant improvements in remission, disease-free survival or overall survival [201]. Later on, some trials demonstrated a good response in patients treated both with GO and standard chemotherapy [156, 202].

Researches directed to the identification of surface markers univocally expressed in leukemic cells has opened the way to the introduction of novel targeted therapies against AML. As an example, targeting of interleukin-3 receptor (CD123) has demonstrated to be an interesting option for the innovative chimeric antigen receptor T-cell engineered (CART) cellular therapy [203]. This strategy consists in the creation of synthetic hybrids between single chain variable fragment (scFv) antibodies and the transmembrane and intracellular domains of a T-cell receptor. A recent work demonstrates the feasibility of this approach: a CAR directed against CD33 has been shown to be active in AML cells [204].

Other studies have focused on targeting another receptor, folate receptor beta (FRβ), physiologically expressed on myeloid cells and overexpressed during tumorigenic transformation [205]. FRβ is frequently present in primary AML and can be overexpressed after all-trans retinoic acid treatment [206]. A possible application of CART cell therapy has been displayed both *in vitro* and *in vivo* in FRβ positive AML cells, without any toxicity in healthy human CD34+ stem cells [207]. These evidences suggest CART as a promising approach for treating relapsed/refractory AML.

Conclusions and Future Perspectives

Acute myeloid leukemia (AML) is the most frequent leukemia in adults and presents a very high incidence all over the world. It is an extremely complex disease with considerable genetic, epigenetic, and phenotypic heterogeneity. Even though recent advances in the prognosis and risk stratification of AML patients have contributed to boost conventional treatments, overall survival remains a major problem especially in older, relapsed or refractory patients. In the last years, most of the researches have focused on investigating the underlying mechanisms of leukemogenesis. Elucidating which mutations are responsible for AML pathogenesis and identifying the signaling molecules involved in the aberrant regulation of downstream pathways has provided potential novel targets for drug development. Today, many different interdisciplinary approaches are available that can help clinicians in refining diagnosis and classifying patients for an adequate therapy. However, the main challenges remain reduction of AML incidence and mortality together with prevention of the development of treatment related resistance. Nowadays, therapy is primarily focused on finding new ways to specifically target leukemic cells without inducing significant toxicity to normal cells and on achieving the best personalized treatment for each patient. New inhibitor compounds designed against precise targets hold great promise for curing AML. Recently, the identification of specific cell surface markers has led to the birth of novel immunotherapeutic approaches for AML: exploiting recombinant monoclonal antibodies and chimeric antigen receptors has shown enhanced selectivity properties and reduced aspecific toxicity.

References

1. Gojo I, Karp JE. New strategies in acute myelogenous leukemia: leukemogenesis and personalized medicine. Clin Cancer Res. 2014;20:6233–41. https://doi.org/10.1158/1078-0432.CCR-14-0900.
2. Estey EH. Acute myeloid leukemia: 2014 update on risk-stratification and management. Am J Hematol. 2014;89:1063–81. https://doi.org/10.1002/ajh.23834.
3. Perry AM, Attar EC. New insights in AML biology from genomic analysis. Semin Hematol. 2014;51:282–97. https://doi.org/10.1053/j.seminhematol.2014.08.005.
4. Sill H, Olipitz W, Zebisch A, Schulz E, Wölfler A. Therapy-related myeloid neoplasms: pathobiology and clinical characteristics. Br J Pharmacol. 2011;162:792–805. https://doi.org/10.1111/j.1476-5381.2010.01100.x.
5. Vardiman JW, Thiele J, Arber DA. The 2008 revision of the WHO classification of myeloid neoplasms and acute leukemia: rationale and important changes. Blood. 2008;114:937–52. https://doi.org/10.1182/blood-2009-03-209262.
6. Arber DA, Orazi A, Hasserjian R, Borowitz MJ, Beau MM, Le Bloomfield CD, Cazzola M, Vardiman JW. The 2016 revision to the World Health Organization classification of myeloid neoplasms and acute leukemia. Blood. 2016;127:2391–406. https://doi.org/10.1182/blood-2016-03-643544.The.
7. Patel JP, Gönen M, Figueroa ME, Fernandez H, Sun Z, Racevskis J, Van Vlierberghe P, Dolgalev I, Thomas S, Aminova O, Huberman K, Cheng J, Viale A, Socci ND, Heguy A,

Cherry A, Vance G, Higgins RR, Ketterling RP, Gallagher RE, Litzow M, van den Brink MRM, Lazarus HM, Rowe JM, Luger S, Ferrando A, Paietta E, Tallman MS, Melnick A, Abdel-Wahab O, Levine RL. Prognostic relevance of integrated genetic profiling in acute myeloid leukemia. N Engl J Med. 2012;366:1079–89. https://doi.org/10.1056/NEJMoa1112304.

8. Voigt P, Reinberg D. Genomic and epigenomic landscapes of adult de novo acute myeloid leukemia The Cancer Genome Atlas Research Network. N Engl J Med. 2013;368:2059–74. https://doi.org/10.1056/NEJMoa1301689.
9. Takahashi S. Current findings for recurring mutations in acute myeloid leukemia. J Hematol Oncol. 2011;4:36. https://doi.org/10.1186/1756-8722-4-36.
10. Kihara R, Nagata Y, Kiyoi H, Kato T, Yamamoto E, et al. Comprehensive analysis of genetic alterations and their prognostic impacts in adult acute myeloid leukemia patients. Leukemia. 2014;28:1586–95.
11. Ghoshal S, Baumann H, Wetzler M. Epigenetic regulation of signal transducer and activator of transcription 3 in acute myeloid leukemia. Leuk Res. 2008;32:1005–14. https://doi.org/10.1016/j.leukres.2007.11.035.
12. Yamada O, Kawauchi K. The role of the JAK-STAT pathway and related signal cascades in telomerase activation during the development of hematologic malignancies. JAK-STAT. 2013;2:e25256. https://doi.org/10.4161/jkst.25256.
13. Schuringa JJ, Wierenga AT, Kruijer W, Vellenga E. Constitutive Stat3, Tyr705, and Ser727 phosphorylation in acute myeloid leukemia cells caused by the autocrine secretion of interleukin-6. Blood. 2000;95:3765–70.
14. Spiekermann K, Bagrintseva K, Schwab R, Schmieja K, Hiddemann W. Overexpression and constitutive activation of FLT3 induces STAT5 activation in primary acute myeloid leukemia blast cells. Clin Cancer Res. 2003;9:2140–50.
15. Steensma DP, McClure RF, Karp JE, Tefferi A, Lasho TL, Powell HL, DeWald GW, Kaufmann SH. JAK2 V617F is a rare finding in de novo acute myeloid leukemia, but STAT3 activation is common and remains unexplained. Leukemia. 2006;20:971–8.
16. Cook AM, Li L, Ho Y, Lin A, Li L, Stein A, Forman S, Perrotti D, Jove R, Bhatia R. Role of altered growth factor receptor-mediated JAK2 signaling in growth and maintenance of human acute myeloid leukemia stem cells. Blood. 2014;123:2826–37. https://doi.org/10.1182/blood-2013-05-505735.
17. Dohner H, Estey EH, Amadori S, Appelbaum FR, Buchner T, Burnett AK, Dombret H, Fenaux P, Grimwade D, Larson RA, Lo-Coco F, Naoe T, Niederwieser D, Ossenkoppele GJ, Sanz MA, Sierra J, Tallman MS, Lowenberg B, Bloomfield CD. Diagnosis and management of acute myeloid leukemia in adults: recommendations from an international expert panel, on behalf of the European LeukemiaNet. Blood. 2010;115:453–74. https://doi.org/10.1182/blood-2009-07-235358.
18. Shah A, Andersson TM-L, Rachet B, Bjorkholm M, Lambert PC. Survival and cure of acute myeloid leukaemia in England, 1971–2006: a population-based study. Br J Haematol. 2013;162:509–16. https://doi.org/10.1111/bjh.12425.
19. Kantarjian H, O'brien S, Cortes J, Giles F, Faderl S, Jabbour E, Garcia-Manero G, Wierda W, Pierce S, Shan J, Estey E. Results of intensive chemotherapy in 998 patients age 65 years or older with acute myeloid leukemia or high-risk myelodysplastic syndrome: predictive prognostic models for outcome. Cancer. 2006;106:1090–8. https://doi.org/10.1002/cncr.21723.
20. Walter RB, Othus M, Borthakur G, Ravandi F, Cortes JE, Pierce SA, Appelbaum FR, Kantarjian HA, Estey EH. Prediction of early death after induction therapy for newly diagnosed acute myeloid leukemia with pretreatment risk scores: a novel paradigm for treatment assignment. J Clin Oncol. 2011;29:4417–23. https://doi.org/10.1200/JCO.2011.35.7525.
21. Hulegardh E, Nilsson C, Lazarevic V, Garelius H, Antunovic P, Rangert Derolf A, Mollgard L, Uggla B, Wennstrom L, Wahlin A, Hoglund M, Juliusson G, Stockelberg D, Lehmann S. Characterization and prognostic features of secondary acute myeloid leukemia in a population-based setting: a report from the Swedish Acute Leukemia Registry. Am J Hematol. 2015;90:208–14. https://doi.org/10.1002/ajh.23908.

22. Mrożek K, Marcucci G, Nicolet D, Maharry KS, Becker H, Whitman SP, Metzeler KH, Schwind S, YZ W, Kohlschmidt J, Pettenati MJ, Heerema NA, Block AW, Patil SR, Baer MR, Kolitz JE, Moore JO, Carroll AJ, Stone RM, Larson RA, Bloomfield CD. Prognostic significance of the European LeukemiaNet standardized system for reporting cytogenetic and molecular alterations in adults with acute myeloid leukemia. J Clin Oncol. 2012;30:4515–23. https://doi.org/10.1200/JCO.2012.43.4738.
23. Qin Y-Z, Zhu H-H, Jiang Q, Jiang H, Zhang L-P, L-P X, Wang Y, Liu Y-R, Lai Y-Y, Shi H-X, Jiang B, Huang X-J. Prevalence and prognostic significance of c-KIT mutations in core binding factor acute myeloid leukemia: a comprehensive large-scale study from a single Chinese center. Leuk Res. 2014;38:1435–40. https://doi.org/10.1016/j.leukres.2014.09.017.
24. Dohner K, Schlenk RF, Habdank M, Scholl C, Rucker FG, Corbacioglu A, Bullinger L, Frohling S, Dohner H. Mutant nucleophosmin (NPM1) predicts favorable prognosis in younger adults with acute myeloid leukemia and normal cytogenetics: interaction with other gene mutations. Blood. 2005;106:3740–6. https://doi.org/10.1182/blood-2005-05-2164.
25. Cagnetta A, Adamia S, Acharya C, Patrone F, Miglino M, Nencioni A, Gobbi M, Cea M. Role of genotype-based approach in the clinical management of adult acute myeloid leukemia with normal cytogenetics. Leuk Res. 2014;38:649–59. https://doi.org/10.1016/j.leukres.2014.03.006.
26. Port M, Bottcher M, Thol F, Ganser A, Schlenk R, Wasem J, Neumann A, Pouryamout L. Prognostic significance of FLT3 internal tandem duplication, nucleophosmin 1, and CEBPA gene mutations for acute myeloid leukemia patients with normal karyotype and younger than 60 years: a systematic review and meta-analysis. Ann Hematol. 2014;93:1279–86. https://doi.org/10.1007/s00277-014-2072-6.
27. Grossmann V, Schnittger S, Kohlmann A, Eder C, Roller A, Dicker F, Schmid C, Wendtner C-M, Staib P, Serve H, Kreuzer K-A, Kern W, Haferlach T, Haferlach C. A novel hierarchical prognostic model of AML solely based on molecular mutations. Blood. 2012;120:2963–72. https://doi.org/10.1182/blood-2012-03-419622.
28. Shivarov V, Gueorguieva R, Stoimenov A, Tiu R. DNMT3A mutation is a poor prognosis biomarker in AML: results of a meta-analysis of 4500 AML patients. Leuk Res. 2013;37:1445–50. https://doi.org/10.1016/j.leukres.2013.07.032.
29. Chen X, Xie H, Wood BL, Walter RB, Pagel JM, Becker PS, Sandhu VK, Abkowitz JL, Appelbaum FR, Estey EH. Relation of clinical response and minimal residual disease and their prognostic impact on outcome in acute myeloid leukemia. J Clin Oncol. 2015;33:1258–64. https://doi.org/10.1200/JCO.2014.58.3518.
30. Walter RB, Kantarjian HM, Huang X, Pierce SA, Sun Z, Gundacker HM, Ravandi F, Faderl SH, Tallman MS, Appelbaum FR, Estey EH. Effect of complete remission and responses less than complete remission on survival in acute myeloid leukemia: a combined Eastern Cooperative Oncology Group, Southwest Oncology Group, and M. D. Anderson Cancer Center Study. J Clin Oncol. 2010;28:1766–71. https://doi.org/10.1200/JCO.2009.25.1066.
31. Hoyos M, Nomdedeu JF, Esteve J, Duarte R, Ribera JM, Llorente A, Escoda L, Bueno J, Tormo M, Gallardo D, de Llano MPQ, Marti JM, Aventin A, Mangues R, Brunet S, Sierra J. Core binding factor acute myeloid leukemia: the impact of age, leukocyte count, molecular findings, and minimal residual disease. Eur J Haematol. 2013;91:209–18. https://doi.org/10.1111/ejh.12130.
32. Yin JAL, O'Brien MA, Hills RK, Daly SB, Wheatley K, Burnett AK. Minimal residual disease monitoring by quantitative RT-PCR in core binding factor AML allows risk stratification and predicts relapse: results of the United Kingdom MRC AML-15 trial. Blood. 2012;120:2826–35. https://doi.org/10.1182/blood-2012-06-435669.
33. Grimwade D, Ivey A, BJP H. Molecular landscape of acute myeloid leukemia in younger adults and its clinical relevance. Blood. 2015;127:29–42. https://doi.org/10.1182/blood-2015-07-604496.have.
34. Grimwade D, Mrozek K. Diagnostic and prognostic value of cytogenetics in acute myeloid leukemia. Hematol Oncol Clin North Am. 2011;25:1135–1161., vii. https://doi.org/10.1016/j.hoc.2011.09.018.

35. Schanz J. Helpful tool or oversimplification? Concept of the monosomal karyotype from the clinical and cytogenetic point of view. Biol Blood Marrow Transplant. 2017;22:191–2. https://doi.org/10.1016/j.bbmt.2015.11.013.
36. Coombs CC, Tallman MS, Levine RL. Molecular therapy for acute myeloid leukaemia. Nat Rev Clin Oncol. 2016;13:305–18.
37. O'Donnell MR, Tallman MS, Abboud CN, et al. National Comprehensive Cancer Network clinical guidelines in oncology (NCCN guidelines). Acute myeloid leukemia 1. Fort Washington, MD: National Comprehensive Cancer Network; 2015.
38. Parcells BW, Ikeda AK, Simms-Waldrip T, Moore TB, Sakamoto KM. FMS-like tyrosine kinase 3 in normal hematopoiesis and acute myeloid leukemia. Stem Cells. 2006;24:1174–84. https://doi.org/10.1634/stemcells.2005-0519.
39. Kottaridis PD, Gale RE, Frew ME, Harrison G, Langabeer SE, Belton AA, Walker H, Wheatley K, Bowen DT, Burnett AK, Goldstone AH, Linch DC. The presence of a FLT3 internal tandem duplication in patients with acute myeloid leukemia (AML) adds important prognostic information to cytogenetic patients from the United Kingdom Medical Research Council AML 10 and 12 .pdf. Blood. 2015;98:1752–60.
40. Whitman SP, Archer KJ, Feng L, Whitman SP, Archer KJ, Feng L, Baldus C, Becknell B, Carlson BD, Carroll AJ, Mro K, Vardiman JW, George SL, Kolitz JE, Larson R, Bloomfield CD, Caligiuri M. Absence of the wild-type allele predicts poor prognosis in adult de novo acute myeloid leukemia with normal cytogenetics and the internal tandem duplication of FLT3 : a cancer and leukemia group B study absence of the wild-type allele predicts poor progno. Cancer Res. 2001;61(19):7233–9.
41. Schlenk RF, Döhner K, Krauter J, Fröhling S, Corbacioglu A, Bullinger L, Habdank M, Späth D, Morgan M, Benner A, Schlegelberger B, Heil G, Ganser A, Döhner H. Mutations and treatment outcome in cytogenetically normal acute myeloid leukemia. N Engl J Med. 2008;358:1909–18. https://doi.org/10.1056/NEJMoa074306.
42. Gilliland DG, Griffin JD. The roles of FLT3 in hematopoiesis and leukemia. Blood. 2002;100:1532 LP–1542.
43. Stirewalt DL, Radich JP. The role of FLT3 in haematopoietic malignancies. Nat Rev Cancer. 2003;3:650–65. https://doi.org/10.1038/nrc1169.
44. Frohling S, Scholl C, Gilliland DG, Levine RL. Genetics of myeloid malignancies: pathogenetic and clinical implications. J Clin Oncol. 2005;23:6285–95. https://doi.org/10.1200/JCO.2005.05.010.
45. Mead AJ, Linch DC, Hills RK, Wheatley K, Burnett AK, Gale RE. FLT3 tyrosine kinase domain mutations are biologically distinct from and have a significantly more favorable prognosis than FLT3 internal tandem duplications in patients with acute myeloid leukemia. Blood. 2007;110:1262–70. https://doi.org/10.1182/blood-2006-04-015826.
46. Choudhary C, Schwable J, Brandts C, Tickenbrock L, Sargin B, Kindler T, Fischer T, Berdel WE, Muller-Tidow C, Serve H. AML-associated Flt3 kinase domain mutations show signal transduction differences compared with Flt3 ITD mutations. Blood. 2005;106:265–73. https://doi.org/10.1182/blood-2004-07-2942.
47. Spiekermann K, Dirschinger RJ, Schwab R, Bagrintseva K, Faber F, Buske C, Schnittger S, Kelly LM, Gilliland DG, Hiddemann W. The protein tyrosine kinase inhibitor SU5614 inhibits FLT3 and induces growth arrest and apoptosis in AML-derived cell lines expressing a constitutively activated FLT3. Blood. 2003;101:1494–504. https://doi.org/10.1182/blood-2002-04-1045.
48. Ozeki K, Kiyoi H, Hirose Y, Iwai M, Ninomiya M, Kodera Y, Miyawaki S, Kuriyama K, Shimazaki C, Akiyama H, Nishimura M, Motoji T, Shinagawa K, Takeshita A, Ueda R, Ohno R, Emi N, Naoe T. Biologic and clinical significance of the FLT3 transcript level in acute myeloid leukemia. Blood. 2004;103:1901–8. https://doi.org/10.1182/blood-2003-06-1845.
49. Zheng R, Levis M, Piloto O, Brown P, Baldwin BR, Gorin NC, Beran M, Zhu Z, Ludwig D, Hicklin D, Witte L, Li Y, Small D. FLT3 ligand causes autocrine signaling in acute myeloid leukemia cells. Blood. 2004;103:267–74. https://doi.org/10.1182/blood-2003-06-1969.

50. Caligiuri MA, Briesewitz R, Yu J, Wang L, Wei M, Arnoczky KJ, Marburger TB, Wen J, Perrotti D, Bloomfield CD, Whitman SP. Novel c-CBL and CBL-b ubiquitin ligase mutations in human acute myeloid leukemia. Blood. 2007;110:1022–4. https://doi.org/10.1182/blood-2006-12-061176.
51. Sargin B, Choudhary C, Crosetto N, Schmidt MHH, Grundler R, Rensinghoff M, Thiessen C, Tickenbrock L, Schwable J, Brandts C, August B, Koschmieder S, Bandi SR, Duyster J, Berdel WE, Muller-Tidow C, Dikic I, Serve H. Flt3-dependent transformation by inactivating c-Cbl mutations in AML. Blood. 2007;110:1004–12. https://doi.org/10.1182/blood-2007-01-066076.
52. Allen C, Hills RK, Lamb K, Evans C, Tinsley S, Sellar R, O'Brien M, Yin JL, Burnett a K, Linch DC, Gale RE. The importance of relative mutant level for evaluating impact on outcome of KIT, FLT3 and CBL mutations in core-binding factor acute myeloid leukemia. Leukemia. 2013;27:1891–901. https://doi.org/10.1038/leu.2013.186.
53. Poiré X, Moser BK, Gallagher RE, Laumann K, Bloomfield CD, Powell BL, Koval G, Gulati K, Holowka N, Larson RA, Tallman MS, Appelbaum FR, Sher D, Willman C, Paietta E, Stock W. Arsenic trioxide in front-line therapy of acute promyelocytic leukemia (C9710): prognostic significance of FLT3 mutations and complex karyotype. Leuk Lymphoma. 2014;55:1523–32. https://doi.org/10.3109/10428194.2013.842985.
54. Paschka P, Marcucci G, Ruppert AS, Mrozek K, Chen H, Kittles RA, Vukosavljevic T, Perrotti D, Vardiman JW, Carroll AJ, Kolitz JE, Larson RA, Bloomfield CD. Adverse prognostic significance of KIT mutations in adult acute myeloid leukemia with inv(16) and t(8;21): a Cancer and Leukemia Group B Study. J Clin Oncol. 2006;24:3904–11. https://doi.org/10.1200/JCO.2006.06.9500.
55. Lennartsson J, Jelacic T, Linnekin D, Shivakrupa R. Normal and oncogenic forms of the receptor tyrosine kinase kit. Stem Cells. 2005;23:16–43. https://doi.org/10.1634/stemcells.2004-0117.
56. Mitin N, Rossman KL, Der CJ. Signaling interplay in Ras superfamily function. Curr Biol. 2005;15:R563–74. https://doi.org/10.1016/j.cub.2005.07.010.
57. Malumbres M, Barbacid M. RAS oncogenes: the first 30 years. Nat Rev Cancer. 2003;3:459–65. https://doi.org/10.1038/nrc1097.
58. Van Meter MEM, Diaz-Flores E, Archard JA, Passegue E, Irish JM, Kotecha N, Nolan GP, Shannon K, Braun BS. K-RasG12D expression induces hyperproliferation and aberrant signaling in primary hematopoietic stem/progenitor cells. Blood. 2007;109:3945–52. https://doi.org/10.1182/blood-2006-09-047530.
59. Martelli AM, Nyakern M, Tabellini G, Bortul R, Tazzari PL, Evangelisti C, Cocco L. Phosphoinositide 3-kinase/Akt signaling pathway and its therapeutical implications for human acute myeloid leukemia. Leukemia. 2006;20:911–28. https://doi.org/10.1038/sj.leu.2404245.
60. Platanias LC. Map kinase signaling pathways and hematologic malignancies. Blood. 2003;101:4667–79. https://doi.org/10.1182/blood-2002-12-3647.
61. Ewings KE, Hadfield-Moorhouse K, Wiggins CM, Wickenden JA, Balmanno K, Gilley R, Degenhardt K, White E, Cook SJ. ERK1/2-dependent phosphorylation of BimEL promotes its rapid dissociation from Mcl-1 and Bcl-xL. EMBO J. 2007;26:2856–67. https://doi.org/10.1038/sj.emboj.7601723.
62. Martelli AM, Evangelisti C, Chiarini F, McCubrey JA. The phosphatidylinositol 3-kinase/Akt/mTOR signaling network as a therapeutic target in acute myelogenous leukemia patients. Oncotarget. 2010;1:89–103. 10.18632/oncotarget.114.
63. Shlush LI, Zandi S, Mitchell A, Chen WC, Brandwein JM, Gupta V, Kennedy JA, Schimmer AD, Schuh AC, Yee KW, McLeod JL, Doedens M, Medeiros JJF, Marke R, Kim HJ, Lee K, McPherson JD, Hudson TJ, Brown AMK, Yousif F, Trinh QM, Stein LD, Minden MD, Wang JCY, Dick JE. Identification of pre-leukaemic haematopoietic stem cells in acute leukaemia. Nature. 2014;506:328–33. https://doi.org/10.1038/nature13038.

64. Welch JS, Ley TJ, Link DC, Miller CA, Larson DE, Koboldt DC, Wartman LD, Lamprecht TL, Liu F, Xia J, Kandoth C, Fulton RS, McLellan MD, Dooling DJ, Wallis JW, Chen K, Harris CC, Schmidt HK, Kalicki-Veizer JM, Lu C, Zhang Q, Lin L, O'Laughlin MD, McMichael JF, Delehaunty KD, Fulton LA, Magrini VJ, McGrath SD, Demeter RT, Vickery TL, Hundal J, Cook LL, Swift GW, Reed JP, Alldredge PA, Wylie TN, Walker JR, Watson MA, Heath SE, Shannon WD, Varghese N, Nagarajan R, Payton JE, Baty JD, Kulkarni S, Klco JM, Tomasson MH, Westervelt P, Walter MJ, Graubert TA, DiPersio JF, Ding L, Mardis ER, Wilson RK. The origin and evolution of mutations in acute myeloid leukemia. Cell. 2012;150:264–78. https://doi.org/10.1016/j.cell.2012.06.023.
65. Laplante M, Sabatini DM. mTOR signaling in growth control and disease. Cell. 2012;149:274–93. https://doi.org/10.1016/j.cell.2012.03.017.
66. Beauchamp EM, Platanias LC. The evolution of the TOR pathway and its role in cancer. Oncogene. 2013;32:3923–32. https://doi.org/10.1038/onc.2012.567.
67. Aoki Y, Matsubara Y. Ras/MAPK syndromes and childhood hemato-oncological diseases. Int J Hematol. 2013;97:30–6. https://doi.org/10.1007/s12185-012-1239-y.
68. Pritchard AL, Hayward NK. Molecular pathways: mitogen-activated protein kinase pathway mutations and drug resistance. Clin Cancer Res. 2013;19:2301–9. https://doi.org/10.1158/1078-0432.CCR-12-0383.
69. Furqan M, Mukhi N, Lee B, Liu D. Dysregulation of JAK-STAT pathway in hematological malignancies and JAK inhibitors for clinical application. Biomark Res. 2013;1:5. https://doi.org/10.1186/2050-7771-1-5
70. Lo M-C, Peterson LF, Yan M, Cong X, Hickman JH, Dekelver RC, Niewerth D, Zhang D-E. JAK inhibitors suppress t(8;21) fusion protein-induced leukemia. Leukemia. 2013;27:2272–9. https://doi.org/10.1038/leu.2013.197.
71. Mizuki M, Fenski R, Halfter H, Matsumura I, Schmidt R, Muller C, Gruning W, Kratz-Albers K, Serve S, Steur C, Buchner T, Kienast J, Kanakura Y, Berdel WE, Serve H. Flt3 mutations from patients with acute myeloid leukemia induce transformation of 32D cells mediated by the Ras and STAT5 pathways. Blood. 2000;96:3907–14.
72. Mendler JH, Maharry K, Radmacher MD, Mrózek K, Becker H, Metzeler KH, Schwind S, Whitman SP, Khalife J, Kohlschmidt J, Nicolet D, Powell BL, Carter TH, Wetzler M, Moore JO, Kolitz JE, Baer MR, Carroll AJ, Larson RA, Caligiuri MA, Marcucci G, Bloomfield CD. RUNX1 mutations are associated with poor outcome in younger and older patients with cytogenetically normal acute myeloid leukemia and with distinct gene and microRNA expression signatures. J Clin Oncol. 2012;30:3109–18. https://doi.org/10.1200/JCO.2011.40.6652.
73. Rücker FG, Schlenk RF, Bullinger L, Kayser S, Teleanu V, Kett H, Habdank M, Kugler CM, Holzmann K, Gaidzik VI, Paschka P, Held G, Von Lilienfeld-Toal M, Lübbert M, Fröhling S, Zenz T, Krauter J, Schlegelberger B, Ganser A, Lichter P, Döhner K, Döhner H. TP53 alterations in acute myeloid leukemia with complex karyotype correlate with specific copy number alterations, monosomal karyotype, and dismal outcome. Blood. 2012;119:2114–21. https://doi.org/10.1182/blood-2011-08-375758.
74. Leroy H, Roumier C, Huyghe P, Biggio V, Fenaux P, Preudhomme C. CEBPA point mutations in hematological malignancies. Leukemia. 2005;19:329–34. https://doi.org/10.1038/sj.leu.2403614.
75. Wouters BJ, Löwenberg B, Erpelinck-Verschueren CAJ, Van Putten WLJ, Valk PJM, Delwel R. Double CEBPA mutations, but not single CEBPA mutations, define a subgroup of acute myeloid leukemia with a distinctive gene expression profile that is uniquely associated with a favorable outcome. Blood. 2009;113:3088–91. https://doi.org/10.1182/blood-2008-09-179895.
76. Shaywitz AJ, Greenberg ME. CREB: a stimulus-induced transcription factor activated by a diverse array of extracellular signals. Annu Rev Biochem. 1999;68:821–61. https://doi.org/10.1146/annurev.biochem.68.1.821.
77. Shankar DB, Cheng JC, Kinjo K, Federman N, Moore TB, Gill A, Rao NP, Landaw EM, Sakamoto KM. The role of CREB as a proto-oncogene in hematopoiesis and in acute myeloid leukemia. Cancer Cell. 2005;7:351–62. https://doi.org/10.1016/j.ccr.2005.02.018.

78. Shankar DB, Cheng JC, Sakamoto KM. Role of cyclic AMP response element binding protein in human leukemias. Cancer. 2005;104:1819–24. https://doi.org/10.1002/cncr.21401.
79. Cho E-C, Mitton B, Sakamoto KM. CREB and leukemogenesis. Crit Rev Oncog. 2011;16:37–46.
80. Esparza SD, Chang J, Shankar DB, Zhang B, Nelson SF, Sakamoto KM. CREB regulates Meis1 expression in normal and malignant hematopoietic cells. Leukemia. 2008;22:665–7. https://doi.org/10.1038/sj.leu.2404933.
81. Pigazzi M, Manara E, Baron E, Basso G. miR-34b targets cyclic AMP-responsive element binding protein in acute myeloid leukemia. Cancer Res. 2009;69:2471–8. https://doi.org/10.1158/0008-5472.CAN-08-3404.
82. Pigazzi M, Manara E, Bresolin S, Tregnago C, Beghin A, Baron E, Giarin E, Cho E-C, Masetti R, Rao DS, Sakamoto KM, Basso G. MicroRNA-34b promoter hypermethylation induces CREB overexpression and contributes to myeloid transformation. Haematologica. 2013;98:602–10. https://doi.org/10.3324/haematol.2012.070664.
83. Schnittger S, Schoch C. Nucleophosmin gene mutations are predictors of favorable prognosis in acute myelogenous leukemia with a normal karyotype. Blood. 2005;106:3733–40. https://doi.org/10.1182/blood-2005-06-2248.Supported.
84. Okuda M, Horn HF, Tarapore P, Tokuyama Y, Smulian AG, Chan P-K, Knudsen ES, Hofmann IA, Snyder JD, Bove KE, Fukasawa K. Nucleophosmin/B23 is a target of CDK2/cyclin E in centrosome duplication. Cell. 2000;103:127–40. https://doi.org/10.1016/S0092-8674(00)00093-3.
85. Negi SS, Olson MOJ. Effects of interphase and mitotic phosphorylation on the mobility and location of nucleolar protein B23. J Cell Sci. 2006;119:3676 LP–3685.
86. Rubbi CP, Milner J. Disruption of the nucleolus mediates stabilization of p53 in response to DNA damage and other stresses. EMBO J. 2003;22:6068 LP–6077.
87. MH W, Yung BYM. UV stimulation of nucleophosmin/B23 expression is an immediate-early gene response induced by damaged DNA. J Biol Chem. 2002;277:48234–40. https://doi.org/10.1074/jbc.M206550200.
88. Li YP, Busch RK, Valdez BC, Busch H. C23 interacts with B23, a putative nucleolar-localization-signal-binding protein. Eur J Biochem. 1996;237:153–8. https://doi.org/10.1111/j.1432-1033.1996.0153n.x.
89. Valdez BC, Perlaky L, Henning D, Saijo Y, Chan PK, Busch H. Identification of the nuclear and nucleolar localization signals of the protein p120: interaction with translocation protein B23. J Biol Chem. 1994;269:23776–83.
90. Colombo E, Marine J-C, Danovi D, Falini B, Pelicci PG. Nucleophosmin regulates the stability and transcriptional activity of p53. Nat Cell Biol. 2002;4:529–33.
91. Kurki S, Peltonen K, Latonen L, Kiviharju TM, Ojala PM, Meek D, Laiho M. Nucleolar protein NPM interacts with HDM2 and protects tumor suppressor protein p53 from HDM2-mediated degradation. Cancer Cell. 2004;5:465–75. https://doi.org/10.1016/S1535-6108(04)00110-2.
92. Itahana K, Bhat KP, Jin A, Itahana Y, Hawke D, Kobayashi R, Zhang Y. Tumor suppressor ARF degrades B23, a nucleolar protein involved in ribosome biogenesis and cell proliferation. Mol Cell. 2003;12:1151–64. https://doi.org/10.1016/S1097-2765(03)00431-3.
93. Korgaonkar C, Hagen J, Tompkins V, Frazier AA, Allamargot C, Quelle FW, Quelle DE. Nucleophosmin (B23) targets ARF to nucleoli and inhibits its function. Mol Cell Biol. 2005;25:1258–71. https://doi.org/10.1128/MCB.25.4.1258-1271.2005.
94. Mariano AR, Colombo E, Luzi L, Martinelli P, Volorio S, Bernard L, Meani N, Bergomas R, Alcalay M, Pelicci PG. Cytoplasmic localization of NPM in myeloid leukemias is dictated by gain-of-function mutations that create a functional nuclear export signal. Oncogene. 2006;25:4376–80.
95. Falini B, Nicoletti I, Martelli MF, Mecucci C. Acute myeloid leukemia carrying cytoplasmic/mutated nucleophosmin (NPMc+ AML): biologic and clinical features. Blood. 2007;109:874–85.
96. Falini B, Bolli N, Shan J, Martelli MP, Liso A, Pucciarini A, Bigerna B, Pasqualucci L, Mannucci R, Rosati R, Gorello P, Diverio D, Roti G, Tiacci E, Cazzaniga G, Biondi A,

Schnittger S, Haferlach T, Hiddemann W, Martelli MF, Gu W, Mecucci C, Nicoletti I. Both carboxy-terminus NES motif and mutated tryptophan(s) are crucial for aberrant nuclear export of nucleophosmin leukemic mutants in NPMc+ AML. Blood. 2006;107:4514–23. https://doi.org/10.1182/blood-2005-11-4745.
97. Morris SW, Kirstein MN, Valentine MB, Dittmer K, Shapiro DN, Look AT, Saltman DL. Fusion of a kinase gene, ALK, to a nucleolar protein gene, NPM, in non-Hodgkin's lymphoma. Science. 1995;267:316–7.
98. Redner RL, Rush EA, Faas S, Rudert WA, Corey SJ. The t(5;17) variant of acute promyelocytic leukemia expresses a nucleophosmin-retinoic acid receptor fusion. Blood. 1996;87:882–6.
99. Redner RL, Chen JD, Rush EA, Li H, Pollock SL. The t(5;17) acute promyelocytic leukemia fusion protein NPM-RAR interacts with co-repressor and co-activator proteins and exhibits both positive and negative transcriptional properties. Blood. 2000;95:2683–90.
100. Okazuka K, Masuko M, Seki Y, Hama H, Honma N, Furukawa T, Toba K, Kishi K, Aizawa Y. Successful all-trans retinoic acid treatment of acute promyelocytic leukemia in a patient with NPM/RAR fusion. Int J Hematol. 2007;86:246–9. https://doi.org/10.1532/IJH97.07036.
101. Yoneda-Kato N, Look AT, Kirstein MN, Valentine MB, Raimondi SC, Cohen KJ, Carroll AJ, Morris SW. The t(3;5)(q25.1;q34) of myelodysplastic syndrome and acute myeloid leukemia produces a novel fusion gene, NPM-MLF1. Oncogene. 1996;12:265–75.
102. Yoneda-Kato N, Kato J-Y. Shuttling imbalance of MLF1 results in p53 instability and increases susceptibility to oncogenic transformation. Mol Cell Biol. 2008;28:422–34. https://doi.org/10.1128/MCB.02335-06.
103. Falini B, Mecucci C, Saglio G, Lo Coco F, Diverio D, Brown P, Pane F, Mancini M, Martelli MP, Pileri S, Haferlach T, Haferlach C, Schnittger S. NPM1 mutations and cytoplasmic nucleophosmin are mutually exclusive of recurrent genetic abnormalities: a comparative analysis of 2562 patients with acute myeloid leukemia. Haematologica. 2008;93:439–42. https://doi.org/10.3324/haematol.12153.
104. Ammatuna E, Noguera NI, Zangrilli D, Curzi P, Panetta P, Bencivenga P, Amadori S, Federici G, Lo-Coco F. Rapid detection of nucleophosmin Flt3 mutations in acute myeloid leukemia by denaturing HPLC. Clin Chem. 2005;51:2165–7.
105. Falini B, Martelli MP, Bolli N, Bonasso R, Ghia E, Pallotta MT, Diverio D, Nicoletti I, Pacini R, Tabarrini A, Galletti BV, Mannucci R, Roti G, Rosati R, Specchia G, Liso A, Tiacci E, Alcalay M, Luzi L, Volorio S, Bernard L, Guarini A, Amadori S, Mandelli F, Pane F, Lo-Coco F, Saglio G, Pelicci P-G, Martelli MF, Mecucci C. Immunohistochemistry predicts nucleophosmin (NPM) mutations in acute myeloid leukemia. Blood. 2006;108:1999–2005. https://doi.org/10.1182/blood-2006-03-007013.
106. Gorello P, Cazzaniga G, Alberti F, Dell'Oro MG, Gottardi E, Specchia G, Roti G, Rosati R, Martelli MF, Diverio D, Lo Coco F, Biondi A, Saglio G, Mecucci C, Falini B. Quantitative assessment of minimal residual disease in acute myeloid leukemia carrying nucleophosmin (NPM1) gene mutations. Leukemia. 2006;20:1103–8. https://doi.org/10.1038/sj.leu.2404149.
107. Martelli MP, Manes N, Liso A, Pettirossi V, Verducci Galletti B, Bigerna B, Pucciarini A, De Marco MF, Pallotta MT, Bolli N, Sborgia M, di Raimondo F, Fabbiano F, Meloni G, Specchia G, Martelli MF, Falini B. A western blot assay for detecting mutant nucleophosmin (NPM1) proteins in acute myeloid leukaemia. Leukemia. 2008;22:2285–8. https://doi.org/10.1038/leu.2008.149.
108. Noguera NI, Ammatuna E, Zangrilli D, Lavorgna S, Divona M, Buccisano F, Amadori S, Mecucci C, Falini B, Lo-Coco F. Simultaneous detection of NPM1 and FLT3-ITD mutations by capillary electrophoresis in acute myeloid leukemia. Leukemia. 2005;19:1479–82. https://doi.org/10.1038/sj.leu.2403846.
109. Bacher U, Badbaran A, Fehse B, Zabelina T, Zander AR, Kroger N. Quantitative monitoring of NPM1 mutations provides a valid minimal residual disease parameter following allogeneic stem cell transplantation. Exp Hematol. 2009;37:135–42. https://doi.org/10.1016/j.exphem.2008.09.014.

110. Papadaki C, Dufour A, Seibl M, Schneider S, Bohlander SK, Zellmeier E, Mellert G, Hiddemann W, Spiekermann K. Monitoring minimal residual disease in acute myeloid leukaemia with NPM1 mutations by quantitative PCR: clonal evolution is a limiting factor. Br J Haematol. 2009;144:517–23. https://doi.org/10.1111/j.1365-2141.2008.07488.x.
111. Ma W, Kantarjian H, Zhang X, Jilani I, Sheikholeslami MR, Donahue AC, Ravandi F, Estey E, O'Brien S, Keating M, Giles FJ, Albitar M. Detection of nucleophosmin gene mutations in plasma from patients with acute myeloid leukemia: clinical significance and implications. Cancer Biomark. 2009;5:51–8. https://doi.org/10.3233/CBM-2009-0583.
112. Gruszka AM, Lavorgna S, Consalvo MI, Ottone T, Martinelli C, Cinquanta M, Ossolengo G, Pruneri G, Buccisano F, Divona M, Cedrone M, Ammatuna E, Venditti A, De Marco A, Lo-Coco F, Pelicci PG. A monoclonal antibody against mutated nucleophosmin 1 for the molecular diagnosis of acute myeloid leukemias. Blood. 2010;116:2096–102. https://doi.org/10.1182/blood-2010-01-266908.
113. Gruszka AM, Martinelli C, Sparacio E, Pelicci PG, De Marco A. The concurrent use of N- and C-terminal antibodies anti-nucleophosmin 1 in immunofluorescence experiments allows for precise assessment of its subcellular localisation in acute myeloid leukaemia patients. Leukemia. 2012;26:159–62. https://doi.org/10.1038/leu.2011.177.
114. Yang L, Rau R, Goodell MA. DNMT3A in haematological malignancies. Nat Rev Cancer. 2015;15:152–65. https://doi.org/10.1038/nrc3895.
115. Ley TJ, Ding L, Walter MJ, McLellan MD, Lamprecht T, Larson DE, Kandoth C, Payton JE, Baty J, Welch J, Harris CC, Lichti CF, Townsend RR, Fulton RS, Dooling DJ, Koboldt DC, Schmidt H, Zhang Q, Osborne JR, Lin L, O'Laughlin M, McMichael JF, Delehaunty KD, McGrath SD, Fulton LA, Magrini VJ, Vickery TL, Hundal J, Cook LL, Conyers JJ, Swift GW, Reed JP, Alldredge PA, Wylie T, Walker J, Kalicki J, Watson MA, Heath S, Shannon WD, Varghese N, Nagarajan R, Westervelt P, Tomasson MH, Link DC, Graubert TA, DiPersio JF, Mardis ER, Wilson RK. DNMT3A mutations in acute myeloid leukemia. N Engl J Med. 2010;363:2424–33. https://doi.org/10.1056/NEJMoa1005143.
116. Gaidzik VI, Schlenk RF, Paschka P, Stolzle A, Spath D, Kuendgen A, von Lilienfeld-Toal M, Brugger W, Derigs HG, Kremers S, Greil R, Raghavachar A, Ringhoffer M, Salih HR, Wattad M, Kirchen HG, Runde V, Heil G, Petzer AL, Girschikofsky M, Heuser M, Kayser S, Goehring G, Teleanu M-V, Schlegelberger B, Ganser A, Krauter J, Bullinger L, Dohner H, Dohner K. Clinical impact of DNMT3A mutations in younger adult patients with acute myeloid leukemia: results of the AML Study Group (AMLSG). Blood. 2013;121:4769–77. https://doi.org/10.1182/blood-2012-10-461624.
117. Marcucci G, Metzeler KH, Schwind S, Becker H, Maharry K, Mrozek K, Radmacher MD, Kohlschmidt J, Nicolet D, Whitman SP, Y-Z W, Powell BL, Carter TH, Kolitz JE, Wetzler M, Carroll AJ, Baer MR, Moore JO, Caligiuri MA, Larson RA, Bloomfield CD. Age-related prognostic impact of different types of DNMT3A mutations in adults with primary cytogenetically normal acute myeloid leukemia. J Clin Oncol. 2012;30:742–50. https://doi.org/10.1200/JCO.2011.39.2092.
118. Gale RE, Lamb K, Allen C, El-Sharkawi D, Stowe C, Jenkinson S, Tinsley S, Dickson G, Burnett AK, Hills RK, Linch DC. Simpson's paradox and the impact of different DNMT3A mutations on outcome in younger adults with acute myeloid leukemia. J Clin Oncol. 2015;33:2072–83. https://doi.org/10.1200/JCO.2014.59.2022.
119. Luskin MR, Lee J-W, Fernandez HF, Abdel-Wahab O, Bennett JM, Ketterling RP, Lazarus HM, Levine RL, Litzow MR, Paietta EM, Patel JP, Racevskis J, Rowe JM, Tallman MS, Sun Z, Luger SM. Benefit of high-dose daunorubicin in AML induction extends across cytogenetic and molecular groups. Blood. 2016;127:1551–8. https://doi.org/10.1182/blood-2015-07-657403.
120. Metzeler KH, Walker A, Geyer S, Garzon R, Klisovic RB, Bloomfield CD, Blum W, Marcucci G. DNMT3A mutations and response to the hypomethylating agent decitabine in acute myeloid leukemia. Leukemia. 2012;26:1106–7. https://doi.org/10.1038/leu.2011.342.

121. Celik H, Mallaney C, Kothari A, Ostrander EL, Eultgen E, Martens A, Miller CA, Hundal J, Klco JM, Challen GA. Enforced differentiation of Dnmt3a-null bone marrow leads to failure with c-Kit mutations driving leukemic transformation. Blood. 2015;125:619–28. https://doi.org/10.1182/blood-2014-08-594564.
122. Mardis ER, Ding L, Dooling DJ, Larson DE, McLellan MD, Chen K, Koboldt DC, Fulton RS, Delehaunty KD, McGrath SD, Fulton LA, Locke DP, Magrini VJ, Abbott RM, Vickery TL, Reed JS, Robinson JS, Wylie T, Smith SM, Carmichael L, Eldred JM, Harris CC, Walker J, Peck JB, Du F, Dukes AF, Sanderson GE, Brummett AM, Clark E, McMichael JF, Meyer RJ, Schindler JK, Pohl CS, Wallis JW, Shi X, Lin L, Schmidt H, Tang Y, Haipek C, Wiechert ME, Ivy JV, Kalicki J, Elliott G, Ries RE, Payton JE, Westervelt P, Tomasson MH, Watson MA, Baty J, Heath S, Shannon WD, Nagarajan R, Link DC, Walter MJ, Graubert TA, DiPersio JF, Wilson RK, Ley TJ. Recurring mutations found by sequencing an acute myeloid leukemia genome. N Engl J Med. 2009;361:1058–66. https://doi.org/10.1056/NEJMoa0903840.
123. Marcucci G, Maharry K, Y-Z W, Radmacher MD, Mrozek K, Margeson D, Holland KB, Whitman SP, Becker H, Schwind S, Metzeler KH, Powell BL, Carter TH, Kolitz JE, Wetzler M, Carroll AJ, Baer MR, Caligiuri MA, Larson RA, Bloomfield CD. IDH1 and IDH2 gene mutations identify novel molecular subsets within de novo cytogenetically normal acute myeloid leukemia: a Cancer and Leukemia Group B study. J Clin Oncol. 2010;28:2348–55. https://doi.org/10.1200/JCO.2009.27.3730.
124. Paschka P, Schlenk RF, Gaidzik VI, Habdank M, Kronke J, Bullinger L, Spath D, Kayser S, Zucknick M, Gotze K, Horst H-A, Germing U, Dohner H, Dohner K. IDH1 and IDH2 mutations are frequent genetic alterations in acute myeloid leukemia and confer adverse prognosis in cytogenetically normal acute myeloid leukemia with NPM1 mutation without FLT3 internal tandem duplication. J Clin Oncol. 2010;28:3636–43. https://doi.org/10.1200/JCO.2010.28.3762.
125. Figueroa ME, Abdel-Wahab O, Lu C, Ward PS, Patel J, Shih A, Li Y, Bhagwat N, Vasanthakumar A, Fernandez HF, Tallman MS, Sun Z, Wolniak K, Peeters JK, Liu W, Choe SE, Fantin VR, Paietta E, Lowenberg B, Licht JD, Godley LA, Delwel R, Valk PJM, Thompson CB, Levine RL, Melnick A. Leukemic IDH1 and IDH2 mutations result in a hypermethylation phenotype, disrupt TET2 function, and impair hematopoietic differentiation. Cancer Cell. 2010;18:553–67. https://doi.org/10.1016/j.ccr.2010.11.015.
126. Ward PS, Patel J, Wise DR, Abdel-Wahab O, Bennett BD, Coller HA, Cross JR, Fantin VR, Hedvat CV, Perl AE, Rabinowitz JD, Carroll M, SM S, Sharp KA, Levine RL, Thompson CB. The common feature of leukemia-associated IDH1 and IDH2 mutations is a neomorphic enzyme activity converting alpha-ketoglutarate to 2-hydroxyglutarate. Cancer Cell. 2010;17:225–34. https://doi.org/10.1016/j.ccr.2010.01.020.
127. Gaidzik VI, Paschka P, Spath D, Habdank M, Kohne C-H, Germing U, von Lilienfeld-Toal M, Held G, Horst H-A, Haase D, Bentz M, Gotze K, Dohner H, Schlenk RF, Bullinger L, Dohner K. TET2 mutations in acute myeloid leukemia (AML): results from a comprehensive genetic and clinical analysis of the AML study group. J Clin Oncol. 2012;30:1350–7. https://doi.org/10.1200/JCO.2011.39.2886.
128. Krivtsov AV, Figueroa ME, Sinha AU, Stubbs MC, Feng Z, Valk PJM, Delwel R, Dohner K, Bullinger L, Kung AL, Melnick AM, Armstrong SA. Cell of origin determines clinically relevant subtypes of MLL-rearranged AML. Leukemia. 2013;27:852–60. https://doi.org/10.1038/leu.2012.363.
129. Whitman SP, Ruppert AS, Marcucci G, Mrozek K, Paschka P, Langer C, Baldus CD, Wen J, Vukosavljevic T, Powell BL, Carroll AJ, Kolitz JE, Larson RA, Caligiuri MA, Bloomfield CD. Long-term disease-free survivors with cytogenetically normal acute myeloid leukemia and MLL partial tandem duplication: a Cancer and Leukemia Group B study. Blood. 2007;109:5164–7. https://doi.org/10.1182/blood-2007-01-069831.
130. Neff T, Armstrong SA. Recent progress toward epigenetic therapies: the example of mixed lineage leukemia. Blood. 2013;121:4847–53. https://doi.org/10.1182/blood-2013-02-474833.
131. de Boer J, Walf-Vorderwulbecke V, Williams O. In focus: MLL-rearranged leukemia. Leukemia. 2013;27:1224–8. https://doi.org/10.1038/leu.2013.78.

132. Ernst T, Chase AJ, Score J, Hidalgo-Curtis CE, Bryant C, Jones AV, Waghorn K, Zoi K, Ross FM, Reiter A, Hochhaus A, Drexler HG, Duncombe A, Cervantes F, Oscier D, Boultwood J, Grand FH, Cross NCP. Inactivating mutations of the histone methyltransferase gene EZH2 in myeloid disorders. Nat Genet. 2010;42:722–6. https://doi.org/10.1038/ng.621.
133. Nikoloski G, Langemeijer SMC, Kuiper RP, Knops R, Massop M, Tonnissen ERLTM, van der Heijden A, Scheele TN, Vandenberghe P, de Witte T, van der Reijden BA, Jansen JH. Somatic mutations of the histone methyltransferase gene EZH2 in myelodysplastic syndromes. Nat Genet. 2010;42:665–7. https://doi.org/10.1038/ng.620.
134. Paschka P, Schlenk RF, Gaidzik VI, Herzig JK, Aulitzky T, Bullinger L, Spath D, Teleanu V, Kundgen A, Kohne C-H, Brossart P, Held G, Horst H-A, Ringhoffer M, Gotze K, Nachbaur D, Kindler T, Heuser M, Thol F, Ganser A, Dohner H, Dohner K. ASXL1 mutations in younger adult patients with acute myeloid leukemia: a study by the German-Austrian Acute Myeloid Leukemia Study Group. Haematologica. 2015;100:324–30. https://doi.org/10.3324/haematol.2014.114157.
135. Izutsu K, Kurokawa M, Imai Y, Maki K, Mitani K, Hirai H. The corepressor CtBP interacts with Evi-1 to repress transforming growth factor beta signaling. Blood. 2001;97:2815–22.
136. Senyuk V, Chakraborty S, Mikhail FM, Zhao R, Chi Y, Nucifora G. The leukemia-associated transcription repressor AML1/MDS1/EVI1 requires CtBP to induce abnormal growth and differentiation of murine hematopoietic cells. Oncogene. 2002;21:3232–40. https://doi.org/10.1038/sj.onc.1205436.
137. Villa R, Morey L, Raker VA, Buschbeck M, Gutierrez A, De Santis F, Corsaro M, Varas F, Bossi D, Minucci S, Pelicci PG, Di Croce L. The methyl-CpG binding protein MBD1 is required for PML-RARalpha function. Proc Natl Acad Sci U S A. 2006;103:1400–5. https://doi.org/10.1073/pnas.0509343103.
138. Santoro F, Botrugno OA, Dal Zuffo R, Pallavicini I, Matthews GM, Cluse L, Barozzi I, Senese S, Fornasari L, Moretti S, Altucci L, Pelicci PG, Chiocca S, Johnstone RW, Minucci S. A dual role for Hdac1: oncosuppressor in tumorigenesis, oncogene in tumor maintenance. Blood. 2013;121:3459–68. https://doi.org/10.1182/blood-2012-10-461988.
139. Spensberger D, Vermeulen M, Le Guezennec X, Beekman R, van Hoven A, Bindels E, Stunnenberg H, Delwel R. Myeloid transforming protein Evi1 interacts with methyl-CpG binding domain protein 3 and inhibits in vitro histone deacetylation by Mbd3/Mi-2/NuRD. Biochemistry. 2008;47:6418–26. https://doi.org/10.1021/bi800267f.
140. Belkina AC, Denis GV. BET domain co-regulators in obesity, inflammation and cancer. Nat Rev Cancer. 2012;12:465–77. https://doi.org/10.1038/nrc3256.
141. S-Y W, Chiang C-M. The double bromodomain-containing chromatin adaptor Brd4 and transcriptional regulation. J Biol Chem. 2007;282:13141–5. https://doi.org/10.1074/jbc.R700001200.
142. Dawson MA, Prinjha RK, Dittmann A, Giotopoulos G, Bantscheff M, Chan W-I, Robson SC, Chung C, Hopf C, Savitski MM, Huthmacher C, Gudgin E, Lugo D, Beinke S, Chapman TD, Roberts EJ, Soden PE, Auger KR, Mirguet O, Doehner K, Delwel R, Burnett AK, Jeffrey P, Drewes G, Lee K, Huntly BJP, Kouzarides T. Inhibition of BET recruitment to chromatin as an effective treatment for MLL-fusion leukaemia. Nature. 2011;478:529–33. https://doi.org/10.1038/nature10509.
143. Filippakopoulos P, Qi J, Picaud S, Shen Y, Smith WB, Fedorov O, Morse EM, Keates T, Hickman TT, Felletar I, Philpott M, Munro S, McKeown MR, Wang Y, Christie AL, West N, Cameron MJ, Schwartz B, Heightman TD, La Thangue N, French CA, Wiest O, Kung AL, Knapp S, Bradner JE. Selective inhibition of BET bromodomains. Nature. 2010;468:1067–73. https://doi.org/10.1038/nature09504.
144. Nicodeme E, Jeffrey KL, Schaefer U, Beinke S, Dewell S, Chung C-W, Chandwani R, Marazzi I, Wilson P, Coste H, White J, Kirilovsky J, Rice CM, Lora JM, Prinjha RK, Lee K, Tarakhovsky A. Suppression of inflammation by a synthetic histone mimic. Nature. 2010;468:1119–23. https://doi.org/10.1038/nature09589.
145. Marteijn JAF, van Emst L, Erpelinck-Verschueren CAJ, Nikoloski G, Menke A, de Witte T, Lowenberg B, Jansen JH, van der Reijden BA. The E3 ubiquitin-protein ligase Triad1 inhibits

clonogenic growth of primary myeloid progenitor cells. Blood. 2005;106:4114–23. https://doi.org/10.1182/blood-2005-04-1450.

146. Marteijn JA, van der Meer LT, Smit JJ, Noordermeer SM, Wissink W, Jansen P, Swarts HG, Hibbert RG, de Witte T, Sixma TK, Jansen JH, van der Reijden BA. The ubiquitin ligase Triad1 inhibits myelopoiesis through UbcH7 and Ubc13 interacting domains. Leukemia. 2009;23:1480–9. https://doi.org/10.1038/leu.2009.57.
147. Pietschmann K, Buchwald M, Muller S, Knauer SK, Kogl M, Heinzel T, Kramer OH. Differential regulation of PML-RARalpha stability by the ubiquitin ligases SIAH1/SIAH2 and TRIAD1. Int J Biochem Cell Biol. 2012;44:132–8. https://doi.org/10.1016/j.biocel.2011.10.008.
148. Johansson B, Billstrom R, Kristoffersson U, Akerman M, Garwicz S, Ahlgren T, Malm C, Mitelman F. Deletion of chromosome arm 3p in hematologic malignancies. Leukemia. 1997;11:1207–13.
149. Shi G, Weh HJ, Martensen S, Seeger D, Hossfeld DK. 3p21 is a recurrent treatment-related breakpoint in myelodysplastic syndrome and acute myeloid leukemia. Cytogenet Cell Genet. 1996;74:295–9.
150. Zhu X, He F, Zeng H, Ling S, Chen A, Wang Y, Yan X, Wei W, Pang Y, Cheng H, Hua C, Zhang Y, Yang X, Lu X, Cao L, Hao L, Dong L, Zou W, Wu J, Li X, Zheng S, Yan J, Zhou J, Zhang L, Mi S, Wang X, Zhang L, Zou Y, Chen Y, Geng Z, Wang J, Zhou J, Liu X, Wang J, Yuan W, Huang G, Cheng T, Wang Q-F. Identification of functional cooperative mutations of SETD2 in human acute leukemia. Nat Genet. 2014;46:287–93. https://doi.org/10.1038/ng.2894.
151. Rhodes DR, Yu J, Shanker K, Deshpande N, Varambally R, Ghosh D, Barrette T, Pandey A, Chinnaiyan AM. ONCOMINE: a cancer microarray database and integrated data-mining platform. Neoplasia. 2004;6:1–6.
152. Wang H, Bei L, Shah CA, Horvath E, Eklund EA. HoxA10 influences protein ubiquitination by activating transcription of ARIH2, the gene encoding triad1. J Biol Chem. 2011;286(19):16832–45. https://doi.org/10.1074/jbc.M110.213975
153. Thorsteinsdottir U, Sauvageau G, Hough MR, Dragowska W, Lansdorp PM, Lawrence HJ, Largman C, Humphries RK. Overexpression of HOXA10 in murine hematopoietic cells perturbs both myeloid and lymphoid differentiation and leads to acute myeloid leukemia. Mol Cell Biol. 1997;17:495–505.
154. Bjornsson JM, Andersson E, Lundstrom P, Larsson N, Xu X, Repetowska E, Humphries RK, Karlsson S. Proliferation of primitive myeloid progenitors can be reversibly induced by HOXA10. Blood. 2001;98:3301–8.
155. Camós M, Esteve J, Jares P, Colomer D, Rozman M, Villamor N, Costa D, Carrió A, Nomdedéu J, Montserrat E, Campo E. Gene expression profiling of acute myeloid leukemia with translocation t(8;16)(p11;p13) and MYST3-CREBBP rearrangement reveals a distinctive signature with a specific pattern of *HOX* gene expression. Cancer Res. 2006;66:6947–54.
156. Hills RK, Castaigne S, Appelbaum FR, Delaunay J, Petersdorf S, Othus M, Estey EH, Dombret H, Chevret S, Ifrah N, Cahn J-Y, Recher C, Chilton L, Moorman AV, Burnett AK. Addition of gemtuzumab ozogamicin to induction chemotherapy in adult patients with acute myeloid leukaemia: a meta-analysis of individual patient data from randomised controlled trials. Lancet Oncol. 2014;15:986–96. https://doi.org/10.1016/S1470-2045(14)70281-5.
157. Armand P, Kim HT, Logan BR, Wang Z, Alyea EP, Kalaycio ME, Maziarz RT, Antin JH, Soiffer RJ, Weisdorf DJ, Rizzo JD, Horowitz MM, Saber W. Validation and refinement of the Disease Risk Index for allogeneic stem cell transplantation. Blood. 2014;123:3664–71. https://doi.org/10.1182/blood-2014-01-552984.
158. Gupta V, Tallman MS, Weisdorf DJ. Allogeneic hematopoietic cell transplantation for adults with acute myeloid leukemia: myths, controversies, and unknowns. Blood. 2011;117:2307–18. https://doi.org/10.1182/blood-2010-10-265603.
159. Sorror ML, Storb RF, Sandmaier BM, Maziarz RT, Pulsipher MA, Maris MB, Bhatia S, Ostronoff F, Deeg HJ, Syrjala KL, Estey E, Maloney DG, Appelbaum FR, Martin PJ, Storer

BE. Comorbidity-age index: a clinical measure of biologic age before allogeneic hematopoietic cell transplantation. J Clin Oncol. 2014;32:3249–56. https://doi.org/10.1200/JCO.2013.53.8157.
160. Koreth J, Schlenk R, Kopecky KJ, Honda S, Sierra J, Djulbegovic BJ, Wadleigh M, DeAngelo DJ, Stone RM, Sakamaki H, Appelbaum FR, Dohner H, Antin JH, Soiffer RJ, Cutler C. Allogeneic stem cell transplantation for acute myeloid leukemia in first complete remission: systematic review and meta-analysis of prospective clinical trials. JAMA. 2009;301:2349–61. https://doi.org/10.1001/jama.2009.813.
161. Gupta V, Tallman MS, He W, Logan BR, Copelan E, Gale RP, Khoury HJ, Klumpp T, Koreth J, Lazarus HM, Marks DI, Martino R, Rizzieri DA, Rowe JM, Sabloff M, Waller EK, DiPersio JF, Bunjes DW, Weisdorf DJ. Comparable survival after HLA-well-matched unrelated or matched sibling donor transplantation for acute myeloid leukemia in first remission with unfavorable cytogenetics at diagnosis. Blood. 2010;116:1839–48. https://doi.org/10.1182/blood-2010-04-278317.
162. McClune BL, Weisdorf DJ, Pedersen TL, Tunes da Silva G, Tallman MS, Sierra J, Dipersio J, Keating A, Gale RP, George B, Gupta V, Hahn T, Isola L, Jagasia M, Lazarus H, Marks D, Maziarz R, Waller EK, Bredeson C, Giralt S. Effect of age on outcome of reduced-intensity hematopoietic cell transplantation for older patients with acute myeloid leukemia in first complete remission or with myelodysplastic syndrome. J Clin Oncol. 2010;28:1878–87. https://doi.org/10.1200/JCO.2009.25.4821.
163. Farag SS, Maharry K, Zhang M-J, Perez WS, George SL, Mrozek K, DiPersio J, Bunjes DW, Marcucci G, Baer MR, Cairo M, Copelan E, Cutler CS, Isola L, Lazarus HM, Litzow MR, Marks DI, Ringden O, Rizzieri DA, Soiffer R, Larson RA, Tallman MS, Bloomfield CD, Weisdorf DJ. Comparison of reduced-intensity hematopoietic cell transplantation with chemotherapy in patients age 60–70 years with acute myelogenous leukemia in first remission. Biol Blood Marrow Transplant. 2011;17:1796–803. https://doi.org/10.1016/j.bbmt.2011.06.005.
164. Litzow MR, Tarima S, Perez WS, Bolwell BJ, Cairo MS, Camitta BM, Cutler CS, de Lima M, Dipersio JF, Gale RP, Keating A, Lazarus HM, Luger S, Marks DI, Maziarz RT, McCarthy PL, Pasquini MC, Phillips GL, Rizzo JD, Sierra J, Tallman MS, Weisdorf DJ. Allogeneic transplantation for therapy-related myelodysplastic syndrome and acute myeloid leukemia. Blood. 2010;115:1850–7. https://doi.org/10.1182/blood-2009-10-249128.
165. Gooley TA, Chien JW, Pergam SA, Hingorani S, Sorror ML, Boeckh M, Martin PJ, Sandmaier BM, Marr KA, Appelbaum FR, Storb R, McDonald GB. Reduced mortality after allogeneic hematopoietic-cell transplantation. N Engl J Med. 2010;363:2091–101. https://doi.org/10.1056/NEJMoa1004383.
166. de Lima M, Porter DL, Battiwalla M, Bishop MR, Giralt SA, Hardy NM, Kroger N, Wayne AS, Schmid C. Proceedings from the National Cancer Institute's second international workshop on the biology, prevention, and treatment of relapse after hematopoietic stem cell transplantation: part III. Prevention and treatment of relapse after allogeneic transplantation. Biol Blood Marrow Transplant. 2014;20:4–13. https://doi.org/10.1016/j.bbmt.2013.08.012.
167. Thol F, Schlenk RF, Heuser M, Ganser A. How I treat refractory and early relapsed acute myeloid leukemia. Blood. 2015;126:319–27. https://doi.org/10.1182/blood-2014-10-551911.
168. Wander SA, Levis MJ, Fathi AT. The evolving role of FLT3 inhibitors in acute myeloid leukemia: quizartinib and beyond. Ther Adv Hematol. 2014;5:65–77. https://doi.org/10.1177/2040620714532123.
169. Furqan M, Akinleye A, Mukhi N, Mittal V, Chen Y, Liu D. STAT inhibitors for cancer therapy. J Hematol Oncol. 2013;6:90. https://doi.org/10.1186/1756-8722-6-90.
170. Geest CR, Coffer PJ. MAPK signaling pathways in the regulation of hematopoiesis. J Leukoc Biol. 2009;86:237–50. https://doi.org/10.1189/jlb.0209097.
171. Volk A, Li J, Xin J, You D, Zhang J, Liu X, Xiao Y, Breslin P, Li Z, Wei W, Schmidt R, Li X, Zhang Z, Kuo PC, Nand S, Zhang J, Chen J, Zhang J. Co-inhibition of NF-kappaB and JNK is synergistic in TNF-expressing human AML. J Exp Med. 2014;211:1093–108. https://doi.org/10.1084/jem.20130990.

172. Altman JK, Sassano A, Kaur S, Glaser H, Kroczynska B, Redig AJ, Russo S, Barr S, Platanias LC. Dual mTORC2/mTORC1 targeting results in potent suppressive effects on acute myeloid leukemia (AML) progenitors. Clin Cancer Res. 2011;17:4378–88. https://doi.org/10.1158/1078-0432.CCR-10-2285.
173. Zeng Z, Shi YX, Tsao T, Qiu Y, Kornblau SM, Baggerly KA, Liu W, Jessen K, Liu Y, Kantarjian H, Rommel C, Fruman DA, Andreeff M, Konopleva M. Targeting of mTORC1/2 by the mTOR kinase inhibitor PP242 induces apoptosis in AML cells under conditions mimicking the bone marrow microenvironment. Blood. 2012;120:2679–89. https://doi.org/10.1182/blood-2011-11-393934.
174. Willems L, Chapuis N, Puissant A, Maciel TT, Green AS, Jacque N, Vignon C, Park S, Guichard S, Herault O, Fricot A, Hermine O, Moura IC, Auberger P, Ifrah N, Dreyfus F, Bonnet D, Lacombe C, Mayeux P, Bouscary D, Tamburini J. The dual mTORC1 and mTORC2 inhibitor AZD8055 has anti-tumor activity in acute myeloid leukemia. Leukemia. 2012;26:1195–202. https://doi.org/10.1038/leu.2011.339.
175. Kampa-Schittenhelm KM, Heinrich MC, Akmut F, Rasp KH, Illing B, Dohner H, Dohner K, Schittenhelm MM. Cell cycle-dependent activity of the novel dual PI3K-MTORC1/2 inhibitor NVP-BGT226 in acute leukemia. Mol Cancer. 2013;12:46. https://doi.org/10.1186/1476-4598-12-46.
176. Chapuis N, Tamburini J, Green AS, Vignon C, Bardet V, Neyret A, Pannetier M, Willems L, Park S, Macone A, Maira S-M, Ifrah N, Dreyfus F, Herault O, Lacombe C, Mayeux P, Bouscary D. Dual inhibition of PI3K and mTORC1/2 signaling by NVP-BEZ235 as a new therapeutic strategy for acute myeloid leukemia. Clin Cancer Res. 2010;16:5424–35. https://doi.org/10.1158/1078-0432.CCR-10-1102.
177. Greger JG, Eastman SD, Zhang V, Bleam MR, Hughes AM, Smitheman KN, Dickerson SH, Laquerre SG, Liu L, Gilmer TM. Combinations of BRAF, MEK, and PI3K/mTOR inhibitors overcome acquired resistance to the BRAF inhibitor GSK2118436 dabrafenib, mediated by NRAS or MEK mutations. Mol Cancer Ther. 2012;11:909–20. https://doi.org/10.1158/1535-7163.MCT-11-0989.
178. Brundage ME, Tandon P, Eaves DW, Williams JP, Miller SJ, Hennigan RH, Jegga A, Cripe TP, Ratner N. MAF mediates crosstalk between Ras-MAPK and mTOR signaling in NF1. Oncogene. 2014;33:5626–36. https://doi.org/10.1038/onc.2013.506.
179. Zhang W, Ruvolo VR, Gao C, Zhou L, Bornmann W, Tsao T, Schober WD, Smith P, Guichard S, Konopleva M, Andreeff M. Evaluation of apoptosis induction by concomitant inhibition of MEK, mTOR, and Bcl-2 in human acute myelogenous leukemia cells. Mol Cancer Ther. 2014;13:1848–59. https://doi.org/10.1158/1535-7163.MCT-13-0576.
180. Kojima K, Kornblau SM, Ruvolo V, Dilip A, Duvvuri S, Davis RE, Zhang M, Wang Z, Coombes KR, Zhang N, Qiu YH, Burks JK, Kantarjian H, Shacham S, Kauffman M, Andreeff M. Prognostic impact and targeting of CRM1 in acute myeloid leukemia. Blood. 2013;121:4166–74. https://doi.org/10.1182/blood-2012-08-447581.
181. ClinicalTrials.gov Identifier: NCT02088541.
182. Martinelli C, Colombo E, Piccini D, Sironi C, Pelicci PG, De Marco A. An intrabody specific for the nucleophosmin carboxy-terminal mutant and fused to a nuclear localization sequence binds its antigen but fails to relocate it in the nucleus. Biotechnol Rep. 2014;3:27–33. https://doi.org/10.1016/j.btre.2014.05.008.
183. Dawson MA, Kouzarides T, Huntly BJP. Targeting epigenetic readers in cancer. N Engl J Med. 2012;367:647–57. https://doi.org/10.1056/NEJMra1112635.
184. Abdel-Wahab O, Levine RL. Mutations in epigenetic modifiers in the pathogenesis and therapy of acute myeloid leukemia. Blood. 2013;121:3563–72. https://doi.org/10.1182/blood-2013-01-451781.
185. Wang F, Travins J, DeLaBarre B, Penard-Lacronique V, Schalm S, Hansen E, Straley K, Kernytsky A, Liu W, Gliser C, Yang H, Gross S, Artin E, Saada V, Mylonas E, Quivoron C, Popovici-Muller J, Saunders JO, Salituro FG, Yan S, Murray S, Wei W, Gao Y, Dang L, Dorsch M, Agresta S, Schenkein DP, Biller SA, SM S, de Botton S, Yen KE. Targeted inhibition of

mutant IDH2 in leukemia cells induces cellular differentiation. Science. 2013;340:622–6. https://doi.org/10.1126/science.1234769.
186. Yen K, Travins J, Wang F, David MD, Artin E, Straley K, Padyana A, Gross S, DeLaBarre B, Tobin E, Chen Y, Nagaraja R, Choe S, Jin L, Konteatis Z, Cianchetta G, Saunders JO, Salituro FG, Quivoron C, Opolon P, Bawa O, Saada V, Paci A, Broutin S, Bernard OA, de Botton S, Marteyn BS, Pilichowska M, Xu Y, Fang C, Jiang F, Wei W, Jin S, Silverman L, Liu W, Yang H, Dang L, Dorsch M, Penard-Lacronique V, Biller SA, S-SM S. AG-221, a first-in-class therapy targeting acute myeloid leukemia harboring oncogenic IDH2 mutations. Cancer Discov. 2017;7:478–93. https://doi.org/10.1158/2159-8290.CD-16-1034.
187. Chan SM, Thomas D, Corces-Zimmerman MR, Xavy S, Rastogi S, Hong W-J, Zhao F, Medeiros BC, Tyvoll DA, Majeti R. Isocitrate dehydrogenase 1 and 2 mutations induce BCL-2 dependence in acute myeloid leukemia. Nat Med. 2015;21:178–84. https://doi.org/10.1038/nm.3788.
188. Navada SC, Steinmann J, Lubbert M, Silverman LR. Clinical development of demethylating agents in hematology. J Clin Invest. 2014;124:40–6. https://doi.org/10.1172/JCI69739.
189. Huls G. Azacitidine in AML: a treatment option? Blood. 2015;126:283 LP–284.
190. Dombret H, Seymour JF, Butrym A, Wierzbowska A, Selleslag D, Jang JH, Kumar R, Cavenagh J, Schuh AC, Candoni A, Recher C, Sandhu I, Bernal del Castillo T, Al-Ali HK, Martinelli G, Falantes J, Noppeney R, Stone RM, Minden MD, McIntyre H, Songer S, Lucy LM, Beach CL, Dohner H. International phase 3 study of azacitidine vs conventional care regimens in older patients with newly diagnosed AML with >30% blasts. Blood. 2015;126:291–9. https://doi.org/10.1182/blood-2015-01-621664.
191. Blum W, Garzon R, Klisovic RB, Schwind S, Walker A, Geyer S, Liu S, Havelange V, Becker H, Schaaf L, Mickle J, Devine H, Kefauver C, Devine SM, Chan KK, Heerema NA, Bloomfield CD, Grever MR, Byrd JC, Villalona-Calero M, Croce CM, Marcucci G. Clinical response and miR-29b predictive significance in older AML patients treated with a 10-day schedule of decitabine. Proc Natl Acad Sci U S A. 2010;107:7473–8. https://doi.org/10.1073/pnas.1002650107.
192. He J, Xiu L, De Porre P, Dass R, Thomas X. Decitabine reduces transfusion dependence in older patients with acute myeloid leukemia: results from a post hoc analysis of a randomized phase III study. Leuk Lymphoma. 2015;56:1033–42. https://doi.org/10.3109/10428194.2014.951845.
193. Wagner JM, Hackanson B, Lübbert M, Jung M. Histone deacetylase (HDAC) inhibitors in recent clinical trials for cancer therapy. Clin Epigenetics. 2010;1:117–36. https://doi.org/10.1007/s13148-010-0012-4.
194. Lubbert M, Kuendgen A. Combining DNA methyltransferase and histone deacetylase inhibition to treat acute myeloid leukemia/myelodysplastic syndrome: achievements and challenges. Cancer. 2015;121:498–501. https://doi.org/10.1002/cncr.29083.
195. Kirschbaum M, Gojo I, Goldberg SL, Bredeson C, Kujawski LA, Yang A, Marks P, Frankel P, Sun X, Tosolini A, Eid JE, Lubiniecki GM, Issa J-P. A phase 1 clinical trial of vorinostat in combination with decitabine in patients with acute myeloid leukaemia or myelodysplastic syndrome. Br J Haematol. 2014;167:185–93. https://doi.org/10.1111/bjh.13016.
196. Strati P, Kantarjian H, Ravandi F, Nazha A, Borthakur G, Daver N, Kadia T, Estrov Z, Garcia-Manero G, Konopleva M, Rajkhowa T, Durand M, Andreeff M, Levis M, Cortes J. Phase I/II trial of the combination of midostaurin (PKC412) and 5-azacytidine for patients with acute myeloid leukemia and myelodysplastic syndrome. Am J Hematol. 2015;90:276–81. https://doi.org/10.1002/ajh.23924.
197. Groschel S, Sanders MA, Hoogenboezem R, de Wit E, Bouwman BAM, Erpelinck C, van der Velden VHJ, Havermans M, Avellino R, van Lom K, Rombouts EJ, van Duin M, Dohner K, Beverloo HB, Bradner JE, Dohner H, Lowenberg B, Valk PJM, Bindels EMJ, de Laat W, Delwel R. A single oncogenic enhancer rearrangement causes concomitant EVI1 and GATA2 deregulation in leukemia. Cell. 2014;157:369–81. https://doi.org/10.1016/j.cell.2014.02.019.

198. Zuber J, Shi J, Wang E, Rappaport AR, Herrmann H, Sison EA, Magoon D, Qi J, Blatt K, Wunderlich M, Taylor MJ, Johns C, Chicas A, Mulloy JC, Kogan SC, Brown P, Valent P, Bradner JE, Lowe SW, Vakoc CR. RNAi screen identifies Brd4 as a therapeutic target in acute myeloid leukaemia. Nature. 2011;478:524–8. https://doi.org/10.1038/nature10334.
199. Herrmann H, Blatt K, Shi J, Gleixner KV, Cerny-Reiterer S, Mullauer L, Vakoc CR, Sperr WR, Horny H-P, Bradner JE, Zuber J, Valent P. Small-molecule inhibition of BRD4 as a new potent approach to eliminate leukemic stem- and progenitor cells in acute myeloid leukemia AML. Oncotarget. 2012;3:1588–99. 10.18632/oncotarget.733.
200. Gasiorowski RE, Clark GJ, Bradstock K, Hart DNJ. Antibody therapy for acute myeloid leukaemia. Br J Haematol. 2014;164:481–95. https://doi.org/10.1111/bjh.12691.
201. Petersdorf SH, Kopecky KJ, Slovak M, Willman C, Nevill T, Brandwein J, Larson RA, Erba HP, Stiff PJ, Stuart RK, Walter RB, Tallman MS, Stenke L, Appelbaum FR. A phase 3 study of gemtuzumab ozogamicin during induction and postconsolidation therapy in younger patients with acute myeloid leukemia. Blood. 2013;121:4854–60. https://doi.org/10.1182/blood-2013-01-466706.
202. Amadori S, Suciu S, Selleslag D, Aversa F, Gaidano G, Musso M, Annino L, Venditti A, Voso MT, Mazzone C, Magro D, De Fabritiis P, Muus P, Alimena G, Mancini M, Hagemeijer A, Paoloni F, Vignetti M, Fazi P, Meert L, Ramadan SM, Willemze R, de Witte T, Baron F. Gemtuzumab ozogamicin versus best supportive care in older patients with newly diagnosed acute myeloid leukemia unsuitable for intensive chemotherapy: results of the randomized phase III EORTC-GIMEMA AML-19 trial. J Clin Oncol. 2016;34:972–9. https://doi.org/10.1200/JCO.2015.64.0060.
203. Gill S, Tasian SK, Ruella M, Shestova O, Li Y, Porter DL, Carroll M, Danet-Desnoyers G, Scholler J, Grupp SA, June CH, Kalos M. Preclinical targeting of human acute myeloid leukemia and myeloablation using chimeric antigen receptor-modified T cells. Blood. 2014;123:2343–54. https://doi.org/10.1182/blood-2013-09-529537.
204. Kenderian SS, Ruella M, Shestova O, Klichinsky M, Aikawa V, Morrissette JJD, Scholler J, Song D, Porter DL, Carroll M, June CH, Gill S. CD33-specific chimeric antigen receptor T cells exhibit potent preclinical activity against human acute myeloid leukemia. Leukemia. 2015;29:1637–47. https://doi.org/10.1038/leu.2015.52.
205. Ross JF, Wang H, Behm FG, Mathew P, Wu M, Booth R, Ratnam M. Folate receptor type beta is a neutrophilic lineage marker and is differentially expressed in myeloid leukemia. Cancer. 1999;85:348–57.
206. Wang H, Zheng X, Behm FG, Ratnam M. Differentiation-independent retinoid induction of folate receptor type beta, a potential tumor target in myeloid leukemia. Blood. 2000;96:3529–36.
207. Lynn RC, Poussin M, Kalota A, Feng Y, Low PS, Dimitrov DS, Powell DJJ. Targeting of folate receptor beta on acute myeloid leukemia blasts with chimeric antigen receptor-expressing T cells. Blood. 2015;125:3466–76. https://doi.org/10.1182/blood-2014-11-612721.

Chapter 14
When the Molecules Start Playing Chess, or How MicroRNAs Acquire Dualistic Activity During Cancer Progression

Krassimira Todorova and Soren Hayrabedyan

Abstract Genomic instability was found to be a major source of a chromosomal rearrangement resulting in multidimensional gene network reprogramming, providing survival benefits to cancer cells. One recently discovered phenomenon caused by these changes is the dualistic microRNA activity, converting tumor suppressor microRNAs into tumor promoting omcomiRs. Understanding mechanics of this dualism will reveal how far tumor progression could go. This chapter will discuss some aspects of the current knowledge and understanding of this side of oncogenesis.

Keywords Dualistic microRNA function · miR-204 · miR-30 · Prostate cancer · Autophagy · Genome instability

Introduction

Cancer development has been investigated for many years, with new molecular players constantly being discovered, shedding light on the mechanics of oncogenesis and cancer progression. One of the few major shifts in our current understanding of cancer evolvement and progression is related to its multidimensional character. Not unlikely, disciplines as systems biology, as well as methods applying high-throughput multidimensional data gathering and analysis helped reveal its nature.

Thus, along with the discovery of several new layers of genome organization and gene regulation, new molecular players came to the scene. Epigenomics currently deals with the complex organization of the genome structure and gene regulation, including additional molecular player such as non-coding RNAs (ncRNAs).

K. Todorova · S. Hayrabedyan (✉)
Institute of Biology and Immunology of Reproduction "Acad. Kiril Bratanov", Bulgarian Academy of Sciences, Laboratory of Reproductive OMICs Technologies, 73, Tsarigradsko shosse blvd, 1113 Sofia, Bulgaria
e-mail: shayrabedyan@ibir.bas.bg

S. Fayyaz, A. A. Farooqi (eds.), *Recent Trends in Cancer Biology: Spotlight on Signaling Cascades and microRNAs*, https://doi.org/10.1007/978-3-319-71553-7_14

The two most prominent types of ncRNAs are microRNAs (miRs) and long non-coding RNAs (lncRNAs). First type (miRs) are short 21–23 nucleotides long ncRNAs, transcribed as longer precursors in nucleus, then exported to cytosol where Dicer enzyme process mature form capable of silencing multiple mRNAs per single miR [1]. They are tightly involved in cell development, signaling, cell cycle [2], but also in carcinogenesis and mutagenesis [3, 4]. Their crucial tumor suppressor role was found through the studies of its abolishment in cancerogenesis and its link to chromosome deletions in leukemia [5]. A miR locus has been found harbored by many chromosomal regions including coding regions for RNA, introns, and intergenic regions. So far over 1881 precursors and 2588 mature miRs are identified in human [6], although far less miRs are real than bioinformatically predicted, suggesting their tertiary structure and real interactors importance. Some of them could act as tumor inducing microRNAs, referred to as oncogenic miRs referred to as oncomiRs. In general, miRs that become dysregulated in cancer are of specific importance, as they are oncogenesis responsible either through targeting and downregulating tumor-suppressor genes or getting downregulated when targeting protooncogenes. For example, the tumor suppressor miR-21 targeting PTEN is upregulated in a several cancers [7], while miR-145 targeting MYC, MUC1, and OCT4 oncogenes is downregulated in many cancers, like colon [8], breast [9] and lung [10] cancer. One important aspect of their physiology is the regulation of their biogenesis by important steroid hormones such as androgens and estrogens [11, 12].

Current Evidence for Dualistic MicroRNA Activity Phenomena

One of the first phenomena of microRNA functional biological dualism have been reported about miR-26a [13], suggesting that miR-26a played a role of tumor suppressor in hepatocellular carcinoma acting restrictively on cyclins D2 and E2 [14], while same miR-26a played a role of tumor promotor (an oncomiR) suppressing PTEN in high-grade glioma [15]. While in hepatocellular carcinoma this miR was reduced (suppressed), in glioma a genome amplification upregulated miR-26a, suggesting a genome dysregulation and rearrangement underlies the biological duality phenomenon.

So far, two types of duality are exhibited by microRNAs—*duality in activity depending on cell type context*, thus specific microRNA could be a tumor suppressor in one cell type, but could act as oncomiR in another cell type, as shown for miR-26a. Several other miRs already have been shown to exhibit such functions, like miR-30 playing role of an oncomiR in glioma, ovarian cancer, renal cell cancer and chronic myeloid leukemia, while acting as a tumor suppressor in colorectal, gastric, lung, prostate and breast cancer [16]. Similarly, miR-23b and miR-7, also act as oncomiRs in glioma [17] and renal cell carcinoma [18], and as tumor suppressors in bladder [19] and breast cancer [19] respectively.

Another microRNA exhibiting such cell context-dependent duality is miR-204. Although it is among the most abundant acute lymphocytic leukemias oncomiRs, miR-204 however acts at the same time as a tumor suppressor in many solid tumors, like melanomas, glioma, non-small cell lung cancer, bladder and gastric cancers, head and neck tumor, and endometrioid endometrial cancer (for Review see [20]).

Surprisingly, this miR was found by several groups, including ours, to act dualistically in the same cell type, during different stages of cancer progression [20–22]. In prostate cancer miR-204 was found to act as tumor suppressor in metastatic, but lacking TMPRSS2:ERG oncofusion (DU145, LNCaP and PC3 cell lines) [20, 21], while miR-204 obtained oncomiR properties in same fusion harboring cells (VCaP, NCI H600 cell lines) [21]. Similar duality was found in breast cancer, where miR-204 acted as oncogenic, targeting several tumor suppressor genes in breast cancer cell lines MDA-MB-231 and MCF-7 [22], but it was also found to act as a tumor suppressor in same MDA-MB-231 cell line by others [23], inducing cell apoptosis inducer through feed-forward cascade between STAT3, BCl-2 and survivin.

This *second mode of duality* is defined by an action of the same microRNA as tumor suppressor, and as an oncomiR *in the same cell type*. Like miR-204, miR-375 behaves as oncogenic in androgen-sensitive 22Rv1 prostate cancer cell line, but as tumor-suppressive in androgen-insensitive PC-3 prostate cancer cell line [24]. While miR-204 activity in prostate cancer changes during genome instability progression, in breast cancer this microRNA and miR-17 exhibit same duality in the same stage of cancer, but in different directions. This way, miR-204 targets both tumor suppressor and oncogenic proteins in MDA-MB-231 cells [24], and so does miR-17 in MCF-7 cells by suppressing cell proliferation via AIB118 targeting, but inhibiting the HBP1 tumor suppressor [25]. A tumor suppressor in many tumors, miR-29, also acts as an oncomiR in breast cancer, facilitating epithelial-to-mesenchymal transition (EMT) and metastasis [26]. Despite exhibiting both oncogenic and tumor suppressive properties, miR-27a, a component of the miR-23a~27a~24-2 polycistronic cluster, also may either facilitate or suppress apoptosis depending on cell type [27, 28].

Thus, all these phenomena converge in and actually coexist with miRNA-targeted genes functional duality as all, as a reflection of the complex non-coding RNAs epigenetic regulation [14]. For example, this complexity at gene and protein level is manifested also by a cell senescence mediated by not-other but oncogene members of RAS, MYC and E2F family [29]. To understand the underlying mechanics of the non-coding microRNAs duality we should further investigate the molecules and biological processes involved that are their downstream targets.

Biological Outcomes of MicroRNA Duality Inferring this Phenomenon Origin

A common feature of microRNAs playing dualistic role in biology of cancer development is the transformation of tumor suppressor to tumor inducing activity [24]. This transformation is usually attained by passing through a phase of bipolar activity where some targets remain pro-oncogenic, hence tumor suppressor activity is preserved, while other new targets are acquired, this time represented by tumor suppressor genes, conferring oncomiR activity to the microRNA. Since most of the studies were initially focused on specific miR target proteins, there have been controversial reports on particular miR activity [24, 25]. Employing more high throughput methods detecting multiple microRNA targets simultaneously allowed for the discovery of this transitional phenomenon of losing tumor suppressor for pro-oncogenic activity of microRNAs [21, 24, 30].

The initial focus on duality research suggested most duality microRNAs affected the expression of major tumor suppressors like PTEN (affected by miR-26a [14], miR-30a-5p [31]); protooncogenes/master transcriptional regulators, involved in cell differentiation and proliferation, such as myc family members MYC, MYB (miR-26a [13, 14, 32], miR-204 [21]), Runt family members RUNX2, RUNX3 (miR-204 [21, 33], miR-30a [34]), Notch1 (miR-30a [34]); epigenome regulators as EZH2 (miR-25a [13], miR-30a [35]); some long non-coding RNAs LncRNA HOTTIP ([36]); epithelial to mesenchymal transformation SOX4 (miR-204 [37], miR-30a [38]), SLUG (miR-204 [39]), SNAI1 [40] and cancer stem-cell like induction pathways such as Wnt/ β-catenin signaling (miR-204 [41], miR-30a [42]); cell cycle regulators Cyclin D2, E2 (miR-26a [14], miR-204 [43]), autophagy genes BECN1, ATG5 (miR-30a [44, 45]), LC3B (miR-204 [46]), invasion and migration PI3K/AKT genes (miR-30a [40]) and also ubiquitination genes like ISOT, UBA5, NED88, TRIP12, E3 Ubiquitin ligases ARIH2 and UBR5 (miR-204 [30]).

As a general common scenario, a multidimensional gene regulatory network of master transcriptional factors and down-stream regulators is affected by the cancerogenesis to confer already known loss of differentiation, cell apoptosis inhibition, epithelial to mesenchymal transformation, along with the acquisition of stem-like properties and higher metastatic potential.

Alike gene regulation, cancerogenesis is not a coordinated activity, but it has a pattern as most of the newer enlightenments in the area suggest, and that pattern is related to specific "weak spots" in genome organization and gene regulation. Every single "weak spot" is always backed up by multiple feedback mechanisms, so only multidirectional gene dysregulation, resulting in simultaneous disorganization of many cellular functions could sustain cancer development and progression.

We will follow a hypothetical scenario driven mostly by the research of microRNA dysregulation in prostate cancer by us and others to depict some aspects of the "peculiar" transition from physiological tumor-suppressed state to oncogene non-differentiated state. A hallmark in every cancer progression is the dysregulation of the epigenome as well, as major mechanism of sustaining the cell differentiated state.

It is to note that prostate epithelial cells are under androgen hormones control through the regulation of the androgen receptor (AR), latter having a highly complex regulatory mechanics on its target downstream genes through androgen response elements [11], including some microRNAs, like miR-21 and others [47]. It has been recently discovered that AR and other steroid receptors (like estrogen one) play a major role in Dicer-mediated microRNA maturation, in a ligand specific manner, hence exhibiting synthesis rate limited regulatory function, as microRNAs are synthesized as primary miRs (pri-miRs) by RNA Pol II and are later converted to mature miRNAs by the RNAse enzymes Drosha and Dicer [48].

It was found that tissue-specific knockout of Dicer completely impairs AR function leading to an androgen-insensitivity syndrome [12]. miRNAs were shown as mediators of AR function with a possible feedback loop between miRNAs, AR, and AR corepressors. Unlike classical hormone action theory postulate that the ligand-bound AR is recruited to the promoter of its target genes, it was found surprisingly that miRNAs are also mediators of the androgen action. Consequently a new hypothesis have been raised, including a 3-step rather than 1-step model of AR activation, where instead of simple ligand-receptor interaction, AR gene regulation is mediated through its action on miRNA maturation via Dicer-AR ligand dependent interaction, an interaction with AR corepressors and coactivators and AR-DNA direct interaction [12]. The importance of this phenomenon becomes clear in the development of cancer-resistant phenotype in prostate cancer, as some of the dualistic micro-RNAs noted so far exert a regulatory function on AR on their own [16, 20, 21]. The following positive feedback loop between AR, miR-204, XRN1 and miR-34a has already been observed, implicating a dual function of the axis between miR-204 and XRN1 in prostate cancer [49]. Androgens, by downregulating miR-204, induce a 5′-3′ exoribonuclease 1 named XRN1, which in turn is a miR-204 target itself, forming a feed-forward regulatory loop. Cancer progression stage-wise, in AR-sensitive prostate cancer cell lines (LNCaP and 22Rv1) miR-204 has been implicated to acts as a tumor suppressor, while in more advanced neuroendocrine-like differentiated prostate cancer cell lines (PC-3 and CL1), miR-204 has been implicated as an oncomiR. In latter case (cell lines) AR is transcribed, but it is not sensitive to androgen hormones signaling. It was shown that AR can induce miR-204 [49], while unlike others [49], we found miR-204 overexpression on other hand to upregulate AR expression in prostate cancer cells [30]. The other dualistic micro-RNA-30a and members of its family were also shown to play a role in AR suppression by targeting its 3′UTR. Those miR-30 members had lowered expression levels in castration-resistant prostate cancer cells [50].

As androgen/AR is able to suppress miR-204 expression [49] in AR-sensitive LNCaP and other prostate cancer cells, it becomes clear why the lower levels of miR-204 in LNCaP and PC3 cells have tumor suppressor activity, while higher levels of miR-204 in TMPRSS2:ERG fusion positive prostate cancer cell lines (VCaP, NCI H660) had pro-oncogenic properties [21]. In first case, lower miR-204 levels coincide with AR hyperstimulation associated with high androgen hormone levels in prostate cancer, and the ability of high AR stimulation to cause chromosomal remodeling and AR dependent TMPRSS2 gene fusion with transcription factor ERG [51–54]. Before TMPRSS2:ERG fusion to occur, the relatively lower miR-204 levels (but still upregulated when compared to normal prostate epithelia) stimulate AR overexpression, but as we have discovered the TMPRSS2:ERG fusion positive VCaP and NCI H660 prostate cancer cell lines exhibited higher miR-204 levels, coinciding with lower or none AR expression as we found [21, 30]. Since this fusion is believed to occur as unnatural splicing intra-chromosomal rearrangement to “enhance” ERG expression and “put it under” AR control, it was not surprising for us why AR inducing and TMPRSS2:ERG suppressing miR-204 is increased in ERG-fusion positive cells [30]. The fusion itself was found to employ a very complex regulatory mechanism, first by regulating the non-fused ERG locus [53–61] and second by utilizing an epigenetic mechanism regulating Polycomb gene EZH2 to change DNA methylation state [51, 52, 55], conferring cancer stem-like and epithelial-to-mesenchymal transition characteristics of prostate cancer cells, and castration resistant phenotype acquisition. The oncogenically induced EZH2 overexpression also exhibited dualistic biological function by inducing in turn an increased cell ageing process [29], as TMPRSS2:ERG fusion transfection resulted in cell proliferation arrest in non-fusion harboring cells [62]. Now, it is not surprising that miR-204 over-expression in ERG-fusion cells could be interpreted as oncomiR activity, as miR-204 is actually promoting an even more malignant phenotype by inducing AR and suppressing TMPRSS2:ERG fusion, helping tumor cell bypass the transitional state and starting to use ERG-fusion Polycomb and other pro-oncogenic pathways [30]. Thus in cells where AR is not sensitive to androgens, but have an internal and reprogrammed signaling as we will talk latter, TMPRSS2:ERG fusion promotes metastasis, but is regulated by former tumor-suppressor miR-204 to reduce its cell senescence activity. Once initiated, these ERG-fusion pathways have other mechanisms for positive feedback loops, and do not need the androgen receptivity and ERG-fusion transcript anymore [63].

What we found in our studies on miR-204 dualistic activity, suggested that in ERG-fusion positive prostate cancer cells, miR-204 participated in several feedback and feed-forward loops with master proliferation and metastasis transcription factors (TFs) as MYB, RUNX2 and miR-204 harboring ETS1. What could explain miR-204 switch form suppressing into inducing miR regarding this TFs is that the regulatory loops are not direct, hence further mutagenesis or perturbations of other regulatory elements could explain the switch.

The feed-forward loops MYB(RUNX2)-miR204-MYB(RUNX2)-SLUG/E-Cadherin and ETS1-miR204-SLUG/E-Cadherin were found in ERG-fusion cells, and while MYB(RUNX2)-miR-204 were working on some of the ETS TFs in EGR-negative cells as well, MYB, RUNX2 and ETS1 were all required for SLUG ETS gene expression in ERG-negative cell lines [39]. This finding demonstrated that ERG-fusion phenomenon is necessary to unlock set of events and feed-forward loops where even some pro-oncogenic master TFs are not necessary for EMT anymore, as Polycomb silencing program is switching off the differentiated cell state. Thus, mir-204 over-expression upregulated EMT-related N-cadherin regardless of ERG-fusion state, while E-cadherin was downregulated only in ERG-positive cells [39], being strongest in NCI-H660 cells, possibly related to an additional malignant tumors associated E-cadherin alternative splicing exclusion of exon-11 [64]. We also observed E-cadherin alternative splicing phenomenon regardless of ERG-status [39].

Additional to Polycomb mechanism, ERG-fusion utilizes another prostate epithelia differentiation disruption mechanism, via SOX9 [65], which is not direct AR target, but being an ERG target is also targeted by AR-induced TMPRSS2:ERG. SOX-9 was fond upregulated by miR-204 in all cell lines, while SOX9 downregulation by inhibiting miR-204 was only possible in ERG-fusion harboring cells [39], suggesting the importance of miR-204 for Sox-9 based epithelial linage dysregulation and an even further "fine-tuning" of miR-204 functionality with the ERG-fusion occurrence. In concordance with effects of miR-204 on EMT changes, miR-204 suppressed cell viability and migration only in TMPRSS2:ERG fusion-negative cells [39].

The bone-specific transcription regulator Runx2 is abnormally expressed in highly metastatic hematopoietic stem cell niche, facilitating local osteolysis [66]. Key event to this process is the induction by RUNX2 of key male/prostate differentiation (SOX9) and EMT molecules (Snail2, SLUG, SNAI2, SMAD3) that promote significantly the invasiveness of cancer cells. SLUG was reported to upregulate RUNX2 [67], but we have also found that RUNX2 upregulated SLUG in positive feedback loop [39]. Indeed, recent data show that RUNX2 is able to bind AR and detach it from its target genes like some tumor suppressors, but at the same time, the two together, RUNX2 and AR, are able to bind the SLUG enhancer region and result in an increased invasiveness [68].

This phenomenon adds yet another layer of complexity in AR signaling where miR-204 participated, helping "reprogram" multiple AR transcriptional targets in again dualistic fashion [30]. We have investigated the interference of miR-204 in AR signaling using high resolution proteomics recently, only to discover that large set proteins encoded by miR-204 target genes were actually upregulated in AR sensitive prostate cancer cells conversely to expected downregulation, and only small part was downregulated [30].

Among the miR-204 significantly overexpressed proteins were Methyl-CpG-binding domain protein 3 (MBD3), downregulation of DNA (cytosine-5)-methyltransferase 1 (DNMT1), enrichment of several histone regulation enzymes—histone methyltransferases, histone arginine methylation, and epigenetic proteins related to DNA methylation, Protein arginine *N*-methyltransferase 5 (PRMT5) methylating, which Arg-3 of histone H3 in glioblastoma and other tumors, as well as tRNA methylation. Dysregulated by miR-204 were also AR direct target, AR co-activators and co-repressors, AR regulators, and AR transcription starting sites related TFs. In parallel, we have found that the actions of miR-204 towards AR and TMPRSS2:ERG expression was mediated by miR-204 mediated regulation of their promotor methylation. Thus, although Histone deacetylases (HDACs) cause transcriptional repression by making histones wrap DNA more tightly, in prostate cancer HDACs are also paradoxically required for the activation of a substantial fraction of AR target genes including TMPRSS2. TMPRSS2 participates in the ERG-fusion, reprograming its transcription under AR control [69, 70]. Using high resolution proteomics, we found histone deacetylase complex of MBD3, MBD2 and HDAC1 to be enriched by miR-204 over-expression [30], further supporting the hypothesis that miR-204 potentiates TMPRSS2:ERG action in prostate cancer, although the fusion itself was downregulated simultaneously by promoter hyper-methylation. In prostate cancer context miR-204 serves as fine tuning regulator of both the ERG-fusion to prevent cell senescence and of epigenetic machinery to promote tumor progress. This notion is supported by the finding that ERG-positive prostate cancers are strongly HDAC1-positive.

We have also another direction of similar bi-polar miR-204 activity of increased AR expression, by miR-204 dependent promotor hypo-methylation, probably related to significant DNMT1 downregulation. This was accompanied by an upregulation of SWI/SNF complex associated SMARCC1 subunit, involved in chromatin remodeling [30]. Remarkably, the SMARCC1 subunit participates in neural progenitor(stem) cell specific chromatin remodeling complexes, important for the self-renewal and proliferative capacity of the multipotent neural stem cells [71]. Overall, we found that miR-204 helps in epigenetic reprograming of AR signaling axis, proliferation and EMT, and in autophagy and ubiquitination reprograming as well [11, 21, 30, 39].

We suspect that part of the action of miR-204 on upregulation of RUNX2 and MYB TFs in ERG-fusion manner is also related to ERG-fusion and AR-sensitivity related ubiquitination reduction, resulting in a decreased protein products turnover [21]. Autophagy is process of non-selective starvation and selective degradation of organelles and misfolded proteins, which is either beneficial or suppressive to cancer development.

About 40% of the human genome is represented by retrotransposons—genetic elements that are ubiquitously presented in the DNA and are able to amplify themselves by something like a "copy-paste" mechanism. The retrotransposone first gets transcribed into RNA, then it is converted to DNA by reverse transcription and inserted into DNA target sites [72]. Their ability to change their position and direction within the genome represents major source of genetic variation in individuals, tissues and cells, as retrotransposons represent 42% of the entire genome in humans [73]. Autophagy prevent genome instability by selective retrotransposon RNA degradation, preventing their genome reinsertion, controlling both long and short interspersed elements. Although normal autophagy is required for NDP52 and p62-mediated selective retrotransposon degradation and genome stability preservation, with the advance of cancer development both autophagy degradation of misfolded proteins is decreased and also the process of genome stability preservation [74]. The inability to selectively clear genome destabilization events eventually result in irregular splicing events and further supports genome reorganization and *de novo* fusion products generation. miR-204 and miR-30 could selectively suppress retrotransposon clearance by inhibiting beclin1 [74] and atg16 [75] from autophagy initiation pathway.

We already found miR-204 involved in upregulation of autophagy initiation ATG16L and maturation important LC3 in AR sensitive LNCaP cells, as without any of these two proteins autophagy process could not propagate to successful degradation. In PC3 AR-resistant cells both factors were downregulated suggesting another oncomiR activity [75]. Similarly, miR-204 was found to regulate another initiation factor—Beclin1 and also LC3 as well, in renal cell carcinoma [46]. In renal cell carcinoma, autophagy, which also exhibits dualistic role in cancer is required for tumor growth. Recently, both miR and autophagy duality were related, as not only miRs regulate autophagy involved proteins at every single stage, but also a complex interaction between the autophagy and miRNA biogenesis and maturation has been recently found. This way autophagosomes degrade selectively miRNA-maturation indispensable DICER1, and the argonaute 2 (AGO2) component of RISC silencing complex [76]. Reciprocally, DICER1 silencing attenuates the induction of autophagy in promyelocytic leukemia differentiation. Similar phenomenon have been observed in ATG5 and ATG16 knockout mouse embryonic fibroblasts, where AGO2 is vastly accumulated [77]. Although it was though initially that only RNA-free DICER1 and AGO2 proteins are degraded by autophagy, new evidence suggested several miRNAs to be directly degraded by autophagy, suggesting that miRNA regulation-autophagy degradation work in both ways. This concept further extended beyond simple miR-regulated autophagy feedback/feedforward loops, rather having an impact on the global cellular non-codding RNA landscape [78].

Bringing all together, miR-204 and other miRs, like miR-30, exhibiting similar directions of duality, interplay in very complex way with processes like epigenetic regulation, transcriptional regulation, autophagy and ubiquitination. They are both regulators and being regulated by these processes. One additional aspect of microRNAs we would like to discuss is a yet another source of multi-dimensionality in their genesis and regulation that could explain their ability for biological and functional duality.

Genome Organization and When Molecules Start Playing Chess, by Converting Small Pawns into a Queen

As we mentioned above, part of the mechanics of acquiring mutants and gene rearrangement is related to changes in RNA splice patterns. One of the outcomes of the genome rearrangement during prostate cancer progression is placing the oncogenic transcription factor ERG under the androgen-regulated TMPRSS2 promoter control thus "reprograming" ERG induction and alternative allele regulation [52, 69]. The newly formed TMPRSS2:ERG fusion gene exhibits different alternative splicing patterns between invasive and localized prostate cancer [70]. The switching among different alternative androgen receptor mRNA isoforms underlies a crucial mechanism of prostate cancer drug resistance generation, enabling prostate cancer cells growth in limited serum androgen conditions [79]. Not only AR, but also AR-regulated genes undergo alternative splicing in prostate cancer, providing additional dimension to AR targets "reprogramming", opening new avenues for even more complex gene regulation. Following this pattern, besides the full-length transcript of tumor suppressor TSC2 (Tuberous Sclerosis 2) gene, after androgen stimulation, another shorter (C-terminal only) mRNA isoform is alternatively spliced. While full-length TSC2 suppresses cell growth by autophagy inhibiting mTOR pathway, the truncated C-terminal only isoform actually promotes cell growth. This is possible, as an additional internal promoter exists allowing direct transcription of its downstream C-terminal exons only [79]. Lots of mRNA isoforms in general could preserve, but also could lose their ability to remain tumor-suppressor miR targets, as is the case with the ones observed in miR-204 overexpressed prostate cancer cells.

Alternative splicing of androgen receptor gene generates new mRNA isoforms enabling prostate cancer hormone resistant state with so far about 20 isoforms already been implicated [reviewed in [79]]. Arv7 is the most frequent hormone resistant splicing isoform. It is produced by the switch of the terminal exon with a new one, referred to as CE3, residing cryptically within intron 3. The alternative splicing generates new truncated AR mRNA, lacking the exons encoding Ligand-Binding Domain (LBD), but also having a new poly(A) tail, subject to different non-coding RNA regulation. Unlike the full-length form, ARv7 shorter form is constitutively active [80]. ARv7 expression is controlled by two regulatory layers—the prostate cancer overexpressed transcription factors Myc and NFκβ2 [81], and also by this TFs targeting miRs—miR-26a and miR-204. Indeed, we have shown recently that miR-204 increased NF-κB transactivation [82].

Prostate cancer splicing changes are associated with changes in splicing regulator proteins, exhibiting altered prostate cancer expression patterns, including the neuro-endocrine castration-resistant form, represented by the PC3 cell line. Thus, the development of neuroendocrine form of prostate cancer was found to be promoted by an increase in the splicing factor SRRM4 (serine/arginine repetitive matrix 4 protein) expression. Another splicing regulator, Sam68 is also upregulated in prostate cancer [83]. Sam68 is directly regulated by miR-204 and promotes breast cancer cells self-renewal potential by Wnt/Beta-Catenin pathway and by miR-30 dual activity miR. Thus miR-204 and Sam68 are involved in splicing of the metastasis important CD44 mRNA isoforms, apoptosis related BCLx and cell cycle related CCND1 splice isoforms [83].

The analysis of human and mouse genome and transcriptome have also revealed remarkable phenomena of both strands overlapping transcription in “transcriptional forests”, separated by “transcriptional deserts”, having only few transcripts [84]. Genome-wide transcription studies revealed a remarkable transcription pattern, where almost both entire DNA strands of the human genome are transcribed, pointing toward an extensive overlap of DNA transcriptional units and DNA regulatory elements. Indeed, genome data indicate that the amount of undergoing transcription cannot be entirely explained by the genome-wide annotations and the genomic architecture was postulated to be not colinear, but instead having an interleaved and modular design, where many genomic sequences are multifunctional. Multifunctional sequences are used to encode for multiple, independently regulated transcripts, and as regulatory regions as well [85]. Thus, a complex hierarchy of overlapping isoforms is defined, as transcriptional foci genomic sequences are often shared within a number of different coding and non-coding transcripts in the *sense* and *antisense* directions [86]. This allowed for 3012 sequences to be recognized as genuine ncRNA variants of protein-coding cDNAs, among 8961 previously annotated as truncated coding sequences by FANTOM2 project [84]. While these interleaved arrangements abundance and conservation do suggest a strong biological relevance, their complexity is almost impossible to be analyzed currently. This multifunctionality of the genome encoding, and the functions of another class of long non-coding RNAs, being able to sponge the micro-RNAs could further explain the mechanics of miR duality. Even more, some non-coding RNAs have been shown to have double functionality, serving both as non-coding and as a protein coding RNAs. miR-204 also repressed the expression of lncRNA HOTTIP by interfering with RISK silencing component AGO2 in hepatocellular carcinoma [36], suggesting how lncRNAs expression could be controlled by short ncRNAs, such as miR-204. The gene encoding for this lncRNA, is located at the 5′ of the HOXA locus, coordinating several 5′ HOXA genes activation by chromosomal looping, leading to a spatial relocation of lncRNA HOTTIP in close proximity with those genes. HOTTIP binds WDR5, forming a complex with a histone methyltransferase resulting in H3K4 methylation and transcriptional activation of the HOXA locus. In breast cancer, miR-204 interacts with another lncRNA [22]. Quantitatively, lncRNAs are ten-times less abundant than mRNAs in cells [87], due to their higher cell-to-cell variation of expression levels [88]. The majority of lncRNAs are highly tissue- and cell-specific, unlike mRNAs, where only ~19% exhibit such specificity [87].

Overall, the nucleosomal occupancy that is regulated by histone methyl transferases controls the "open" chromatin states and provides means together with other organizing proteins for bringing together in three-dimensional space of distinct genomic loci. Thus cell-dependent context of miRs action could be preserved, as in different cell types, this epigenome regulatory landscape differs, conferring distinct targets to distinct microRNAs.

Overall, the multidimensionality of the genome organization underlay the multidimensionality of genome perturbation and its many manifestations.

In conclusion, if we imagine that the cell is a chessboard and the molecules such as microRNAs and transcriptional factors are the figures on that board, then for the microRNAs we could look at as pawns, that although being the smallest non-coding RNAs, have the power of a Queen once they reach the end of the board and got matured.

Acknowledgements The authors would like to acknowledge the financial support of the Bulgarian National Research Fund (Grant # DCOST 01/23, 2016).

References

1. Bartel DP. MicroRNAs: genomics, biogenesis, mechanism, and function. Cell. 2004;116(2):281–97.
2. Harfe BD. MicroRNAs in vertebrate development. Curr Opin Genet Dev. 2005;15(4):410–5.
3. Bommer GT, Gerin I, Feng Y, Kaczorowski AJ, Kuick R, Love RE, et al. p53-Mediated activation of miRNA34 candidate tumor-suppressor genes. Curr Biol. 2007;17(15):1298–307.
4. Ma L, Teruya-Feldstein J, Weinberg RA. Tumour invasion and metastasis initiated by microRNA-10b in breast cancer. Nature. 2007;449(7163):682–8.
5. Calin GA, Dumitru CD, Shimizu M, Bichi R, Zupo S, Noch E, et al. Frequent deletions and down-regulation of micro- RNA genes miR15 and miR16 at 13q14 in chronic lymphocytic leukemia. Proc Natl Acad Sci U S A. 2002;99(24):15524–9.
6. Kozomara A, Griffiths-Jones S. MiRBase: annotating high confidence microRNAs using deep sequencing data. Nucleic Acids Res. 2014;42(Database issue):D68–73.
7. Meng F, Henson R, Wehbe-Janek H, Ghoshal K, Jacob ST, Patel T. MicroRNA-21 regulates expression of the PTEN tumor suppressor gene in human hepatocellular cancer. Gastroenterology. 2007;133(2):647–58.

8. Sachdeva M, Zhu S, Wu F, Wu H, Walia V, Kumar S, et al. p53 represses c-Myc through induction of the tumor suppressor miR-145. Proc Natl Acad Sci U S A. 2009;106(9):3207–12.
9. Sachdeva M, Mo YY. MicroRNA-145 suppresses cell invasion and metastasis by directly targeting mucin 1. Cancer Res. 2010;70(1):378–87.
10. Yin R, Zhang S, Wu Y, Fan X, Jiang F, Zhang Z, et al. microRNA-145 suppresses lung adenocarcinoma-initiating cell proliferation by targeting OCT4. Oncol Rep. 2011;25(6): 1747–54.
11. Todorova K, Mincheff M, Hayrabedyan S, Mincheva J, Zasheva D, Kuzmanov A, Fernández N. Fundamental role of microRNAs in androgen-dependent male reproductive biology and prostate cancerogenesis. Am J Reprod Immunol. 2013;69(2):100–4.
12. Narayanan R, Jiang J, Gusev Y, Jones A, Kearbey JD, Miller DD, et al. MicroRNAs are mediators of androgen action in prostate and muscle. PLoS One. 2010;5(10):e13637.
13. Sander S, Bullinger L, Klapproth K, Fiedler K, Kestler HA, Barth TFE, et al. MYC stimulates EZH2 expression by repression of its negative regulator miR-26a. Blood. 2008;112(10):4202–12.
14. Huse JT, Holland EC. Yin and yang: cancer-implicated miRNAs that have it both ways. Cell Cycle. 2009;8(22):3611–2.
15. Kota J, Chivukula RR, O'Donnell KA, Wentzel EA, Montgomery CL, Hwang H-W, et al. Therapeutic microRNA delivery suppresses tumorigenesis in a murine liver cancer model. Cell. 2009;137(6):1005–17.
16. Yang X, Chen Y, Chen L. The versatile role of microRNA-30a in human cancer. Cell Physiol Biochem. 2017;41(4):1616–32.
17. Chen L, Han L, Zhang K, Shi Z, Zhang J, Zhang A, et al. VHL regulates the effects of miR-23b on glioma survival and invasion via suppression of HIF-1α/VEGF and β-catenin/Tcf-4 signaling. Neuro-Oncology. 2012;14(8):1026–36.
18. Majid S, Dar AA, Saini S, Deng G, Chang I, Greene K, et al. MicroRNA-23b functions as a tumor suppressor by regulating Zeb1 in bladder cancer. PLoS One. 2013;8(7):e67686.
19. Kong X, Li G, Yuan Y, He Y, Wu X, Zhang W, et al. MicroRNA-7 inhibits epithelial-to-mesenchymal transition and metastasis of breast cancer cells via targeting FAK expression. PLoS One. 2012;7(8):e41523.
20. Li T, Pan H, Li R. The dual regulatory role of miR-204 in cancer. Tumor Biol. 2016;37(9): 11667–77.
21. Todorova K, Metodiev MV, Metodieva G, Zasheva D, Mincheff M, Hayrabedyan S. miR-204 is dysregulated in metastatic prostate cancer in vitro. Mol Carcinog. 2016;55(2):131–47.
22. Lee H, Lee S, Bae H, Kang H-S, Kim SJ. Genome-wide identification of target genes for miR-204 and miR-211 identifies their proliferation stimulatory role in breast cancer cells. Sci Rep. 2016;6:25287.
23. Wang X, Qiu W, Zhang G, Xu S, Gao Q, Yang Z. MicroRNA-204 targets JAK2 in breast cancer and induces cell apoptosis through the STAT3/BCl-2/survivin pathway. Int J Clin Exp Pathol. 2015;8(5):5017–25.
24. Costa-Pinheiro P, Ramalho-Carvalho J, Vieira FQ, Torres-Ferreira J, Oliveira J, Gonçalves CS, et al. MicroRNA-375 plays a dual role in prostate carcinogenesis. Clin Epigenetics. 2015;7(1):42.
25. Li H, Bian C, Liao L, Li J, Zhao RC. miR-17-5p promotes human breast cancer cell migration and invasion through suppression of HBP1. Breast Cancer Res Treat. 2011;126(3):565–75.
26. Gebeshuber CA, Zatloukal K, Martinez J. miR-29a suppresses tristetraprolin, which is a regulator of epithelial polarity and metastasis. EMBO Rep. 2009;10(4):400–5.
27. Mertens-Talcott SU, Chintharlapalli S, Li X, Safe S. The oncogenic microRNA-27a targets genes that regulate specificity protein transcription factors and the G2-M checkpoint in MDA-MB-231 breast cancer cells. Cancer Res. 2007;67(22):11001–11.
28. Guttilla IK, White BA. Coordinate regulation of FOXO1 by miR-27a, miR-96, and miR-182 in breast cancer cells. J Biol Chem. 2009;284(35):23204–16.
29. Mooi WJ, Peeper DS. Oncogene-induced cell senescence – halting on the road to cancer. N Engl J Med. 2006;355(10):1037–46.

30. Todorova K, Metodiev MV, Metodieva G, Mincheff M, Fernández N, Hayrabedyan S. MicroRNA-204 participates in TMPRSS2/ERG regulation and androgen receptor reprogramming in prostate cancer. Horm Cancer. 2017;8(1):28–48.
31. Li WF, Dai H, Ou Q, Zuo GQ, Liu CA. Overexpression of microRNA-30a-5p inhibits liver cancer cell proliferation and induces apoptosis by targeting MTDH/PTEN/AKT pathway. Tumor Biol. 2016;37(5):5885–95.
32. Chang T-C, Yu D, Lee Y-S, Wentzel EA, Arking DE, West KM, et al. Widespread microRNA repression by Myc contributes to tumorigenesis. Nat Genet. 2008;40(1):43–50.
33. Huang J, Zhao L, Xing L, Chen D. MicroRNA-204 regulates Runx2 protein expression and mesenchymal progenitor cell differentiation. Stem Cells. 2010;28(2):357–64.
34. Liu Z, Chen L, Zhang X, Xu X, Xing H, Zhang Y, et al. RUNX3 regulates vimentin expression via miR-30a during epithelial-mesenchymal transition in gastric cancer cells. J Cell Mol Med. 2014;18(4):610–23.
35. Zhang P, Yang X, Ma X, Ingram DR, Lazar AJ, Torres KE, Pollock RE. Antitumor effects of pharmacological EZH2 inhibition on malignant peripheral nerve sheath tumor through the miR-30a and KPNB1 pathway. Mol Cancer. 2015;14:55.
36. Ge Y, Yan X, Jin Y, Yang X, Yu X, Zhou L, et al. fMiRNA-192 and miRNA-204 directly suppress lncRNA HOTTIP and interrupt GLS1-mediated glutaminolysis in hepatocellular carcinoma. PLoS Genet. 2015;11(12):e1005726.
37. Ying Z, Li Y, Wu J, Zhu X, Yang Y, Tian H, et al. Loss of miR-204 expression enhances glioma migration and stem cell like phenotype. Cancer Res. 2012.; Retrieved from http://cancerres.aacrjournals.org/content/early/2012/12/01/0008-5472.CAN-12-2895.abstract
38. Lu N, Lin T, Wang L, Qi M, Liu Z, Dong H, et al. Association of SOX4 regulated by tumor suppressor miR-30a with poor prognosis in low-grade chondrosarcoma. Tumour Biol. 2015;36(5):3843–52.
39. Todorova K, Zasheva D, Kanev K, Hayrabedyan S. miR-204 shifts the epithelial to mesenchymal transition in concert with the transcription factors RUNX2, ETS1, and cMYB in prostate cancer cell line model. J Cancer Res. 2014;2014(840906):1–14.
40. Liu Z, Tu K, Liu Q. Effects of microRNA-30a on migration, invasion and prognosis of hepatocellular carcinoma. FEBS Lett. 2014;588(17):3089–97.
41. He H, Chen K, Wang F, Zhao L, Wan X, Wang L, Mo Z. miR-204-5p promotes the adipogenic differentiation of human adipose-derived mesenchymal stem cells by modulating DVL3 expression and suppressing Wnt/β-catenin signaling. Int J Mol Med. 2015;35(6):1587–95.
42. Zhao J-J, Lin J, Zhu D, Wang X, Brooks D, Chen M, et al. miR-30-5p functions as a tumor suppressor and novel therapeutic tool by targeting the oncogenic Wnt/β-catenin/BCL9 pathway. Cancer Res. 2014;74(6):1801–13.
43. Wu X, Zeng Y, Wu S, Zhong J, Wang Y, Xu J. MiR-204, down-regulated in retinoblastoma, regulates proliferation and invasion of human retinoblastoma cells by targeting CyclinD2 and MMP-9. FEBS Lett. 2015;589(5):645–50.
44. Zheng B, Zhu H, Gu D, Pan X, Qian L, Xue B, et al. MiRNA-30a-mediated autophagy inhibition sensitizes renal cell carcinoma cells to sorafenib. Biochem Biophys Res Commun. 2015;459(2):234–9.
45. Yu Y, Cao L, Yang L, Kang R, Lotze M, Tang D. microRNA 30A promotes autophagy in response to cancer therapy. Autophagy. 2012;8(5):853–5.
46. Mikhaylova O, Stratton Y, Hall D, Kellner E, Ehmer B, Drew AF, et al. VHL-regulated MiR-204 suppresses tumor growth through inhibition of LC3B-mediated autophagy in renal clear cell carcinoma. Cancer Cell. 2012;21(4):532–46.
47. Shi X-B, Xue L, Yang J, Ma A-H, Zhao J, Xu M, et al. An androgen-regulated miRNA suppresses Bak1 expression and induces androgen-independent growth of prostate cancer cells. Proc Natl Acad Sci U S A. 2007;104(50):19983–8.
48. Kai ZS, Pasquinelli AE. MicroRNA assassins: factors that regulate the disappearance of miRNAs. Nat Struct Mol Biol. 2010;17(1):5–10.
49. Ding M, Lin B, Li T, Liu Y, Li Y, Zhou X, et al. A dual yet opposite growth-regulating function of miR-204 and its target XRN1 in prostate adenocarcinoma cells and neuroendocrine-like prostate cancer cells. Oncotarget. 2015;6(10):7686–700.

50. Kumar B, Khaleghzadegan S, Mears B, Hatano K, Kudrolli TA, Chowdhury WH, et al. Identification of miR-30b-3p and miR-30d-5p as direct regulators of androgen receptor signaling in prostate cancer by complementary functional microRNA library screening. Oncotarget. 2016;7(45):72593–607.
51. Leshem O, Madar S, Kogan-Sakin I, Kamer I, Goldstein I, Brosh R, et al. TMPRSS2/ERG promotes epithelial to mesenchymal transition through the ZEB1/ZEB2 axis in a prostate cancer model. PLoS One. 2011;6(7):e21650.
52. Yu J, Yu J, Mani R-S, Cao Q, Brenner CJ, Cao X, et al. An integrated network of androgen receptor, polycomb, and TMPRSS2-ERG gene fusions in prostate cancer progression. Cancer Cell. 2010;17(5):443–54.
53. Tomlins SA, Laxman B, Varambally S, Cao X, Yu J, Helgeson BE, et al. Role of the *TMPRSS2–ERG* gene fusion in prostate cancer. Neoplasia. 2008;10(2):177–88.
54. Wang J, Cai Y, Ren C, Ittmann M. Expression of variant TMPRSS2/ERG fusion messenger RNAs is associated with aggressive prostate cancer. Cancer Res. 2006;66(17):8347–51.
55. Alumkal JJ, Herman JG. Distinct epigenetic mechanisms distinguishTMPRSS2-ERG fusion-positive and -negative prostate cancers. Cancer Discov. 2012;2(11):979–81.
56. Narod SA, Seth A, Nam R. Fusion in the ETS gene family and prostate cancer. Br J Cancer. 2008;99(6):847–51.
57. Yin L, Rao P, Elson P, Wang J, Ittmann M, Heston WDW. Role of TMPRSS2-ERG gene fusion in negative regulation of PSMA expression. PLoS One. 2011;6(6):e21319.
58. Barwick BG, Abramovitz M, Kodani M, Moreno CS, Nam R, Tang W, et al. Prostate cancer genes associated with TMPRSS2-ERG gene fusion and prognostic of biochemical recurrence in multiple cohorts. Br J Cancer. 2010;102(3):570–6.
59. Hu Y, Dobi A, Sreenath T, Cook C, Tadase AY, Ravindranath L, et al. Delineation of TMPRSS2-ERG splice variants in prostate cancer. Clin Cancer Res. 2008;14(15):4719–25.
60. Attard G, Clark J, Ambroisine L, Fisher G, Kovacs G, Flohr P, et al. Duplication of the fusion of TMPRSS2 to ERG sequences identifies fatal human prostate cancer. Oncogene. 2008;27(3):253–63.
61. Burdova A, Bouchal J, Tavandzis S, Kolar Z. TMPRSS2-erg gene fusion in prostate cancer. Biomed Papers. 2014;158(4):502–10.
62. Bastus NC, Boyd LK, Mao X, Stankiewicz E, Kudahetti SC, Oliver RTD, et al. Androgen-induced TMPRSS2:ERG fusion in nonmalignant prostate epithelial cells. Cancer Res. 2010;70(23):9544–8.
63. Fayyaz S, Farooqi AA. miRNA and TMPRSS2-ERG do not mind their own business in prostate cancer cells. Immunogenetics. 2013;65(5):315–32.
64. Sharma S, Lichtenstein A. Aberrant splicing of the E-cadherin transcript is a novel mechanism of gene silencing in chronic lymphocytic leukemia cells. Blood. 2009;114(19):4179–85.
65. Huang Z, Hurley PJ, Simons BW, Marchionni L, Berman DM, Ross AE, Schaeffer EM. Sox9 is required for prostate development and prostate cancer initiation. Oncotarget. 2012;3(6):651–63.
66. Akech J, Wixted JJ, Bedard K, van der Deen M, Hussain S, Guise TA, et al. Runx2 association with progression of prostate cancer in patients: mechanisms mediating bone osteolysis and osteoblastic metastatic lesions. Oncogene. 2010;29(6):811–21.
67. Lambertini E, Franceschetti T, Torreggiani E, Penolazzi L, Pastore A, Pelucchi S, et al. SLUG: a new target of lymphoid enhancer factor-1 in human osteoblasts. BMC Mol Biol. 2010;11:13.
68. Little GH, Baniwal SK, Adisetiyo H, Groshen S, Chimge N-O, Kim SY, et al. Differential effects of RUNX2 on the androgen receptor in prostate cancer: synergistic stimulation of a gene set exemplified by SNAI2 and subsequent invasiveness. Cancer Res. 2014;74(10):2857–68.
69. Farooqi AA, Hou M-F, Chen C-C, Wang C-L, Chang H-W. Androgen receptor and gene network: micromechanics reassemble the signaling machinery of TMPRSS2-ERG positive prostate cancer cells. Cancer Cell Int. 2014;14:34.
70. Lin X, Qureshi MZ, Romero MA, Yaylim I, Arif S, Ucak I, et al. Signaling networks in TMPRSS2-ERG positive prostate cancers: do we need a Pied Piper or sharpshooter to deal with "at large" fused oncoprotein. Cell Mol Biol (Noisy-le-Grand). 2017;63(2):1–8. Retrieved from http://www.ncbi.nlm.nih.gov/pubmed/28364793

71. Kadam S, McAlpine GS, Phelan ML, Kingston RE, Jones KA, Emerson BM. Functional selectivity of recombinant mammalian SWI/SNF subunits. Genes Dev. 2000;14(19):2441–51. Retrieved from http://www.ncbi.nlm.nih.gov/pubmed/11018012
72. Sanmiguel P. Evidence that a recent increase in maize genome size was caused by the massive amplification of intergene retrotransposons. Ann Bot. 1998;82:37–44.
73. Lander ES, Linton LM, Birren B, Nusbaum C, Zody MC, Baldwin J, et al. Initial sequencing and analysis of the human genome. Nature. 2001;409(6822):860–921.
74. Guo H, Chitiprolu M, Gagnon D, Meng L, Perez-Iratxeta C, Lagace D, Gibbings D. Autophagy supports genomic stability by degrading retrotransposon RNA. Nat Commun. 2014;5:1–11.
75. Hayrabedyan S, Mincheff M, Zasheva D, Manolova N, Todorova K. Autophagy signalling is differentially modulated by miR-204 in context of innate immunity induction. Comptes Rendus de L'Academie Bulgare des Sciences. 2013;66(1):127–32.
76. Gibbings D, Mostowy S, Jay F, Schwab Y, Cossart P, Voinnet O. Selective autophagy degrades DICER and AGO2 and regulates miRNA activity. Nat Cell Biol. 2012;14(12):1314–21.
77. Sibony M, Abdullah M, Greenfield L, Raju D, Wu T, Rodrigues DM, et al. Microbial disruption of autophagy alters expression of the RISC component AGO2, a critical regulator of the miRNA silencing pathway. Inflamm Bowel Dis. 2015;21(12):2778–86.
78. Gozuacik D, Akkoc Y, Ozturk DG, Kocak M. Autophagy-regulating microRNAs and cancer. Front Oncol. 2017;7:65.
79. Lu C, Luo J. Decoding the androgen receptor splice variants. Transl Androl Urol. 2013;2(3):178–86.
80. Cao B, Qi Y, Zhang G, Xu D, Zhan Y, Alvarez X, et al. Androgen receptor splice variants activating the full-length receptor in mediating resistance to androgen-directed therapy. Oncotarget. 2014;5(6):1646–56.
81. Nadiminty N, Tummala R, Liu C, Lou W, Evans CP, Gao AC. NF-κB2/p52:c-Myc:hnRNPA1 pathway regulates expression of androgen receptor splice variants and enzalutamide sensitivity in prostate cancer. Mol Cancer Ther. 2015;14(8):1884–95.
82. Todorova K, Zasheva D, Hayrabedyan S. Innate immunity challenge differently modulates inflammatory and apoptosis regulation in lymph node and bone marrow metastatic cell line models, favouring higher metastatic phenotype. Comptes Rendus de L'Academie Bulgare des Sciences. 2014;67(11):1575–82.
83. Munkley J, Livermore K, Rajan P, Elliott DJ. RNA splicing and splicing regulator changes in prostate cancer pathology. Hum Genet. 2017. https://doi.org/10.1007/s00439-017-1792-9.
84. Carninci P, Kasukawa T, Katayama S, Gough J, Frith MC, Maeda N, et al. The transcriptional landscape of the mammalian genome. Science (New York, NY). 2005;309(5740):1559–63.
85. Kapranov P, Willingham AT, Gingeras TR. Genome-wide transcription and the implications for genomic organization. Nat Rev Genet. 2007;8(6):413–23.
86. ENCODE Project Consortium, Birney E, Stamatoyannopoulos JA, Dutta A, Guigó R, Gingeras TR, et al. Identification and analysis of functional elements in 1% of the human genome by the ENCODE pilot project. Nature. 2007;447(7146):799–816.
87. Cabili MN, Trapnell C, Goff L, Koziol M, Tazon-Vega B, Regev A, Rinn JL. Integrative annotation of human large intergenic noncoding RNAs reveals global properties and specific subclasses. Genes Dev. 2011;25(18):1915–27.
88. Yunusov D, Anderson L, DaSilva LF, Wysocka J, Ezashi T, Roberts RM, Verjovski-Almeida S. HIPSTR and thousands of lncRNAs are heterogeneously expressed in human embryos, primordial germ cells and stable cell lines. Sci Rep. 2016;6:32753.

Chapter 15
The Crosstalk Between miRNAs and Autophagy in Cancer Progression

Bayraktar Oznur and Gozuacik Devrim

Abstract MicroRNAs (miRNAs) are small (17–25 nucleotide-long), non-coding RNAs that modulate and repress the expression of their target mRNAs. Aberrant expression of miRNAs is linked to many human diseases including cancer. Impaired levels of miRNAs may also result in defected autophagy. Autophagy is a double-edged sword during cancer initiation and progression. At the beginning of tumorigenesis, autophagy suppresses tumor formation by removing defective organelles such as mitochondria; restricting oxidative stress and protecting genome stability. However, in the later stages of tumor formation, autophagy is a survival pathway for tumor cells under the low levels of oxygen (hypoxia), deprivation of growth factors and glucose. The main focus of this chapter is the interplay between miRNAs and autophagy during initiation and progression of cancer.

Keywords Autophagy · Cancer · MicroRNA · miRNA · Anti-autophagic miRNA · Pro-autophagic miRNA · Tumor · Tumor suppressive miRNA · Oncogenic miRNA · mTOR · Beclin1 · Atg5

B. Oznur (✉)
Faculty of Medicine, Medical Biology and Genetics Department, Okan University, Istanbul, Turkey
e-mail: oznur.bayraktar@okan.edu.tr

G. Devrim
Molecular Biology, Genetics and Bioengineering Program, Faculty of Engineering and Natural Sciences, Sabanci University, Istanbul, Turkey

S. Fayyaz, A. A. Farooqi (eds.), *Recent Trends in Cancer Biology: Spotlight on Signaling Cascades and microRNAs*, https://doi.org/10.1007/978-3-319-71553-7_15

Autophagy and Its Involvement in Cancer Progression

Autophagic Machinery

Autophagy is a cellular degradation system for misfolded/unfolded and long-lived proteins as well as for damaged or old organelles (e.g. ER, mitochondria). It is highly conserved from yeast to man, indicating its importance for the organisms. Autophagy is already active under basal conditions but stimuli such as serum, aminoacid or glucose deprivation as well as hypoxia and certain chemicals upregulate the autophagic machinery to enable cell survival under stress conditions [1].

At the initiation phase of autophagy, double or multi membrane vesicles, called autophagosomes, are formed in the cytoplasm around the target molecule. Later, these autophagosomes fuse with lysosomes, forming 'autolysosomes'. Finally, hydrolytic enzymes in the lysosomes degrade autophagic content in the autolysosomes and the end products are recycled for further use [2].

At the molecular level, pre-initiation complex (formed by ULK1/2, ATG13 and FIP200) and initiation complex (formed by Beclin1, Vps34, Vps15 and Atg14L) induce the elongation of a double or multi-membrane structure. The source of membrane might be variable such as ER, mitochondria or Golgi. Two ubiquitin-like conjugation systems regulate the elongation of the autophagosomal membrane. In the first system, Atg12 and Atg5 are covalently attached first and then binds to Atg16. Atg5-Atg12-Atg16 complex has E3-like enzymatic activity catalyzing phosphatidylethanolamine (PE) conjugation to LC3 [3]. The conjugation of LC3 to PE also requires the activity of Atg4, Atg3, Atg7. LC3 protein conjugated to PE is named as LC3-II and directed to autophagosome membrane. Therefore, the ratio of LC3-II to LC3-I is well-accepted indication of autophagic activity.

The most important autophagy-regulating protein is mTOR. It joins two complexes, namely mTORC1 and mTORC2. Six proteins from mTORC1; mTOR, Deptor, Raptor, Tti1/Tel2 complex, PRAS40 and mLST8. On the other hand, mTORC2 is made up by seven proteins; mTOR, Deptor, Rictor, mLST8, mSin1, Tti1/Tel2 complex and Protor1/2 [4].

Under nutrient-rich conditions, mTORC1 complex prevents initiation of autophagy via deactivation of autophagy inducer ULK1. It also upregulates the synthesis of lipids and proteins through phosphorylation of its downstream targets, 4EBP1 and S6K1. Several stimuli such as deprivation of aminoacid, glucose and growth factors block mTORC1. Subsequent dephosphorylation of ULK1 results in activation of autophagic machinery. Although mTORC2 can be induced by growth factors, it is unresponsive to nutrients. The biological role of mTORC2 is less elucidated. Of note, involvement of mTORC2 in the reorganization of cytoskeleton, cell migration and cell survival have been shown [4]. mTORC1 prevents autophagosome formation via inhibition of autophagosome formation, while mTORC2 limits the expression of autophagy related genes.

The Role of Autophagy in Cancer Progression

Autophagy can have both oncogenic and tumor-suppressive roles depending on the stage and type of the tumor. Downregulation of autophagy due to deficiency of some autophagy genes causes abnormal accumulation of p62 as well as increased mitochondrial defects, oxidative stress, DNA damage and finally cell death. These changes destroy genomic stability and tumor formation is initiated. Therefore, functional autophagic machinery has a tumor-suppressive role at the beginning of tumorigenesis. On the other hand, elevated levels of autophagy due to constitutively active EGFR pathway,triggered by KRAS and BRAF mutations, restrain cellular stress responses and maintain mitochondrial health. Therefore, increased levels of autophagy promotes cell survival and tumor progression [5].

Autophagy as a Tumor Suppressive Mechanism

Deletion of one allele of Beclin1 is detected in 40–75% of cases of human sporadic breast, prostate and ovarian cancer. Heterozygous disruption of Beclin1 in mice accelerated formation of premalignant lesions and elevated the frequency of spontaneous malignancies in mice. Moreover, autophagy was diminished while proliferation rate of tumor cells was elevated in these mice. In addition, Beclin1-mediated autophagy emerged as a controller of cell-growth and a repressor of the tumor formation [6].

Beclin-1 was not the only autophagy gene which has been found to have tumor suppressor roles. In mice with Atg7 deletion in liver and systemic mosaic deletion of Atg5, autophagy defect triggers mitochondrial abnormalities, accumulation of p62, increased ROS levels and formation of benign adenomas in liver [7]. In another study, direct interaction of LC3 with LaminB1 (nuclear lamina protein) leads to collapse of nuclear lamina by autophagy under 'aberrant' stress due to oncogenic and genotoxic insults. LMNB1 degradation through autophagy triggers cellular senescence and limits carcinogenesis [8].

Mutations or abnormal expression of oncogenes might alter carcinogenesis via dysregulation of autophagy. Oncogenic RAS upregulates the expression of Beclin1 and Noxa, and elevated levels of Noxa induces autophagy via preventing Mcl-1 binding to Beclin-1. Consequently, a reduction in survival of tumor cell colonies is observed. Furthermore, knockdown of Beclin1, Atg5 or Noxa inhibits Ras-induced cell death. These findings suggest a way to control oncogenic capacity of misregulated Ras signals via autophagy [9].

Dysregulation of cell growth and survival pathways such as PI3K/AKT/mTOR associates with different types of cancer such as hepatocellular carcinoma (HCC), breast and endometrial cancer [10–12] and regulators of mTOR are subjects for cancer studies.

Via modulation of autophagy, Deptor (DEP-domain containing mTOR interacting protein) can both promote or suppress tumor formation. Interaction of Deptor with mTOR reduces the kinase activity of mTOR and activates autophagy. In addition, Deptor induces proliferation and survival of cells via activating Akt pathway. Reduced levels of Deptor are correlated with massive mTOR activation in particular tumor models. Deptor primarily suppresses tumor formation in subsets of multiple myeloma (MM), pancreatic, colorectal and liver cancer due to activation of Akt/mTOR pathway [4]. However, Deptor might also function as an oncogene by inducing cancer cell proliferation and survival in a few exceptional types of cancer such as lung, thyroid carcinoma or a subset of multiple myeloma. In these types of cancer cells, the expression levels of Deptor are higher than normal [4].

Autophagy as Tumor Promoter

Abnormally high proliferation rate of tumor cells causes deprivation of glucose, serum, hormones as well as oxygen. Functional autophagic machinery is vital for the survival of cancer cells under these restricting conditions.

High levels of p62, an indicator of defective autophagy, is associated with poor prognosis in HCC. Moreover, p62 is an inducer of mTORC1 and without any other oncogenic stimuli, overexpression of p62 initiates HCC formation [13]. RNA silencing of ULK1, Atg7, or Atg13 limits tumorigenesis, induces senescence and increases the life span in mice model of KRAS-driven glioblastoma [14]. In tumor cells, ablation of FIP200 results in defective autophagy and accumulation of abnormal mitochondria. Moreover, in mammary cells and RAS-transformed mouse embryonic fibroblasts (MEFs), tumor proliferation is decreased although apoptosis levels are not changed [15].

Pancreatic cancer cells and melanomas upregulate autophagy to achieve optimum rate of proliferation both in vitro and in vivo [16, 17]. Atg7 induces melanomagenesis in BrafV600Edriven, PTEN-null mice model. Deletion of Atg7 results in defective autophagy, elevated oxidative stress and senescence as well as extension in life span of mice [18]. In addition, mice with Atg7 deficient tumors accumulate defective mitochondria and trigger p53 expression untimely. This, in turn, results in cell cycle arrest, high levels of apoptotic cell death and finally restrained tumor size [19].

RAS mutations have been found in 33% of human cancer and this oncogene can both upregulate or downregulate the autophagic activity depending on the context. RAS can trigger autophagy via activation of Atg5/Atg7 by JNK, inhibition of Bcl2/Mcl1 via binding to Beclin 1, or GAIP. In RAS-transformed cancer cells, a higher level of autophagy is connected to cancer cell growth, survival and metastasis [5]. In oncogenic RAF- or RAS-driven fast-growing tumors, autophagy controls mitochondrial quality and metabolic energy levels to support the proliferation and survival of tumors [21].

Control of Autophagy Associated Proteins by miRNAs to Modulate Cancer Progression

Biogenesis of MicroRNAs (miRNAs)

MicroRNAs (miRNAs) are small, non-coding RNA which are 17–25 nucleotides long. They bind to 3′ UTR of their targets, with either partial or perfect complementarity. They downregulate or completely silence the expression of target gene(s) post-transcriptionally. There might be several targets of a single miRNA and miRNAs function in a broad range of molecular pathways [22].

Maturation process of miRNAs begins at the nucleus. RNA pol II or III is responsible for the transcription of primary miRNAs (pri-miRNAs) from endogenous miRNA genes in the nucleus. First cleavage of stem-loop structure of pri-miRNAs is performed by Drosha/DGCR8 complex in the nucleus. After first cleavage, they are called as 'pre-miRNA'. Translocation of pre-miRNAs to cytoplasm is carried out by Exportin-5 complex and then further cleavage of hairpin-structure is performed by Dicer and TRBP (transactivation-response RNA-binding protein). This additional cleavage is crucial for the formation of a miRNA-miRNA* duplex. While thermodynamically more stable miRNA strand is cargo for RNA Induced Silencing Complex (RISC) and driven to its target mRNAs by Argonaute (AGO), the other miRNA strand is degraded in the cytoplasm [23]. If the complementarity between seed sequence of miRNA and target mRNA is perfect or nearly perfect, target mRNA is cleaved and degraded. If the complementarity is partial, translational repression occurs. Therefore, binding of miRNA to its target, either perfectly or partially, leads to a decrease in the level of target mRNA.

The Involvement of in mRNAs in Cancer Via Modulation of Autophagy

Transformation of normal cells into cancer cells used to be characterized by six hallmarks previously by Hanahan and Weinberg. However, in the light of recent findings, hallmarks of cancer now are increased to nine: (1) independency on growth and (2) anti-growth signals, (3) evading from apoptosis, (4) unlimited capacity for replication, (5) increased angiogenesis (6) metastasis capacity, (7) metabolic reprogramming (8) tumor-inducing inflammation and (9) genomic instability [24]. miRNAs are able to regulate many of these hallmarks of cancer, however, we will mainly focus on the modulation of cancer via miRNAs that affect autophagy in this chapter.

Early studies firstly indicate diminished levels of miRNAs in tumors suggesting the role of miRNAs as being tumor suppressors. On the other hand, later studies discovered oncogenic miRNAs as well, such as miR-30a targeting Beclin1, miR-376b targeting Atg4C and Beclin1 and miR-155 targeting mTOR pathway [25].

Modulation of tumor suppressor genes and oncogenes by miRNAs are crucial for both initiation and the progression of cancer. Downregulation of Let-7 family is detected in many cancer types including breast, lung, ovarian and prostate. Tumor supressor miRNA Let-7 represses the expression of crucial oncogenes such as CDK-6, TGFBR1, c-MYC and BCL-xL [26]. Oncogenic miR-17-92 cluster is overexpressed in several cancer types and its overexpression decreases the level of crucial tumor supressors such as p53, MAPK1, RBL1, p15, SMAD2 and SMAD4 [27, 28].

There is also a feedback loop between miRNAs, tumor supressor genes and oncogenes. Therefore, not only miRNAs modulate tumor supressor genes and oncogenes but expression of several miRNA genes is also controlled by oncogenic or tumor-suppressive proteins. To exemplify, the expression of miR-34a, miR-17-92 and miR-21 is controlled by TP53, MYC and RAS, respectively [29].

The dual role of miRNAs in cancer is partly attributed to the interaction of miRNAs with autophagy genes. miRNAs might target directly autophagy genes or upstream signaling pathways of autophagy such as AKT/mTOR [30] and ERK [4]. The consequences of inhibition or promotion of autophagy by miRNAs might be either oncogenic or tumor suppressive depending on the cell-type and the context. In addition, miRNAs can alter the cancer cells' response to chemotherapy and radiotherapy via modulation of autophagy.

Anti-autophagic miRNAs in Cancer Modulation

Two variants from same miRNA family, miR-30a and miR-30d both reduce the expression of Atg5 and Beclin1 and have oncogenic effects [25]. miR-30a attenuates rapamycin-induced autophagy in human glioblastoma cells through targeting Beclin1 [31]. miR-30d targets other autophagy genes such as Atg2 and Atg12 in many cancer tissues including ovarian, breast and prostate [32]. Oncogenic miR-183 targets UVRAG and this interaction leads to a decrease in autophagy stimulated with either starvation or rapamycin [32]. Anti-autophagic and oncogenic miR-376b targets Atg4C and Beclin1 in Huh-7 and MCF-7 cancer cell line [33].

miR-17-92a family has anti-autophagic and oncogenic role in prostate cancer. mir-20a and miR-17 belong to miR-17-92a cluster. In LNCaP cells, these two miRNAs prevent celasterol-induced autophagy via suppression of Atg7 [34]. Elevated level of miR-20a triggers DNA damage response and oxidative stress and impedes autophagy via targeting p62, Beclin1 and Atg16L1 in human breast cancer tissues. In addition, initiation and growth of tumor are accelerated with miR-20a expression in xenograft mice models [35].

The expression of miR-290-295 is found to be upregulated in different melanoma cell lines and this expression pattern is correlated with increased levels of malignancy. Elevated level of malignancy is due to the inhibition of starvation-induced autophagy via suppression of Atg7 and ULK1 by miR-290-295 [36].

Inhibition of autophagy via miR-21, which targets PI3K/AKT/mTOR pathway, causes an attenuation in the radiosensitivity of malignant human gliomas and chemosensitivity of breast cancer cells [30, 37]. In cervical cancer cells, upregulation of miR-21 is positively correlated with HIF1α, and PTEN is targeted by miR-21. These alterations diminish autophagic activity and radiosensitivity through the interaction between PTEN/Akt/HIF1α and PI3K/AKT/mTOR pathways [38].

On the other hand, tumor suppressive and anti-autophagic miR-885-3p and miR-26b both target ULK2 in squamous carcinoma and prostate cancer cells, respectively. The repression of Rictor via binding to miR-15a/miR-16 causes a decrease in the phosphorylation levels of mTORC1 and p70S6K and a consequent decline in cell proliferation [32]. miR-130a prevents autophagy by targeting Atg2B and DICER1 [39] and increases cell death in chronic lymphocytic leukemia cells. miR-224-3p suppresses hypoxia-induced autophagy and enhances apoptosis via targeting Atg5 and FIP200 [40]. Inhibition of autophagy via suppression of Atg12 by miR-140-5p results in decreased survival of colorectal cancer stem cell (CSC) [41]. miR-214 that targets UCP2, a mitochondrial protein involved in ROS production, enhances fulvestrant and tamoxifen sensitivity of breast cancer through activation of PI3K/AKT/mTOR pathway and suppression of autophagy [42].

The targets of miR-34a (e.g. Bcl2, SIRT1) include genes that associate with apoptosis, cell cycle, differentiation as well as metastasis and cancer stemness [43]. In quercetin-treated HepG2 cells, elevated expression of miR-34a leads to a decline in SIRT1 expression and a rise in acetylated p53 levels. Consequently, induced apoptosis is observed, indicating the tumor-suppressive role of miR-34a [44]. In acute myeloid leukemia cells, miR-34a impedes autophagy and induces apoptosis via HMGB1, which is a protein that involves in DNA damage repair [45]. miR-372 blocks cell survival and autophagy via targeting Ulk1 in human pancreatic adenocarcinoma cells [46].

There are multiple miRNAs that modulate the sensitivity of cancer cells to therapy via alteration of autophagy. miR-181a targeting Atg5 diminishes autophagic activity triggered by rapamycin and starvation in breast, liver cancer as well as leukemia cell lines and causes cisplatin sensitivity in gastric cancer cell line [47, 48]. miR-17 downregulates autophagy via targeting Atg7 in human glioblastoma cells. Furthermore, miR-17 increases sensitivity of glioblastomas to radiotherapy and chemotherapy [49]. Diminished levels of miR-199a-5p, of which target is Atg7, reduce cisplatin sensitivity of HCC cells via upregulation of autophagy [50]. Via regulating Atg12 and HMG2B, miR-23b-3p sensitizes gastric cancer cells to chemotherapy [51]. Repression of Atg12 by miR-23b leads to a reduction in radiation-induced autophagy in pancreatic cancer cells and overexpression of miR-23b increases radiosensitivity of cancer cells [52]. Anti-autophagic miR-101 targets multiple autophagy-related genes, namely Rab5A, STMN1 and Atg4D, and induces cisplatin-triggered apoptosis in HCC cell line [53].

Pro-autophagic miRNAs in Cancer Modulation

Autophagy promotes the survival of cancer cells under hypoxia. In human prostate cancer cells, hypoxic conditions induce miR-96 expression which enhances autophagy levels via targeting mTOR. However, ectopic expression of miR-96 up to certain level prevents autophagy via targeting Atg7 [54] unraveling the dual role of miR-96 in cancer progression. Upregulated miR-155 expression in cervical and human nasopharyngeal cancer cells under hypoxia upregulates autophagy by targeting multiple participants of mTOR pathway, such as RICTOR, RPS6KB2 and RHEB [55].

Elevated levels of miR-210 is associated with metastasis, angiogenesis and epithelial to mesenchymal transition [56]. The expression level of miR-210 is induced in endometriotic cells under hypoxia. Upon elevated levels of miR-210, expression of HIF-1α and Bcl2 is incresed. In addition, autophagy is induced upon elevated levels of miR-210 while miR-210 overexpression and Bcl2 silencing cause significant elevation in autophagic activity and in survival of cancer cells [57]. Similarly under hypoxia, repression of Bcl-2 by miR-210 makes colon cancer cells radioresistant due to upregulation of autophagy [58].

miR-193a-5p, miR-503, miR-663, miR-148b and miR-30b target autophagy inhibitor Class I PI3K [59]. miR-193a-5p also directly targets mTOR and inhibits metastasis in non-small cell lung cancer (NSCLC) [60]. miR-99a impedes tumor cell migration, invasion and growth via targeting mTOR in breast cancer [61]. miR-423-5p induces autophagy via increasing Atg7 and LC3-II levels in HCC cells. Upon sorafenib addition, elevated levels of pro-autophagic miR-423-5p causes cell cycle arrest through reduced pErk1/2 activity which is an indicator of better chemoresponse [62].

Conclusion

Despite the presence of growing literature about cancer, there are still missing links due to the existence of variable types and subtypes of tumor cells and complex signaling pathways involved in disease formation and progression.

Autophagy is a well conserved pathway from yeast to human. It functions as a cellular recycling machinery, degrading damaged or old organelles, misfolded proteins to provide energy for cell survival and to avoid toxicity. It has crucial role in regulating the homeostasis through modulation of cellular pathways and therefore autophagic dysregulation results in many pathological conditions, including cancer. Autophagy might act as suppressor or promoter of tumorigenesis depending on the stage and type of cancer. It also modulates sensitivity of tumor cells to radiotherapy and chemotherapy.

Studies performed so far identified the existence of ~2000 miRNAs and these miRNAs affect the expression of 60% of protein coding genes [63] demonstrating the regulation of many cellular pathways through particular miRNAs. miRNAs affect growth, proliferation, viability, metastasis, malignancy, angiogenesis and genomic instability of tumor cells via binding to their targets. Similar to the role of autophagy in cancer, miRNAs can either suppress or promote tumorigenesis depending on the context. Furthermore, they might be either pro-autophagic or anti-autophagic. As a consequence, interaction between autophagy and miRNAs during the onset and progression of cancer should be considered carefully in a tissue- and context-dependent manner.

Acknowledgements This work was supported by Faculty of Medicine, Okan University. I would like to thank Prof. Dr. Semih Baskan and Prof. Dr. Ender Altiok for sharing their precious knowledge and experience.

References

1. Karakas HE, Gözüaçik D. Autophagy and cancer. Turk J Biol. 2014;38:720–39. https://doi.org/10.3906/biy-1408-16.
2. Oral O, Akkoc Y, Bayraktar O, Gozuacik D. Physiological and pathological significance of the molecular cross-talk between autophagy and apoptosis. Histol Histopathol. 2016;31(5):479–98. 10.14670/HH-11-714. PubMed PMID: 26680630
3. Kuma A, Mizushima N, Ishihara N, Ohsumi Y. Formation of the approximately 350-kDa Apg12-Apg5.Apg16 multimeric complex, mediated by Apg16 oligomerization, is essential for autophagy in yeast. J Biol Chem. 2002;277(21):18619–25. https://doi.org/10.1074/jbc.M111889200. PubMed PMID: 11897782
4. Catena V, Fanciulli M. Deptor: not only a mTOR inhibitor. J Exp Clin Cancer Res. 2017;36(1):12. https://doi.org/10.1186/s13046-016-0484-y. PubMed PMID: 28086984; PubMed Central PMCID: PMCPMC5237168
5. White E. The role for autophagy in cancer. J Clin Invest. 2015;125(1):42–6. https://doi.org/10.1172/JCI73941. PubMed PMID: 25654549; PubMed Central PMCID: PMCPMC4382247
6. Qu X, Yu J, Bhagat G, Furuya N, Hibshoosh H, Troxel A, et al. Promotion of tumorigenesis by heterozygous disruption of the beclin 1 autophagy gene. J Clin Invest. 2003;112(12):1809–20. https://doi.org/10.1172/JCI20039. PubMed PMID: 14638851; PubMed Central PMCID: PMCPMC297002
7. Takamura A, Komatsu M, Hara T, Sakamoto A, Kishi C, Waguri S, et al. Autophagy-deficient mice develop multiple liver tumors. Genes Dev. 2011;25(8):795–800. https://doi.org/10.1101/gad.2016211. PubMed PMID: 21498569; PubMed Central PMCID: PMCPMC3078705
8. Dou Z, Xu C, Donahue G, Shimi T, Pan JA, Zhu J, et al. Autophagy mediates degradation of nuclear lamina. Nature. 2015;527(7576):105–9. https://doi.org/10.1038/nature15548. PubMed PMID: 26524528; PubMed Central PMCID: PMCPMC4824414
9. Elgendy M, Sheridan C, Brumatti G, Martin SJ. Oncogenic Ras-induced expression of Noxa and Beclin-1 promotes autophagic cell death and limits clonogenic survival. Mol Cell. 2011;42(1):23–35. https://doi.org/10.1016/j.molcel.2011.02.009. PubMed PMID: 21353614
10. Dong P, Konno Y, Watari H, Hosaka M, Noguchi M, Sakuragi N. The impact of microRNA-mediated PI3K/AKT signaling on epithelial-mesenchymal transition and cancer stemness in endometrial cancer. J Transl Med. 2014;12:231. https://doi.org/10.1186/s12967-014-0231-0. PubMed PMID: 25141911; PubMed Central PMCID: PMCPMC4145234
11. Liu L, Liao JZ, He XX, Li PY. The role of autophagy in hepatocellular carcinoma: friend or foe. Oncotarget. 2017. 10.18632/oncotarget.17202. PubMed PMID: 28476020
12. Zhang HD, Jiang LH, Sun DW, Li J, Ji ZL. The role of miR-130a in cancer. Breast Cancer. 2017. https://doi.org/10.1007/s12282-017-0776-x. PubMed PMID: 28477068
13. Umemura A, He F, Taniguchi K, Nakagawa H, Yamachika S, Font-Burgada J, et al. p62, upregulated during preneoplasia, induces hepatocellular carcinogenesis by maintaining survival of stressed HCC-initiating cells. Cancer Cell. 2016;29(6):935–48. https://doi.org/10.1016/j.ccell.2016.04.006. PubMed PMID: 27211490; PubMed Central PMCID: PMCPMC4907799

14. Gammoh N, Fraser J, Puente C, Syred HM, Kang H, Ozawa T, et al. Suppression of autophagy impedes glioblastoma development and induces senescence. Autophagy. 2016;12(9):1431–9. https://doi.org/10.1080/15548627.2016.1190053. PubMed PMID: 27304681; PubMed Central PMCID: PMCPMC5082770
15. Wei H, Wei S, Gan B, Peng X, Zou W, Guan JL. Suppression of autophagy by FIP200 deletion inhibits mammary tumorigenesis. Genes Dev. 2011;25(14):1510–27. https://doi.org/10.1101/gad.2051011. PubMed PMID: 21764854; PubMed Central PMCID: PMCPMC3143941
16. Yang S, Wang X, Contino G, Liesa M, Sahin E, Ying H, et al. Pancreatic cancers require autophagy for tumor growth. Genes Dev. 2011;25(7):717–29. https://doi.org/10.1101/gad.2016111. PubMed PMID: 21406549; PubMed Central PMCID: PMCPMC3070934
17. Ma XH, Piao S, Wang D, McAfee QW, Nathanson KL, Lum JJ, et al. Measurements of tumor cell autophagy predict invasiveness, resistance to chemotherapy, and survival in melanoma. Clin Cancer Res. 2011;17(10):3478–89. https://doi.org/10.1158/1078-0432.CCR-10-2372. PubMed PMID: 21325076; PubMed Central PMCID: PMCPMC3096713
18. Amaravadi R, Kimmelman AC, White E. Recent insights into the function of autophagy in cancer. Genes Dev. 2016;30(17):1913–30. https://doi.org/10.1101/gad.287524.116. PubMed PMID: 27664235; PubMed Central PMCID: PMCPMC5066235
19. Guo JY, Karsli-Uzunbas G, Mathew R, Aisner SC, Kamphorst JJ, Strohecker AM, et al. Autophagy suppresses progression of K-ras-induced lung tumors to oncocytomas and maintains lipid homeostasis. Genes Dev. 2013;27(13):1447–61. https://doi.org/10.1101/gad.219642.113. PubMed PMID: 23824538; PubMed Central PMCID: PMCPMC3713426
20. Schmukler E, Kloog Y, Pinkas-Kramarski R. Ras and autophagy in cancer development and therapy. Oncotarget. 2014;5(3):577–86. 10.18632/oncotarget.1775. PubMed PMID: 24583697; PubMed Central PMCID: PMCPMC3996671
21. Gozuacik D, Akkoc Y, Ozturk DG, Kocak M. Autophagy-regulating microRNAs and cancer. Front Oncol. 2017;7:65. https://doi.org/10.3389/fonc.2017.00065. PubMed PMID: 28459042; PubMed Central PMCID: PMCPMC5394422
22. Bartel DP. MicroRNAs: genomics, biogenesis, mechanism, and function. Cell. 2004;116(2):281–97. PubMed PMID: 14744438
23. Tekirdag KA, Akkoc Y, Kosar A, Gozuacik D. MIR376 family and cancer. Histol Histopathol. 2016;31(8):841–55. 10.14670/HH-11-752. PubMed PMID: 26940843
24. Fouad YA, Aanei C. Revisiting the hallmarks of cancer. Am J Cancer Res. 2017;7(5):1016–36. PubMed PMID: 28560055; PubMed Central PMCID: PMCPMC5446472
25. Chen Y, Fu LL, Wen X, Liu B, Huang J, Wang JH, et al. Oncogenic and tumor suppressive roles of microRNAs in apoptosis and autophagy. Apoptosis. 2014;19(8):1177–89. https://doi.org/10.1007/s10495-014-0999-7. PubMed PMID: 24850099
26. Naidu S, Magee P, Garofalo M. MiRNA-based therapeutic intervention of cancer. J Hematol Oncol. 2015;8:68. https://doi.org/10.1186/s13045-015-0162-0. PubMed PMID: 26062952; PubMed Central PMCID: PMCPMC4465004
27. Li Y, Lauriola M, Kim D, Francesconi M, D'Uva G, Shibata D, et al. Adenomatous polyposis coli (APC) regulates miR17-92 cluster through beta-catenin pathway in colorectal cancer. Oncogene. 2016;35(35):4558–68. https://doi.org/10.1038/onc.2015.522. PubMed PMID: 26804172; PubMed Central PMCID: PMCPMC4960006
28. Tavakoli R, Vakilian S, Langroudi L, Arefian E, Sahmani M, Soleimani M, et al. The role of miR-17-92 cluster in the expression of tumor suppressor genes in unrestricted somatic stem cells. Biologicals. 2017;46:143–7. https://doi.org/10.1016/j.biologicals.2017.02.006. PubMed PMID: 28222938
29. Farazi TA, Hoell JI, Morozov P, Tuschl T. MicroRNAs in human cancer. Adv Exp Med Biol. 2013;774:1–20. https://doi.org/10.1007/978-94-007-5590-1_1. PubMed PMID: 23377965; PubMed Central PMCID: PMCPMC3704221
30. Yu X, Li R, Shi W, Jiang T, Wang Y, Li C, et al. Silencing of MicroRNA-21 confers the sensitivity to tamoxifen and fulvestrant by enhancing autophagic cell death through inhibition of the PI3K-AKT-mTOR pathway in breast cancer cells. Biomed Pharmacother. 2016;77:37–44. https://doi.org/10.1016/j.biopha.2015.11.005. PubMed PMID: 26796263

31. Zhu H, Wu H, Liu X, Li B, Chen Y, Ren X, et al. Regulation of autophagy by a beclin 1-targeted microRNA, miR-30a, in cancer cells. Autophagy. 2009;5(6):816–23. PubMed PMID: 19535919; PubMed Central PMCID: PMCPMC3669137
32. Yao D, Jiang Y, Gao S, Shang L, Zhao Y, Huang J, et al. Deconvoluting the complexity of microRNAs in autophagy to improve potential cancer therapy. Cell Prolif. 2016;49(5):541–53. https://doi.org/10.1111/cpr.12277. PubMed PMID: 27436709
33. Korkmaz G, le Sage C, Tekirdag KA, Agami R, Gozuacik D. miR-376b controls starvation and mTOR inhibition-related autophagy by targeting ATG4C and BECN1. Autophagy. 2012;8(2):165–76. https://doi.org/10.4161/auto.8.2.18351. PubMed PMID: 22248718
34. Guo J, Mei Y, Li K, Huang X, Yang H. Downregulation of miR-17-92a cluster promotes autophagy induction in response to celastrol treatment in prostate cancer cells. Biochem Biophys Res Commun. 2016;478(2):804–10. https://doi.org/10.1016/j.bbrc.2016.08.029. PubMed PMID: 27501757
35. Liu L, He J, Wei X, Wan G, Lao Y, Xu W, et al. MicroRNA-20a-mediated loss of autophagy contributes to breast tumorigenesis by promoting genomic damage and instability. Oncogene. 2017. https://doi.org/10.1038/onc.2017.193. PubMed PMID: 28628113
36. Chen Y, Liersch R, Detmar M. The miR-290-295 cluster suppresses autophagic cell death of melanoma cells. Sci Rep. 2012;2:808. https://doi.org/10.1038/srep00808. PubMed PMID: 23150779; PubMed Central PMCID: PMCPMC3496171
37. Gwak HS, Kim TH, Jo GH, Kim YJ, Kwak HJ, Kim JH, et al. Silencing of microRNA-21 confers radio-sensitivity through inhibition of the PI3K/AKT pathway and enhancing autophagy in malignant glioma cell lines. PLoS One. 2012;7(10):e47449. https://doi.org/10.1371/journal.pone.0047449. PubMed PMID: 23077620; PubMed Central PMCID: PMCPMC3471817
38. Song L, Liu S, Zhang L, Yao H, Gao F, Xu D, et al. MiR-21 modulates radiosensitivity of cervical cancer through inhibiting autophagy via the PTEN/Akt/HIF-1alpha feedback loop and the Akt-mTOR signaling pathway. Tumour Biol. 2016;37(9):12161–8. https://doi.org/10.1007/s13277-016-5073-3. PubMed PMID: 27220494
39. Kovaleva V, Mora R, Park YJ, Plass C, Chiramel AI, Bartenschlager R, et al. miRNA-130a targets ATG2B and DICER1 to inhibit autophagy and trigger killing of chronic lymphocytic leukemia cells. Cancer Res. 2012;72(7):1763–72. https://doi.org/10.1158/0008-5472.CAN-11-3671. PubMed PMID: 22350415
40. Guo X, Xue H, Guo X, Gao X, Xu S, Yan S, et al. MiR224-3p inhibits hypoxia-induced autophagy by targeting autophagy-related genes in human glioblastoma cells. Oncotarget. 2015;6(39):41620–37. 10.18632/oncotarget.5871. PubMed PMID: 26536662; PubMed Central PMCID: PMCPMC4747177
41. Zhai H, Fesler A, Ba Y, Wu S, Ju J. Inhibition of colorectal cancer stem cell survival and invasive potential by hsa-miR-140-5p mediated suppression of Smad2 and autophagy. Oncotarget. 2015;6(23):19735–46. 10.18632/oncotarget.3771. PubMed PMID: 25980495; PubMed Central PMCID: PMCPMC4637317
42. Yu X, Luo A, Liu Y, Wang S, Li Y, Shi W, et al. MiR-214 increases the sensitivity of breast cancer cells to tamoxifen and fulvestrant through inhibition of autophagy. Mol Cancer. 2015;14:208. https://doi.org/10.1186/s12943-015-0480-4. PubMed PMID: 26666173; PubMed Central PMCID: PMCPMC4678692
43. Misso G, Di Martino MT, De Rosa G, Farooqi AA, Lombardi A, Campani V, et al. Mir-34: a new weapon against cancer? Mol Ther Nucleic Acids. 2014;3:e194. https://doi.org/10.1038/mtna.2014.47. PubMed PMID: 25247240; PubMed Central PMCID: PMCPMC4222652
44. Lou G, Liu Y, Wu S, Xue J, Yang F, Fu H, et al. The p53/miR-34a/SIRT1 positive feedback loop in quercetin-induced apoptosis. Cell Physiol Biochem. 2015;35(6):2192–202. https://doi.org/10.1159/000374024. PubMed PMID: 25896587
45. Liu L, Ren W, Chen K. MiR-34a promotes apoptosis and inhibits autophagy by targeting HMGB1 in acute myeloid leukemia cells. Cell Physiol Biochem. 2017;41(5):1981–92. https://doi.org/10.1159/000475277. PubMed PMID: 28478444

46. Chen H, Zhang Z, Lu Y, Song K, Liu X, Xia F, et al. Down-regulation of ULK1 by microRNA-372 inhibits the survival of human pancreatic adenocarcinoma cells. Cancer Sci. 2017. https://doi.org/10.1111/cas.13315. PubMed PMID: 28677209
47. Zhao J, Nie Y, Wang H, Lin Y. MiR-181a suppresses autophagy and sensitizes gastric cancer cells to cisplatin. Gene. 2016;576(2 Pt 2):828–33. https://doi.org/10.1016/j.gene.2015.11.013. PubMed PMID: 26589846
48. Tekirdag KA, Korkmaz G, Ozturk DG, Agami R, Gozuacik D. MIR181A regulates starvation- and rapamycin-induced autophagy through targeting of ATG5. Autophagy. 2013;9(3):374–85. https://doi.org/10.4161/auto.23117. PubMed PMID: 23322078; PubMed Central PMCID: PMCPMC3590257
49. Comincini S, Allavena G, Palumbo S, Morini M, Durando F, Angeletti F, et al. microRNA-17 regulates the expression of ATG7 and modulates the autophagy process, improving the sensitivity to temozolomide and low-dose ionizing radiation treatments in human glioblastoma cells. Cancer Biol Ther. 2013;14(7):574–86. https://doi.org/10.4161/cbt.24597. PubMed PMID: 23792642; PubMed Central PMCID: PMCPMC3742487
50. Xu N, Zhang J, Shen C, Luo Y, Xia L, Xue F, et al. Cisplatin-induced downregulation of miR-199a-5p increases drug resistance by activating autophagy in HCC cell. Biochem Biophys Res Commun. 2012;423(4):826–31. https://doi.org/10.1016/j.bbrc.2012.06.048. PubMed PMID: 22713463
51. An Y, Zhang Z, Shang Y, Jiang X, Dong J, Yu P, et al. miR-23b-3p regulates the chemoresistance of gastric cancer cells by targeting ATG12 and HMGB2. Cell Death Dis. 2015;6:e1766. https://doi.org/10.1038/cddis.2015.123. PubMed PMID: 25996293; PubMed Central PMCID: PMCPMC4669702
52. Wang P, Zhang J, Zhang L, Zhu Z, Fan J, Chen L, et al. MicroRNA 23b regulates autophagy associated with radioresistance of pancreatic cancer cells. Gastroenterology. 2013;145(5):1133–43.e12. https://doi.org/10.1053/j.gastro.2013.07.048. PubMed PMID: 23916944
53. Xu Y, An Y, Wang Y, Zhang C, Zhang H, Huang C, et al. miR-101 inhibits autophagy and enhances cisplatin-induced apoptosis in hepatocellular carcinoma cells. Oncol Rep. 2013;29(5):2019–24. https://doi.org/10.3892/or.2013.2338. PubMed PMID: 23483142
54. Ma Y, Yang HZ, Dong BJ, Zou HB, Zhou Y, Kong XM, et al. Biphasic regulation of autophagy by miR-96 in prostate cancer cells under hypoxia. Oncotarget. 2014;5(19):9169–82. 10.18632/oncotarget.2396. PubMed PMID: 25333253; PubMed Central PMCID: PMCPMC4253426
55. Wan G, Xie W, Liu Z, Xu W, Lao Y, Huang N, et al. Hypoxia-induced MIR155 is a potent autophagy inducer by targeting multiple players in the MTOR pathway. Autophagy. 2014;10(1):70–9. https://doi.org/10.4161/auto.26534. PubMed PMID: 24262949; PubMed Central PMCID: PMCPMC4389881
56. Yu P, Fan S, Huang L, Yang L, Du Y. MIR210 as a potential molecular target to block invasion and metastasis of gastric cancer. Med Hypotheses. 2015;84(3):209–12. https://doi.org/10.1016/j.mehy.2014.12.024. PubMed PMID: 25618442
57. Xu TX, Zhao SZ, Dong M, Yu XR. Hypoxia responsive miR-210 promotes cell survival and autophagy of endometriotic cells in hypoxia. Eur Rev Med Pharmacol Sci. 2016;20(3):399–406. PubMed PMID: 26914112
58. Sun Y, Xing X, Liu Q, Wang Z, Xin Y, Zhang P, et al. Hypoxia-induced autophagy reduces radiosensitivity by the HIF-1alpha/miR-210/Bcl-2 pathway in colon cancer cells. Int J Oncol. 2015;46(2):750–6. https://doi.org/10.3892/ijo.2014.2745. PubMed PMID: 25385144
59. Su Z, Yang Z, Xu Y, Chen Y, Yu Q. MicroRNAs in apoptosis, autophagy and necroptosis. Oncotarget. 2015;6(11):8474–90. 10.18632/oncotarget.3523. PubMed PMID: 25893379; PubMed Central PMCID: PMCPMC4496162
60. Yu T, Li J, Yan M, Liu L, Lin H, Zhao F, et al. MicroRNA-193a-3p and -5p suppress the metastasis of human non-small-cell lung cancer by downregulating the ERBB4/PIK3R3/mTOR/S6K2 signaling pathway. Oncogene. 2015;34(4):413–23. https://doi.org/10.1038/onc.2013.574. PubMed PMID: 24469061
61. Yang Z, Han Y, Cheng K, Zhang G, Wang X. miR-99a directly targets the mTOR signalling pathway in breast cancer side population cells. Cell Prolif. 2014;47(6):587–95. https://doi.org/10.1111/cpr.12146. PubMed PMID: 25348507

62. Stiuso P, Potenza N, Lombardi A, Ferrandino I, Monaco A, Zappavigna S, et al. MicroRNA-423-5p promotes autophagy in cancer cells and is increased in serum from hepatocarcinoma patients treated with sorafenib. Mol Ther Nucleic Acids. 2015;4:e233. https://doi.org/10.1038/mtna.2015.8. PubMed PMID: 25782064
63. Friedman RC, Farh KK, Burge CB, Bartel DP. Most mammalian mRNAs are conserved targets of microRNAs. Genome Res. 2009;19(1):92–105. https://doi.org/10.1101/gr.082701.108. PubMed PMID: 18955434; PubMed Central PMCID: PMCPMC2612969

Chapter 16
Role of CEACAM in Different Cancers

Ilhan Yaylim, Ghazala Butt, Sumbul Khalid, and Ammad Ahmad Farooqi

Abstract Researchers have developed a better understanding of deregulation of spatio-temporally controlled signaling cascades which contribute in cancer development and progression. The greatest stumbling block in improving the clinical outcome of cancer patients is that tumors continuously adapt to their micro-environment and evolve ways to develop resistance against molecular therapeutics. We are developing systems-level understanding of tumors about their interactions with micro-environment, how they respond to the immunological system and impair the ligand-receptor interaction of killer cells and cancer cells, how they sequentially develop resistance against therapeutics, and how they evolve over time. In this chapter, we have attempted to provide an overview of role of CEACAMs in different cancers. It is relevant to mention that we still have incompletely studied various facets related to CEACAMs and how microRNAs modulate different CEACAMs. Future studies must converge on unraveling more sophisticated biology of CEACAMs, if they are to be targeted effectively.

Keywords CEACAM · Cancer · Signaling · Oncology · Therapy

I. Yaylim
Aziz Sancar Institute of Experimental Medicine, İstanbul University, İstanbul, Turkey

G. Butt (✉)
Department of Botany, GCU, Lahore, Pakistan
e-mail: dr.ghazalayasmeen@gcu.edu.pk

S. Khalid
Department of Bioinformatics and Biotechnology, International Islamic University, Islamabad, Pakistan

A. A. Farooqi
Institute of Biomedical and Genetic Engineering (IBGE), Islamabad, Pakistan

S. Fayyaz, A. A. Farooqi (eds.), *Recent Trends in Cancer Biology: Spotlight on Signaling Cascades and microRNAs*, https://doi.org/10.1007/978-3-319-71553-7_16

Introduction

Carcinoembryonic-antigen-related cell-adhesion molecule (CEACAM) family of proteins is reportedly involved in intercellular binding interactions for modulation of many biological mechanisms related to growth and differentiation of the cells. Data obtained through high-throughput technologies has considerably improved our understanding of characteristically unique features of CEACAM and it is now known that these proteins consist of different family members (CEACAM1, CEACAM 8 and CEACAM 3–7) and several secretory molecules (pregnancy-specific glycoproteins). Mechanistically it has been shown that these molecules are connected to membrane either through a transmembrane anchor (CEACAM1, CEACAM 3, and CEACAM 4) or via glycophosphatidyl anchor (CEACAM 5, CEACAM 6, CEACAM 7, and CEACAM 8) [1, 2]. These molecules adhere homophilically and heterophilically through intercellular interaction site present in N-terminal domains. N-domain of CEACAM (CEA) interacted efficiently with CD8 molecule on T lymphocytes. CEACAM1 is expressed by endothelial, epithelial, myeloid and lymphoid cells. Structural studies had shown that it consisted of an NH_2-terminal, distally located IgV-like domain, and followed by up to three proximally located IgC2-like domains. Extracellularly located domains of all CEACAMs are heavily glycosylated and it has been investigated that more than half of molecular weight of receptor consists of carbohydrate [3].

Efficacy of Labetuzumab Govitecan (Anti-CEACAM5/SN-38 Antibody-drug conjugate) has recently been tested in relapsing and refractory metastatic colorectal cancer patients [4]. Monotherapy with labetuzumab govitecan was effective and had manageable safety profile. Future studies must converge on the use of labetuzumab govitecan with different combinations [4]. In this chapter we have summarized how different CEACAMs play role in regulation of cancer.

CEACAM1

CEACAM1 is also involved in cancer and it has been shown that full-length long tail variant enhances proliferation of melanoma cells. Surprisingly, CEACAM1 promoter analysis revealed 2 SNPs that considerably increased promoter's activity. Results revealed that rs8102519 and rs8103285 located in the 5'UTR of CEACAM1 showed variation in allele distribution [5]. Germ line genotype carrying the double SNPs in promoter region of CEACAM1 enhanced the risk of development of melanoma [5]. Previous investigation also suggested that either loss of expression or genetic alteration of CEACAM1 contributed to the development of colorectal cancer [6].

G-2 cells derived from primary mammary adenocarcinomas grown in WAP-T mice exhibited cancer stem cell like features. These cells were composed of a mixture of epithelial and mesenchymal subpopulations [7]. These two cellular subsets were complementary and a notable feature was reciprocal switching between a mesenchymal-like and an epithelial-like phenotype. There was a co-expression of epithelial markers EpCAM and CEACAM1 on surface of Epithelial-like subpopulation. Cytoplasmic tail of human CEACAM1-L contained a β-catenin binding motif. Increase in cellular plasticity in CEACAM1low cells co-incided with a decrease in phosphorylated β-catenin (S33/S37/T41) levels. Surprisingly, instead of S33/S37/T41, the levels of phosphorylated β-catenin (Y86) were notably increased in CEACAM1low cell populations. S33/S37/T41 phosphorylated β-catenin enhanced dramatically after enforced expression of CEACAM1 [7]. CEACAM1 deficient mice were crossbred with WAP-T mice to study spontaneous growth of mammary tumors, their progression and metastasizing potential. WAP-T/CEACAM1null mice exhibited considerably high rate of metastatic spread. Pulmonary metastasis was detected in 40% of WAP-T/CEACAM1null mice xenografted with tumors [7].

miR-342 gene was present within intron of the EVL gene whose protein product enhanced actin nucleation, bundling and polymerization [8]. EVL gene was frequently methylated in colon cancer which consequently resulted in simultaneous downregulation of EVL and the miR-342 genes. ID4 transfection into CEACAM1null MCF-7 cells enabled lumen formation, but ID4 knockdown in CEACAM1-SF expressing MCF-7 cells significantly impaired lumen formation [8].

Significantly upregulated CEACAM1 was noted in both medullary thyroid carcinoma cells and neoplastic mast cells [9]. Knockdown of CEACAM1 enhanced cellular growth and adhesive features of LAD2, HMC1.2 and HMC1.1 cells, but suppressed cell growth of thyroid cancer TT cells. CEACAM1 exerted its inhibitory effects by activation of SHP-1 after its binding to the phosphorylated cytoplasmic tail of CEACAM1. Interaction of SHP-1 with CEACAM1 resulted in inhibition of critical tyrosine kinase mediated signaling events in cells. Knockdown of CEACAM1 reduced phosphorylated-SHP-1 in HMC1.2 cells [9].

Microphthalmia-associated transcription factor (MITF) has previously been shown to transcriptionally regulate CEACAM1. Detailed mechanistic insights provided clues of presence of M-box motifs located within promoter region of CEACAM1 [10]. MITF overexpression in Ma-Mel-63a melanoma cells induced an increase in protein level of CEACAM1 [10].

Glioblastoma-initiating cells (GIC) had much stronger self-renewal capacity, expressed neural stem cell (NSC) markers, such as CD133, CD15 and showed resistance against radio- and chemotherapies. CEACAM1L overexpression in NSCs induced nuclear translocation of the phosphorylated STAT3 [11]. CEACAM1L existed as a monomer, cis-/trans homo-dimer or cis-/trans hetero-dimer. Cytoplasmic tail of monomeric CEACAM1L activated c-Src–mediated STAT3 transduction cascade however, oligomerically assembled CEACAM markedly inhibited signaling pathway in glioblastoma-initiating cells [11].

Colon carcinogenesis was noted to be correlated with the emergence of a CEACAM1-negative stem cell population. To solve the puzzle related to role of CEACAM1-reduced stem cell population in increasing the tumor-forming potential of these cells, Lgr5-EGFP/ApcMin/+ mice were treated with dextran sulfate sodium to generate colon tumors [12]. The generated tumors were used for isolation of cells that expressed higher levels of CEACAM, GFP, and EPCAM (CEACAM1high/GFP+/EPCAM+) and those that expressed high levels of GFP and EPCAM but reduced levels of CEACAM1 (CEACAM1–/low/GFP+/EPCAM+). Formation of organoids by CEACAM1–/low/GFP+/EPCAM+ cells was at a higher frequency as compared to CEACAM1high/GFP+/EPCAM+ cells. Tumors developed in mice subcutaneously injected with 1 × 105 of CEACAM1–/low/GFP+/EPCAM+ cells [12].

CEACAM1 is composed of four extracellularly located domains and a long (CEACAM1-4L) or short (CEACAM1-4S) cytoplasmically located domain [13]. Serine residues present within cytoplasmic domains of CEACAM1 are phosphorylated and generate binding site for calmodulin. CEACAM1-4L domain contained two tyrosine residues which recruited and activated pp60c-src, a nonreceptor protein tyrosine kinase or the phosphatases SHP1 and −2 upon phosphorylation. Carcinoembryonic antigen induced CEACAM1-dependent initiation of apoptotic cell death in HT29 cells [13]. Camptothecin significantly reduced cell surface expression of CEACAM1-4L (−35%) and CEACAM1-4S (−22%) in Jurkat cells as evidenced by flow cytometry. Actinomycin D or Cycloheximide induced 24% decrease in CEACAM1 staining on the surface of colon HT29 cancer cells which indicated that programmed cell death consequently resulted in shedding of a significant extracellular portion of CEACAM1 [13].

Significant positive correlation was found between expression of CEACAM1 on cells of the primary tumor, hematogenous metastases and lymph node metastases [14]. There was no correlation between CEACAM1 expression with stage, gender, grading or patients' age. Compared to patients having CEACAM1-negative tumors, patients with a CEACAM1-expressing tumor had a shorter progression-free survival and median overall survival [14].

CEACAM5 and CEACAM6

CEA, CEACAM6, CEACAM7, CEACAM19 immunoreactivity in EGFR (epidermal growth factor receptor) mutation-positive cases was considerably higher as compared to patients who did not have mutations in EGFR [15]. Surprisingly, status of CEACAMs between responders and non-responders was not statistically significant in EGFR mutation-positive lung adenocarcinoma patients. Correlation between CEACAM status and the progression-free-survival (PFS) of cancer patients was analyzed critically by log-rank test and Kaplan–Meier survival curves [15]. Data clearly suggested that CEACAM6-positive status correlated with an enhanced PFS in patients treated with tyrosine kinase inhibitors [15].

HT29p cells were treated with Hyper-Interleukin-6 (15 ng/ml) or IL-6 (100 ng/ml) and results revealed that expression of CEACAM5 and CEACAM6 increased dramatically after 24 h by Hyper-IL-6 stimulation [16]. Detailed mechanistic insights revealed that STAT3 phosphorylation was crucial for the Hyper-IL-6-induced upregulation of CEACAM5 and CEACAM6. Trans-signaling pathway of IL-6 stabilized hypoxia-inducible factor 1-alpha and chemical stability of HIF-1α upregulated CEACAM5 and CEACAM6 [16].

CEACAM6 silencing sensitized tamoxifen-resistant breast cancer MMU1 cells to 4-hydroxytamoxifen and 17β-estradiol and reduced anchorage-independent growth and clonogenicity of MMU1 cells [17].

Binding and cross-linking of CEACAM6 by cytotoxic T cells exerted inhibitory effects on their activation which consequently resulted in T-cell non-responsiveness [18]. Blocking of CEACAM6 on the surface of myeloma cells by monoclonal antibodies (mAbs) or CEACAM6 gene silencing by RNA interference strategy induced restoration of T-cell reactivity against malignant plasma cells [18].

BxPC3 cells markedly overexpressed CEACAM6, as compared to Capan2 cells and were resistant to gemcitabine [19]. Gemcitabine (1 μm) induced apoptotic rate was low in CEACAM6 overexpressing Capan2 cells. Moreover, intriguingly, Gemcitabine (1 μm) induced apoptotic rate was significantly higher in CEACAM6 silenced BxPC3 cells. AKT (PKB) activity was dramatically enhanced in activity in CEACAM6 overexpressing Capan2 cells. Conversely, significant loss of AKT activity was noted in CEACAM6 silenced BxPC3 cells. Tumor growth was considerably reduced in mice implanted with CEACAM6 silenced BxPC3 cells [19].

CEABAC10 transgenic mice contained genomic DNA insert of a bacterial artificial chromosome (BAC) which encoded CEA family gene cluster including the CEACAM6 [20]. CEABAC10 mice were infected with B2 *E. coli* strain 11G5 isolates from patient of colon cancer. Infiltration rate of polynuclear cells in crypts was markedly higher in mice infected with *E. coli* strains 11G5 and LF82. Furthermore, these infected mice had higher number of crypt abscesses, larger and multi-focal erosion plates [20]. 11G5-infected mice displayed significant increase in proliferation rate of epithelial cells in crypts as compared to mice infected with LF82 strain and control mice [20]. Findings clearly suggested that mucosa cells of colon rapidly proliferated after infection by 11G5 (*E. coli* strain) associated with colon cancer. Another important clue which advocated the effective role of Colon cancer-associated *E. coli* strains in CEABAC mice was the involvement of CEACAM6. It has previously been reported that CEACAM6 served as a receptor for modulation of adherence or route for the entry of harmful bacteria [21].

CEACAM19

CEACAM19 expression status had a negative association with the estrogen receptors and menopausal status of breast cancer patients to a statistically significant degree [22]. More importantly, regarding ER-status, significantly higher levels of CEACAM19 were noted in tumors with ER-staining as compared to ER+ tumors. CEACAM19-positivity was recorded frequently in ER- tumors (65.2%) as compared to ER+ tumors [22, 23].

CEACAMs and TRAIL Induced Apoptosis

Riproximin (Rpx) is a member of cytotoxic type II ribosome inactivating proteins (RIP) family and noted to significantly improve TRAIL induced apoptosis in pancreatic ductal adenocarcinoma (PDAC) cells [24]. CEACAM overexpressing cell lines were more sensitive to Riproximin as compared to those cancer cell lines which had low CEACAM expression. Tumor necrosis factor receptor superfamily members (TNFRSF1A//TNFR1), TNFRSF1B//TNFR2) and their ligand, tumor necrosis factor (TNF-α) increased in a concentration- and time-dependent manner. Likewise, tumor necrosis factor receptor superfamily members (DR4/TRAILR1) and (TNFRSF10b/DR5) increased but astonishingly 2-fold downregulation of TRAIL was noted at all-time points [24]. In ASML cells, TRAIL inhibited cell growth (50% to 70%) at 1 to 100 nM concentrations and 20% growth inhibition in Suit2–007 cells at 10 nM concentration. The combination of these concentrations of TRAIL with Rpx was synergistically active. Riproximin dose dependently reduced tumor growth in BDX rats implanted intra-portally with ASML cells [24].

Conclusion

Monoclonal antibodies against Tim-3/CEACAM1 have recently been tested in mice with intracranial gliomas [25]. Mice combinatorially treated with antibodies against Tim-3/CEACAM1 demonstrated an increase in the infiltrating CD4+ and CD8+ T cells and simultaneous reduction in immuno-suppressive regulatory T Cells (Tregs) [25]. Exhaustion of reactive lymphocytes particularly cytotoxic T cells drastically abrogates the process of clearance of transformed tumor cells. Production of IFN-γ by CEACAM1+CD8+ TILs was notably reduced which pointed towards T cell exhaustion. It is clear that CEACAM1 plays instrumental role in tumorigenesis. Therefore it will be really important to focus on development of therapeutics against CEACAM1 and CEACAM6. Astonishingly, CEACAMs often behave context dependently and we need to focus on detailed mechanisms through which CEACAMs regulated cancer development and progression. MicroRNA regulation of CEACAMs is insufficiently studied and future studies must converge on unexplored mysteries related to various miRNAs which modulate CEACAMs. Reports also shed light on the use of Salmonella-based CEACAM6 and 4-1BBL vaccines against chemically-induced colorectal tumors [26]. Animal model based study revealed that vaccine treatment increased CD45RO+ memory T cells, decreased FOXP3+ cells and promoted polarization of Th1 cells [26].

References

1. Beauchemin N, Arabzadeh A. Carcinoembryonic antigen-related cell adhesion molecules (CEACAMs) in cancer progression and metastasis. Cancer Metastasis Rev. 2013;32(3–4): 643–71. https://doi.org/10.1007/s10555-013-9444-6.

2. Kuespert K, Pils S, Hauck CR. CEACAMs: their role in physiology and pathophysiology. Curr Opin Cell Biol. 2006;18(5):565–71.
3. Dankner M, Gray-Owen SD, Huang YH, Blumberg RS, Beauchemin N. CEACAM1 as a multi-purpose target for cancer immunotherapy. Oncoimmunology. 2017;6(7):e1328336. https://doi.org/10.1080/2162402X.2017.1328336.
4. Dotan E, Cohen SJ, Starodub AN, Lieu CH, Messersmith WA, Simpson PS, Guarino MJ, Marshall JL, Goldberg RM, Hecht JR, Wegener WA, Sharkey RM, Govindan SV, Goldenberg DM, Berlin JD. Phase I/II trial of Labetuzumab Govitecan (anti-CEACAM5/SN-38 antibody-drug conjugate) in patients with refractory or relapsing metastatic colorectal cancer. J Clin Oncol. 2017;17:JCO2017739011. https://doi.org/10.1200/JCO.2017.73.9011.
5. Ortenberg R, Galore-Haskel G, Greenberg I, Zamlin B, Sapoznik S, Greenberg E, Barshack I, Avivi C, Feiler Y, Zan-Bar I, Besser MJ, Azizi E, Eitan F, Schachter J, Markel G. CEACAM1 promotes melanoma cell growth through Sox-2. Neoplasia. 2014;16(5):451–60.
6. Song JH, Cao Z, Yoon JH, Nam SW, Kim SY, Lee JY, Park WS. Genetic alterations and expression pattern of CEACAM1 in colorectal adenomas and cancers. Pathol Oncol Res. 2011;17(1):67–74. https://doi.org/10.1007/s12253-010-9282-6.
7. Wegwitz F, Lenfert E, Gerstel D, von Ehrenstein L, Einhoff J, Schmidt G, Logsdon M, Brandner J, Tiegs G, Beauchemin N, Wagener C, Deppert W, Horst AK. CEACAM1 controls the EMT switch in murine mammary carcinoma in vitro and in vivo. Oncotarget. 2016;7(39):63730–46. 10.18632/oncotarget.11650.
8. Weng C, Nguyen T, Shively JE. miRNA-342 regulates CEACAM1-induced lumen formation in a three-dimensional model of mammary gland morphogenesis. J Biol Chem. 2016;291(32):16777–86. https://doi.org/10.1074/jbc.M115.710152.
9. Ueshima C, Kataoka TR, Takei Y, Hirata M, Sugimoto A, Hirokawa M, Okayama Y, Blumberg RS, Haga H. CEACAM1 long isoform has opposite effects on the growth of human mastocytosis and medullary thyroid carcinoma cells. Cancer Med. 2017;6(4):845–56. https://doi.org/10.1002/cam4.1050.
10. Ullrich N, Löffek S, Horn S, Ennen M, Sánchez-Del-Campo L, Zhao F, Breitenbuecher F, Davidson I, Singer BB, Schadendorf D, Goding CR, Helfrich I. MITF is a critical regulator of the carcinoembryonic antigen-related cell adhesion molecule 1 (CEACAM1) in malignant melanoma. Pigment Cell Melanoma Res. 2015;28(6):736–40. https://doi.org/10.1111/pcmr.12414.
11. Kaneko S, Nakatani Y, Takezaki T, Hide T, Yamashita D, Ohtsu N, Ohnishi T, Terasaka S, Houkin K, Kondo T. Ceacam1L modulates STAT3 signaling to control the proliferation of glioblastoma-initiating cells. Cancer Res. 2015;75(19):4224–34. https://doi.org/10.1158/0008-5472.CAN-15-0412.
12. Shiokawa D, Sato A, Ohata H, Mutoh M, Sekine S, Kato M, Shibata T, Nakagama H, Okamoto K. The induction of selected Wnt target genes by Tcf1 mediates generation of tumorigenic colon stem cells. Cell Rep. 2017;19(5):981–94. https://doi.org/10.1016/j.celrep.2017.04.017.
13. Nittka S, Böhm C, Zentgraf H, Neumaier M. The CEACAM1-mediated apoptosis pathway is activated by CEA and triggers dual cleavage of CEACAM1. Oncogene. 2008;27(26):3721–8. https://doi.org/10.1038/sj.onc.1211033.
14. Thöm I, Schult-Kronefeld O, Burkholder I, Schuch G, Andritzky B, Kastendieck H, Edler L, Wagener C, Bokemeyer C, Schumacher U, Laack E. Expression of CEACAM-1 in pulmonary adenocarcinomas and their metastases. Anticancer Res. 2009;29(1):249–54.
15. Kobayashi M, Miki Y, Ebina M, Abe K, Mori K, Narumi S, Suzuki T, Sato I, Maemondo M, Endo C, Inoue A, Kumamoto H, Kondo T, Yamada-Okabe H, Nukiwa T, Sasano H. Carcinoembryonic antigen-related cell adhesion molecules as surrogate markers for EGFR inhibitor sensitivity in human lung adenocarcinoma. Br J Cancer. 2012;107(10):1745–53. https://doi.org/10.1038/bjc.2012.422.
16. Holmer R, Wätzig GH, Tiwari S, Rose-John S, Kalthoff H. Interleukin-6 trans-signaling increases the expression of carcinoembryonic antigen-related cell adhesion molecules 5 and 6 in colorectal cancer cells. BMC Cancer. 2015;15:975. https://doi.org/10.1186/s12885-015-1950-1.
17. Maraqa L, Cummings M, Peter MB, Shaaban AM, Horgan K, Hanby AM, Speirs V. Carcinoembryonic antigen cell adhesion molecule 6 predicts breast cancer recurrence following

adjuvant tamoxifen. Clin Cancer Res. 2008;14(2):405–11. https://doi.org/10.1158/1078-0432.CCR-07-1363.

18. Witzens-Harig M, Hose D, Jünger S, Pfirschke C, Khandelwal N, Umansky L, Seckinger A, Conrad H, Brackertz B, Rème T, Gueckel B, Meißner T, Hundemer M, Ho AD, Rossi JF, Neben K, Bernhard H, Goldschmidt H, Klein B, Beckhove P. Tumor cells in multiple myeloma patients inhibit myeloma-reactive T cells through carcinoembryonic antigen-related cell adhesion molecule-6. Blood. 2013;121(22):4493–503. https://doi: 10.1182/blood-2012-05-429415.
19. Duxbury MS, Ito H, Benoit E, Waseem T, Ashley SW, Whang EE. A novel role for carcinoembryonic antigen-related cell adhesion molecule 6 as a determinant of gemcitabine chemoresistance in pancreatic adenocarcinoma cells. Cancer Res. 2004;64(11):3987–93.
20. Raisch J, Buc E, Bonnet M, Sauvanet P, Vazeille E, de Vallée A, Déchelotte P, Darcha C, Pezet D, Bonnet R, Bringer MA, Darfeuille-Michaud A. Colon cancer-associated B2 Escherichia coli colonize gut mucosa and promote cell proliferation. World J Gastroenterol. 2014;20(21):6560–72. https://doi.org/10.3748/wjg.v20.i21.6560.
21. Bos MP, Grunert F, Belland RJ. Differential recognition of members of the carcinoembryonic antigen family by Opa variants of Neisseria gonorrhoeae. Infect Immun. 1997;65(6):2353–61.
22. Michaelidou K, Tzovaras A, Missitzis I, Ardavanis A, Scorilas A. The expression of the CEACAM19 gene, a novel member of the CEA family, is associated with breast cancer progression. Int J Oncol. 2013;42(5):1770–7. https://doi.org/10.3892/ijo.2013.1860.
23. Estiar MA, Esmaeili R, Zare AA, Farahmand L, Fazilaty H, Zekri A, Jafarbeik-Iravani N, Majidzadeh- AK. High expression of CEACAM19, a new member of carcinoembryonic antigen gene family, in patients with breast cancer. Clin Exp Med. 2016;17(4):547–53. https://doi.org/10.1007/s10238-016-0442-1.
24. Adwan H, Murtaja A, Kadhim Al-Taee K, Pervaiz A, Hielscher T, Berger MR. Riproximin's activity depends on gene expression and sensitizes PDAC cells to TRAIL. Cancer Biol Ther. 2014;15(9):1185–97. https://doi: 10.4161/cbt.29503.
25. Li J, Liu X, Duan Y, Liu Y, Wang H, Lian S, Zhuang G, Fan Y. Combined blockade of T cell immunoglobulin and mucin domain 3 and carcinoembryonic antigen-related cell adhesion molecule 1 results in durable therapeutic efficacy in mice with intracranial gliomas. Med Sci Monit. 2017;23:3593–602.
26. Jin C, Duan X, Liu Y, Zhu J, Zhang K, Zhang Y, Xia T, Fei Y, Ye J. T cell immunity induced by a bivalent Salmonella-based CEACAM6 and 4-1BBL vaccines in a rat colorectal cancer model. Oncol Lett. 2017;13(5):3753–9. https://doi.org/10.3892/ol.2017.5938.

Index

S. Fayyaz, A. A. Farooqi (eds.), *Recent Trends in Cancer Biology: Spotlight on Signaling Cascades and microRNAs*, https://doi.org/10.1007/978-3-319-71553-7

The manufacturer's authorised representative in the EU is Springer Nature Customer Service Centre GmbH, Europaplatz 3, 69115 Heidelberg, Germany. If you have any concerns regarding our products, please contact ProductSafety@springernature.com

Printed and bound by CPI Group (UK) Ltd, Croydon, CR0 4YY
05/10/2026
02228141-0008